Crop Safeners for Herbicides

Development, Uses, and Mechanisms of Action

Crop Safeners for Herbicides

Development, Uses, and Mechanisms of Action

Edited by

Kriton K. Hatzios
Department of Plant Pathology,
Physiology and Weed Science
Virginia Polytechnic Institute and State University
Blacksburg, Virginia

Robert E. Hoagland
Southern Weed Science Laboratory
United States Department of Agriculture
Agricultural Research Service
Stoneville, Mississippi

Academic Press, Inc.
Harcourt Brace Jovanovich, Publishers
San Diego New York Berkeley Boston
London Sydney Tokyo Toronto

ACADEMIC PRESS, INC.
San Diego, California 92101

United Kingdom Edition published by
ACADEMIC PRESS, INC. (LONDON) LTD.
24-28 Oval Road, London NW1 7DX

Library of Congress Cataloging-in-Publication Data

Crop safeners for herbicides.

Includes bibliographies and index.
1. Herbicide safeners. I. Hatzios, Kriton K.
II. Hoagland, Robert E.
SB951.45.C76 1988 632′.954 88-3432
ISBN 0-12-332910-8 (alk. paper)

PRINTED IN THE UNITED STATES OF AMERICA
88 89 90 91 9 8 7 6 5 4 3 2 1

Contents

Part Two. Physiological and Biochemical Mechanisms of Action of Herbicide Safeners

Part Four. Summary

Contributors

MICHAEL BARRETT Department of Agronomy, University of Kentucky, Lexington, Kentucky 40546.

E. J. BREAUX[1] Monsanto Agricultural Company, St. Louis, Missouri 63167.

FERENC DUTKA Department of Pesticide Research, Central Research Institute for Chemistry of the Hungarian Academy of Sciences, Budapest II, Pusztaszeri UT 59-67, H-1525 Budapest, Hungary.

EDITH EBERT Ciba-Geigy Ltd., Agricultural Division, CH-4002 Basle, Switzerland.

GAIL EZRA[2] PPG Industries, Inc., P.O. Box 31, Barberton, Ohio 44203.

HANS R. GERBER Ciba-Geigy Ltd., Agricultural Division, CH-4002 Basle, Switzerland.

JOHN W. GRONWALD USDA, ARS, Plant Science Research Unit and Department of Agronomy and Plant Genetics, University of Minnesota, St. Paul, Minnesota 55108.

KRITON K. HATZIOS Department of Plant Pathology, Physiology and Weed Science, Virginia Polytechnic Institute and State University, Blacksburg, Virginia 24061-0330.

ROBERT E. HOAGLAND USDA, ARS, Southern Weed Science Laboratory, Stoneville, Mississippi 38776.

M. A. HOOBLER Monsanto Agricultural Company, St. Louis, Missouri 63167.

JEFFREY S. KARNS USDA, ARS, Beltsville Agricultural Research Center–West, Beltsville, Maryland 20705.

OSAMU KIRINO[3] Pesticide Research Laboratory, Takarazuka Research Center, Sumitomo Chemical Co. Ltd., 4-2-1 Takatsukasa, Takarazuka Hyogo 665, Japan.

TAMAS KOMIVES Department of Pesticide Research, Central Research Institute for Chemistry of the Hungarian Academy of Sciences, Budapest II, Pusztaszeri UT 59-67, H-1525 Budapest, Hungary.

[1]Present address: Rhone Poulenc Agrochimie, 14-20 rue Pierre Baizet, 69009 Lyon, France.

[2]Present address: American Cyanamid Company, P.O. Box 400, Princeton, New Jersey, 08540.

[3]Deceased.

G. A. Leyes Monsanto Agricultural Company, St. Louis, Missouri 63167.

Shooichi Matsunaka Faculty of Agriculture, Kobe University, Nada-ku, Kobe 657, Japan.

J. E. Patanella Monsanto Agricultural Company, St. Louis, Missouri 63167.

Donald Penner Department of Crop and Soil Sciences, Michigan State University, East Lansing, Michigan 48824.

Sharad C. Phatak Department of Horticulture, Coastal Plain Experiment Station, University of Georgia, Tifton, Georgia 31793.

Bruce D. Riggle ICI Americas, Inc., Eastern Research Center, Goldsboro, North Carolina 27533.

Baruch Rubin Department of Field and Vegetable Crops, Faculty of Agriculture, The Hebrew University of Jerusalem, Rehovot 76100, Israel.

Gerald R. Stephenson Department of Environmental Biology, University of Guelph, Guelph, Ontario N1G 2W1, Canada.

Charles S. Vavrina Department of Horticulture, Cooperative Extension Service, University of Georgia, Statesboro, Georgia 30460.

Ko Wakabayashi Mitsubishi Kasei Corporation, Research Center, Kamoshida-cho, Midori-ku, Yokohama 227, Japan.

Robert E. Wilkinson Department of Agronomy, Georgia Station, Agricultural Experiment Stations, University of Georgia, Experiment, Georgia.

Preface

Manipulation of the tolerance of crop plants to herbicides has challenged agronomists and weed scientists since the establishment of chemical weed control as an important facet of agricultural crop production 40 years ago. Several approaches based on mechanical, genetic, and chemical concepts have been tried, with varying degrees of success, to overcome the problem of limited crop selectivity of several important herbicides.

The objective of this book is to cover the subject of the chemical manipulation of crop tolerance to herbicides. Chemical safening agents are commonly referred to as "herbicide safeners" and are used to protect crops against herbicide injury. These agents are also termed "herbicide antidotes" or "crop protectants"; several such chemicals are currently marketed for use with specific crop/herbicide combinations under field conditions. The first attempt to summarize selected aspects of herbicide safeners in book form was "Chemistry and Action of Herbicide Antidotes," edited by Ferenc M. Pallos and John E. Casida and published by Academic Press in 1978. Since that time, several new classes of safeners have been developed, and additional information on the mechanisms of action of old and new herbicide safeners has been published in journal articles but not covered in book form.

The idea for this book originated from a successful symposium, "Chemical Manipulation of Crop Tolerance to Herbicides," that we had the pleasant task of organizing as part of the 26th annual meeting of the Weed Science Society of America in Houston, Texas, in 1986. Although this symposium brought together several scientists, it was somewhat limited in scope since the primary topic discussed was the mechanisms of action of herbicide safeners. After this meeting, invitations were extended to other researchers from several countries who are active in the area of crop safening against herbicide injury. We are indeed pleased that many of these experts agreed to contribute comprehensive chapters for this book.

The purpose of this book is not only to bring the reader up-to-date on the present status of the agronomic uses, development, chemistry, and mechanisms of action of herbicide safeners, but also to assess the impact of safeners throughout the world, to discuss alternative approaches that could be used for increasing herbicide selectivity, and to speculate a bit on future trends in the development of herbicide safeners. The book contains 16 chapters that are divided into four sections. The two chapters in Part 1 offer an extensive and critical review of academic and industrial perspectives that play a key role in the development of herbicide safeners in the United States, Japan, and other parts of the world. The first chapter of Part 2 provides a comprehensive overview of the physiological, biochemical, and molecular aspects of the mechanisms of action of herbicide safeners. The next seven chapters discuss the most prevalent mechanisms of action of selected classes of herbicide safeners. Part 3 provides information on the potential use of alternative approaches for the manipulation of crop tolerance to herbicides. Approaches discussed include the use of adsorbents such as activated charcoal, controlled release formulations of herbicides, prosafeners, microbial safeners, and the potential of antagonistic interactions of herbicides with plant growth regulators, fungicides, and other pesticides. Part 4 contains a single chapter on the progress and prospects of the concept of crop safening against herbicide injury. Many of the contributed chapters are comprehensive reviews on selected aspects of herbicide safeners. Some chapters, however, present and discuss results of research not previously published, and these add considerably to the strength of this book.

We hope this book will serve as a valuable source of basic scientific information to meet the needs of weed scientists, agronomists, pesticide chemists, plant biochemists and physiologists, and those who are responsible for recommending as well as regulating the use of herbicides in agriculture. An additional purpose is to motivate further research in this field, which will undoubtedly lead to the development and commercialization of new and effective crop safeners for herbicides.

Throughout this book, we have tried to be consistent in the use of terms, definitions, and names of plants and chemicals. Whenever possible, we have followed the nomenclature of the Weed Science Society of America (WSSA). Herbicides, safeners, and other agrochemicals are referred to in the text by their common names. A complete list of the common and corresponding chemical names of all herbicides, safeners, and other pesticides is presented in the Appendix. In referring to plants—crops and weeds—we chose to use the WSSA acceptable common names. The Latin binomial names of each plant are given in parentheses or brackets following the first mention of the common name in each chapter. The use of common names was judged more suitable since several weeds that are botanically related to certain crops are known by the same Latin binomial name as those crops (i.e., are conspecific). For example, the weed shattercane and the cultivated grain sorghum are known by the same Latin binomial [*Sorghum bicolor* (L.) Moench].

The editors are indebted to all the contributors for their sincere and painstaking efforts in bringing this book to publication. We also acknowledge the cooperation and guidance of the technical staff of Academic Press. Finally, to Maria and Rebecca—thank you for your patience, encouragement, and support throughout this project.

Kriton K. Hatzios

Robert E. Hoagland

Part One

Development and Uses of Herbicide Safeners

Chapter 1

Development of Herbicide Safeners: Industrial and University Perspectives

Kriton K. Hatzios
Department of Plant Pathology, Physiology and Weed Science
Virginia Polytechnic Institute and State University
Blacksburg, Virginia 24061-0330

> *"It appeared so simple, but getting to this result (practical use of safeners) entailed many trials over the time span of 21 years."*
>
> Otto Hoffman, 1978

I. INTRODUCTION AND TERMINOLOGY

The idea of enhancing crop tolerance to nonselective herbicides with the use of chemical treatments was conceived by Otto Hoffman in the late 1940s (Hoffman, 1978). As a result of his pioneering efforts this concept became a reality and chemical crop safening against herbicide injury is today an accepted agricultural practice.

In analogy to the situation in pharmacology, where the use of antidotes for drug-poisoning is a widely accepted practice, Hoffman (1962) introduced the term "herbicide antidotes" to describe the function of chemical agents which protect crop plants from herbicide injury. However, although widely used, the appropriateness of the term "herbicide antidotes" has been questioned (Pallos and Casida, 1978; Parker, 1983). Much of the criticism stems from the notion that, whereas in medicine and pharmacology the term "antidote" describes a medication used to reverse human intoxication; "herbicide antidotes" prevent herbicide injury on crop plants, but fail to reverse it (Donald and Fawcett, 1975, 1976; Hatzios, 1983; Parker, 1983). The timing of antidote and herbicide applications is very critical for the action of these chemicals as crop protectants. All currently available antidotes are most effective when applied prior to or simultaneously with the herbicides whose injury they prevent (Blair *et al.*, 1976; Hatzios, 1983).

To avoid the existing confusion created by the traditional use of the term "antidote" in medicine, "herbicide antidotes" are now more commonly termed "herbicide safeners." In spite of concerns raised as to the correct spelling or the existence of the word "safeners" in the English language (Parker, 1983), the term "herbicide safeners" has become quite popular among scientists working in this field. It should always be kept in mind, however, that these chemicals do not safen herbicides as the name "herbicide safeners" might imply, but they safen crop plants against herbicide injury. Alternative terms such as "antagonists," "protectants," "modifiers," and "agriregulators" have been proposed and they are used interchangeably with the terms "herbicide antidotes" or "herbicide safeners" in the literature. All currently available chemical safeners appear to act internally (inside the protected plant). Chemicals protecting crops against herbicide injury by acting externally are also available and they are commonly classified as "herbicide adsorbents." Such protectants include primarily activated carbon (charcoal) and selected lignin by-products and they will be discussed later in this this book.

The term "herbicide safener," as used by weed scientists, agronomists, and pesticide chemists, refers to a chemical compound which has limited phytotoxicity on its own and selectively protects crop plants against herbicide injury, without protecting any weeds. The selectivity of herbicide safeners originates either from a selective placement such as the dressing of crop seeds with the safener or it is the result of some very specific crop–herbicide–safener interactions (Hatzios, 1983; Parker, 1983). Figure 1 illustrates pictorially the concept of crop safening against herbicide injury. Naphthalic anhydride (NA) applied as seed dressing at 1.0% by seed weight provided satisfactory protection to corn (*Zea mays* L.) exposed to a lethal dose of chlorsulfuron (0.1 kg/ha), applied preemergence.

The principles underlying the commercial development, chemistry, and practical applications of herbicide safeners are reviewed briefly in this chapter. Herbicide safeners have been discussed in several earlier reviews (Blair *et al.*, 1976; Gray *et al.*, 1982; Hatzios, 1983; Hoffman, 1962, 1978; Pallos and Casida, 1978; Parker, 1983; Peek *et al.*, 1981; Stephenson and Ezra, 1983; Stephenson and Pallos, 1983).

Fig. 1. Amelioration of chlorsulfuron injury to corn by the safener naphthalic anhydride (NA).

II. NEED FOR HERBICIDE SAFENERS

The discovery and development of phenoxyalkanoic acid herbicides such as 2,4-D and MCPA during the years of World War II revolutionized our approach to weed management in crop production (Kirby, 1980). For the first time it was possible to selectively control broadleaf weeds in grass crops with the use of chemical herbicides. Following this milestone, an impressive number of organic herbicides with varying degrees of crop selectivity have been introduced and developed for use in modern agriculture (Adler *et al.*, 1977; Fedtke, 1982; Sanders, 1981). The crop selectivity of any herbicide, however, is a relative rather than absolute characteristic determined by the complex interactions that occur between crops and weeds, their environment, and the given herbicide (Ashton and Harvey, 1971). Many of the currently available herbicides which are effective in controlling problem weeds are either nonselective (e.g., glyphosate, paraquat, glufosinate, etc.) or they possess marginal crop selectivity (e.g.,chloroacetanilides, thiocarbamates, dinitroanilines, etc.).

On the other hand, in spite of the plethora of selective herbicides that are available today, chemical control of weeds in botanically related crops is still a challenge confronting weed technologists. On a worldwide basis, the problem is exemplified by the difficulty in chemically controlling wild oats (*Avena fatua* L.) in cultivated oats (*Avena sativa* L.), shattercane in grain sorghum [both *Sorghum bicolor* (L.) Moench], red rice in cultivated rice (both *Oryza sativa* L.), wild mustard [*Brassica kaber* (DC.) L.C. Wheeler] in rapeseed (*Brassica campestris* L.), wild beet in sugar beet (both *Beta vulgaris* L.), and itchgrass (*Rottboelia exaltata* L.) in corn (Hoffman, 1978; Parker, 1983). Similar problems have also been experienced in crop rotation systems, from the appearance of cultivated crops produced in the previous season as "volunteer weeds" during the next growing season. "Volunteer" wheat (*Triticum aestivum* L.) or barley (*Hordeum vulgare* L.) are presently serious weeds in a barley–wheat rotation system used in many European countries and the western United States (Parker, 1983). "Volunteer" corn is often encountered as a weed problem in corn–soybean [*Glycine max* (L.) Merr.] rotations used in the midwestern United States (Paul and Knake, 1984).

Apart from the aforementioned problems and limitations of the crop selectivity of many of the currently marketed herbicides, there is a continuous need for updating herbicide technology because of changes and developments in agricultural practices. Crop rotations, changes in row widths, monocultural or multiple cropping systems, and crop production with minimum or no tillage represent current facets of modern crop production which have imposed changes in crop protection technology (Wiese, 1985). In addition, natural or chemically induced shifts in weed populations present new challenges that need to be addressed. The weed complexes affecting any given crop are in continual change and new weed problems develop as existing problems are solved. As a result, the need for alternative herbicides to deal with new weed problems continues. The development of weed biotypes which are resistant to triazine herbicides represents a good example of chemically induced shifts in weed populations and the need for alternative herbicides to control these weeds (LeBaron and Gressel, 1982). The most recent list of weed species which have developed resistance to triazine herbicides includes 49 species distributed in many countries all over the world. (LeBaron, 1986). Thirty-six of them are broadleaf weeds and thirteen are monocotyledonous weeds. In addition, selected weeds have developed resistance to many other herbicides including 2,4-D, diclofop-methyl, paraquat, trifluralin, and other dinitroanilines (Gressel, 1986; LeBaron and Gressel, 1982).

It is evident, therefore, that our war against weeds is endless and in order to solve the aforementioned problems we have to rely on two main approaches. The traditional approach, of course, is the development of new herbicides possessing good crop selectivity and strong activity against new weed problems that need to be solved. This approach has been quite successful in the past. However, in recent years, several issues facing the development of new herbicides have imposed serious limitations on the feasibility of this approach. Rapidly escalating research and development costs, increased difficulty in registration, and increased concerns for our environment have been responsible for a decreased rate of discovering new herbicides with sufficient activity and economic viability to justify commercialization. Welcomed exceptions to this trend have been the recent introductions of several new herbicides with excellent crop selectivity and high activity on weeds such as the sulfonylureas (e.g., chlorsulfuron), imidazolinones (e.g., imazaquin), arylophenoxyalkanoic acids (e.g. fluazifop-butyl), and cyclohexenones (e.g,. sethoxydim) (Sanders, 1981).

A second approach to overcoming the weed management problems discussed earlier is to make better use of herbicides that are already available. This objective could be achieved by several means, including improvements in formulation technology of herbicides, improvements in application technology, circumvention of weed resistance to herbicides, and enhancement of crop selectivity of herbicides with the use of herbicide safeners.

Recent improvements in the formulation technology of herbicides are exemplified by the development of granules, microencapsulations, and other controlled-release formulations which are currently the subject of intensive research (Kydonieus, 1980). Directed sprays or critical timing of herbicide applications to weeds prior to crop emergence (preplant incorporated or preemergence treatments) have been used for many years as mechanical means to increase the selectivity of many soil-applied herbicides (Adler *et al.*, 1977). More recently, the use of recirculating sprayers (e.g., ropewick applicator) have been instrumental in increasing the selectivity of postemergence-applied nonselective herbicides such as glyphosate (Dale, 1979).

Circumventing the development of weed resistance to specific herbicides could be achieved primarily by delaying its onset. Use of crop and chemical rotations as well as the use of herbicide combinations would facilitate this objective. A thorough understanding of the biochemical and genetic bases of weed resistance to herbicides will allow us to genetically manipulate crop resistance to the same herbicides (LeBaron and Gressel, 1982). Presently, this is a very active area of research and with recent advances in recombinant DNA technology it is expected that the development and commercialization of crop plants with engineered resistance to herbicides will become a reality in the immediate future (Comai and Stalker, 1984; Glass, 1986; Hatzios, 1987; Netzer, 1984; Zaitlin *et al.*, 1986).

The concept of using herbicide safeners offers an attractive and promising alternative for increasing the crop selectivity of many of the currently available herbicides. Herbicide safeners offer new opportunities for progress in chemical weed control and they are currently the subject of research at both academic and industrial establishments. Chemical manipulation of crop tolerance to herbicides with the use of herbicide safeners is appealing for many reasons. The most important of them are summarized in Table I.

So far, the major advantage provided by the safener concept has been the achievement of selective control of weeds closely related to the crop for which selectivity is otherwise impossible. Thus, it might be argued that the answer to the commonly asked question "Are completely weedless crops possible?" could be yes only with the use of the appropriate herbicide–safener combinations.

Table I Benefits of Safener Concept

Selective chemical control of weeds in botanically related crops
Nonselective herbicides could be used selectively
Counteraction of the residual activity of soil-applied persistent herbicides in crop rotation systems
Increased spectrum of herbicides available for weed control in "minor" crops
Expansion and/or extension of uses and marketability of "out-of-patent" herbicides
Reduction of cost of chemical weed control
Safeners could be used as biochemical probes to elucidate sites and mechanisms of action of herbicides

The counteraction of the residual activity of persistent soil-applied herbicides with safeners may provide flexibility in some crop rotations. For example, the amelioration of chlorsulfuron injury on corn with the safener NA (Frear *et al.*, 1987; Hatzios, 1984a; Parker *et al.*, 1980; Sweetser, 1985) would allow corn to be planted safely in a wheat/corn rotation system. This interaction is illustrated in Fig. 1. Research on the use of safeners for increasing the spectrum of herbicides available for use in "minor" crops has been limited. However, the recent report by Mersie and Parker (1983) demonstrating the protection of the small cereal teff (*Eragrostris teff* Zucc. Trotter) from herbicide injury points out the potential of this concept.

The role of safeners in "post-patent strategy" is expected to become more important as the patents of many of the currently available herbicides expire. Establishment of successful crop–herbicide–safener combinations would offer new market advantages to many currently available herbicides and improve their sales. If current trends are an indication, successful herbicides whose patents expire, decrease in price, and combinations of such herbicides with safeners would enhance their uses. Theoretically then, safeners may reduce the cost of chemical weed control. However, this would depend on the cost of the safeners combined with the "out-of-patent" inexpensive herbicides.

Finally, safeners could be used as useful probes for elucidating sites and mechanisms of herbicide action. Selected aspects of the mechanism of action of safeners and their interactions with herbicides are discussed in later chapters of this book.

From the above summary of the main benefits of the safener concept, it becomes obvious that successful herbicide–safener combinations would remain competitive with new herbicides introduced for use in the same markets. The selection of an inexpensive herbicide that is effective on a hard-to-control weed problem, plus a chemical safener for the crop, would provide a very competitive alternative to a new herbicide that might be developed to answer the same problem.

III. DEVELOPMENT OF HERBICIDE SAFENERS

A. Search for Herbicide Safeners

1. Historical Aspects

Serendipity combined with Hoffman's vision was instrumental for the development of the safener concept (Hoffman, 1978). It all began in the summer of 1947, when, because of a malfunction in the ventilation system of a greenhouse, tomatoes (*Lycopersicon esculentum* Mill.) tested for their responses to structural analogs of the herbicide 2,4-D were exposed accidentally to fumes of 2,4-D. Hoffman observed that the epinastic responses caused by a sublethal dose of 2,4-D on most tomatoes were absent in plants treated with the inactive analog 2,4,6-T. Later, Hoffman *et al.* (1960) reported that foliar treatments of 2,4-D antagonized the injurious effects of the carbamate herbicide barban on wheat. Based on these observations of antagonistic herbicide interactions, Hoffman evaluated several chemicals, capable of causing formative effects similar to 2,4-D, as candidate safeners and in 1962 he introduced the compound S-449 as an effective safener of wheat against injury from the herbicide barban (Hoffman, 1962).

Hoffman (1978) is also credited with the development and use of the first systematic screening program directed at identifying chemical crop safeners for selected herbicides. Research based on this screening program proved quite successful and resulted in the discovery and development of NA as an effective protectant of corn against injury from thiocarbamate herbicides such as EPTC (Hoffman, 1969). NA was patented in 1971 (Hoffman, 1971) and its commercialization epitomized the successful evolution of the safener concept from a scientific curiosity to a practical realization. The practical success of NA, however, was never matched by its performance as a saleable commodity in the corn herbicide market. The reasons for this are many and will be discussed later in this chapter. Nevertheless, the commercialization of NA sparked the initiation of intensive research on herbicide safeners and competition among industry for new products and discoveries. A recent count of companies and research institutions engaged in patented safener research included more than seventeen companies and institutions worldwide (Gray *et al.*, 1982).

These efforts have resulted in the discovery and commercialization of several herbicide safeners during the last decade. Among them, the chloroacetamide safeners of corn against thiocarbamate or chloroacetanilide herbicide injury have been the most successful (Dutka and Komives, 1987; Görög *et al.*, 1982; Gray *et al.*, 1982; Nagy and Balogh, 1985; Pallos *et al.*, 1975, 1977; Wuertzer *et al.*, 1983). Oxime ether derivatives and disubstituted thiazole carboxylates are also important milestones in safener research (Ellis *et al.*, 1980; Rufener *et al.*, 1982; Sacher *et al.*, 1983). These safeners protect grain sorghum from chloroacetanilide herbicide injury. More recently, fenclorim (CGA-123407) was introduced as an effective safener of rice against injury from the chloroacetanilide herbicide pretilachlor (Quadranti and Ebner, 1984).

2. Discovery and Screening of Candidate Safeners

a. General Considerations In the search for new herbicide safeners, three key components need to interact in a desirable manner to make any screening program successful. They include the crop to be safened, the herbicide to be antidoted, and the potential safener. To be economical, any screening program for candidate safeners has to be selective as to the crops and herbicides that should be considered. So far, commercial success in crop safening against herbicides has been limited to three major grass crops: corn, grain sorghum, and rice. These crops can be successfully safened against injury from thiocarbamate and chloroacetanilide herbicides. Results on the safening of other grass crops such as wheat, oats, barley, and ryegrass (*Lolium perenne* L.) against injury from several herbicides have been encouraging (Hatzios, 1983; Parker, 1983). Commercial success in the safening of broadleaf crops against injury from any herbicide or the protection of any crop against photosynthesis-inhibiting herbicides has been rather limited. Chemical adsorbents such as lignin or activated charcoal have been more successful than chemical safeners in achieving these objectives (see discussion in Chapters 11 and 12, this volume).

The development of new herbicide safeners is a costly process justified by the economic incentives offered by this practice. As a result, research on the discovery and development of safeners is conducted primarily by the major agrochemical companies worldwide. However, active research programs in this field are also maintained by a number of public institutions (e.g., universities, research institutes, etc.) and they have resulted in the discovery and introduction of several promising safeners (Chang and Merkle, 1983; Dutka and Komives, 1987; Hilton and Pillai, 1986; Nagy and Balogh, 1985).

From an industrial point of view, a major incentive for engaging in safener research is to expand or extend the use of a proprietary herbicide. Thus, a thorough analysis of current and potential new markets and opportunities is imperative for the initiation of a research program on herbicide safeners. Regardless of its effectiveness, a candidate safener would be developed and reach commercialization only if the range of its uses is limited to proprietary herbicides of a single company and avoids favoring the products of other companies. The experience of NA illustrated clearly the insurmountable difficulties faced by a company (Gulf Oil Company) trying to market a safener designed to expand the sales of the proprietary herbicide of another company (Stauffer Chemical Company). The marketing potential of NA was severely hampered by the discovery and commercialization of the dichloroacetamide safener dichlormid by Stauffer for use with its own proprietary herbicides. Therefore, it is not surprising that all safener–herbicide–crop combinations that have been exploited commercially after the introduction of NA include proprietary safeners and herbicides developed by the same company. Alternatively, a particular company or public institution succeeding in developing an effective safener for another company's proprietary herbicides could postpone its commercialization until the proprietary herbicide goes out of patent. An example illustrating this approach is the development and commercialization of the safener PPG-1292 by the PPG Industries for safening corn against Stauffer's thiocarbamate herbicides EPTC and butylate, which are currently out of patent. The development of the safeners DKA-24 and MG-191 by state-run institutions in Hungary provides additional examples. Both of these safeners protect corn against injury from thiocarbamate herbicides (Dutka and Komives, 1987; Nagy and Balogh, 1985).

b. Screening for Candidate Safeners The discovery of herbicide safeners is usually pursued through the selection of candidate chemicals included in screening tests designed to evaluate crop safener activity against injury from selected herbicides. Three main methods, known as empirical, imitative, and rational methods, have been traditionally used for this purpose (Saggers, 1976). Random screening techniques based primarily on the empirical and to a lesser extent on the imitative methods of chemical selection have been used successfully for the discovery and development of herbicide safeners. The use of the rational (also known as biorational) approach in safener screening has been limited.

Candidate chemicals screened as safeners are either products of targeted synthesis following the identification of chemical leads or they are acquired from a large pool of chemicals with known or unknown biological properties available at the archives of any major pesticide company. Candidate safeners that are effective in protecting one or more major crops against injury from one or more important herbicides are identified in the primary safener screen. This process involves the selection of active compounds from a large number of candidate safeners in empirical biological assays conducted under laboratory or greenhouse conditions. In addition to being very useful for the characterization of safening activity of candidate chemicals, the primary screen provides information that could be used in structure–activity relationship (SAR) studies. Correlations of structures with activity are very important in the agrochemical industry because they provide organic chemists and biologists with useful information about the chemical substituents that are necessary for safening activity or other functions of each chemical molecule. Safeners that show promise in the primary screen are evaluated further under field conditions (secondary screen). The most critical characteristics of any potential safener which are considered at both screens of safener development are listed in Table II.

The importance of an early identification of new markets or opportunities for herbicide–safener combinations was discussed earlier. The degree of crop

Table II Critical Considerations in Screening for Candidate Safeners

Market needs/opportunities for new safeners or herbicide/safener packages
Degree of crop safening (efficacy)
Chemical and botanical specificity of candidate safeners
Ratio of safener-to-herbicide doses
Reliability of safening effect under field conditions
Suitability of active ingredient for proper formulation
Potential phytotoxicity of candidate safeners
Toxicology of candidate safeners
Environmental fate and residual effects of candidate safeners
Interactions of candidate safeners with other biological systems
Registration requirements of EPA and FDA

safening is a very important consideration for selecting candidate safeners for herbicides. The protection afforded by safeners to crop plants against herbicide injury should be either complete or economically acceptable. Our current knowledge on crop safening indicates that it is easier to find crop safeners for herbicides which are marginally selective on the protected crop rather than herbicides which are completely nonselective (Pallos and Casida, 1978). Protection far in excess of fourfold would probably be required when the crop to be safened is extremely sensitive to a given herbicide (Parker, 1982). As mentioned earlier, safening has been successful only with certain crops such as corn, sorghum, and rice and certain classes of soil-applied herbicides such as chloroacetanilides and thiocarbamates.

A successful safener should not only provide a consistent and acceptable degree of protection to a desirable crop. At the same time it should exhibit selectivity by not providing any safening to undesirable weeds. The selectivity of some safeners seems to result from a unique interaction which is inherent to specific crop–herbicide–safener combinations. An example of such a unique interaction is the counteraction of the toxicity of the herbicide EPTC by the safener dichlormid on corn (Pallos *et al.*, 1977; Stephenson and Chang, 1978). Grass weeds such as foxtails (*Setaria* spp.) treated with the prepackaged mixture of EPTC plus dichlormid are not safened (Stephenson and Chang, 1978). In other situations, however, some safeners provide protection to weeds that are botanically related to the crop. In such cases, selectivity is induced by the selective placement of the safener only to the desirable crop. This is accomplished by the dressing or coating of crop seeds with the successful safener. In practice, coating of corn or grain sorghum seeds with the safeners NA, cyometrinil (CGA-43089), oxabetrinil (CGA-92194), CGA-133205, and flurazole makes them sufficiently tolerant to thiocarbamate or chloroacetanilide herbicides (Hatzios, 1983; Parker, 1983).

To be marketable, safeners have also to be effective at low doses. The ratio of safener-to-herbicide doses is a useful parameter measuring the effectiveness of a given safener. Basic economics dictate that this ratio should be kept as low as possible. The ratio of safener-to-herbicide doses for the currently available safeners which are marketed as prepackaged mixtures with herbicides ranges from 1:4 to 1:24 (Hatzios, 1983). Furthermore, a desirable safener should also be reliable under field conditions, easy to formulate, and not cause any adverse effects on plants or other biological systems.

It is evident, then, that the development of proprietary crop safeners for specific herbicides is a lengthy and expensive process achieved through team efforts aimed at characterization, field research and development, patent protection, and registration of these auxiliary chemicals (Schafer, 1985). The major steps of the discovery and industrial development of herbicide safeners are illustrated in Fig. 2.

c. University's Role in Safener Development For the reasons outlined earlier, much of the research on crop safeners for herbicides is conducted by the private sector (agrochemical industry) to secure the patentability and marketing of new compounds. The role of the public sector, however, in the development and use of herbicide safeners is also significant. Similar to the case of herbicides or other crop protection chemicals, university or government scientists provide an

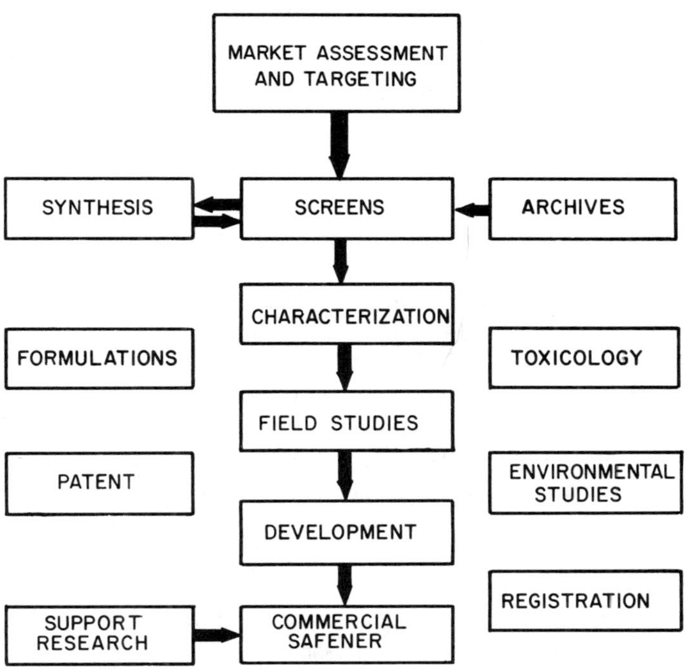

Fig. 2. Flowchart of processes involved in the commercial development of herbicide safeners. (Courtesy of Dr. D.E. Schafer, Monsanto Chemical Company, St. Louis, MO.)

invaluable service by conducting independent and unbiased analyses of the effectiveness of crop safeners for herbicides at the user's level. A healthy cooperation between industry and university or government scientists has been quite successful in advancing crop protection technology in the past and it will undoubtedly continue to be fruitful in the future.

University scientists contribute to the development of herbicide safeners by conducting either applied or basic research. From a practical point of view, university scientists test and evaluate the efficacy of marketed or experimental safeners under a variety of climatic or soil conditions and establish recommendations for the use of these chemicals within each state. Then, the extension network spreads this information throughout the state. At the level of basic research, the contributions of university scientists include studies on the synthesis or discovery of new safeners based primarily on the identification of antagonistic interactions of herbicides with other agrochemicals or the use of structure–activity relationships (Hatzios and Penner, 1985; Stephenson *et al.*, 1978, 1979). Major contributions by university and government scientists have also been made in the area of the mechanisms of action of herbicides and safeners as well as in the metabolic detoxication of herbicides by higher plants. Additional contributions may also result from innovations in the development of rapid and effective screening systems for discovering potential safeners or other chemicals with biological activity or studying the mechanisms of action of these chemicals (Ashton *et al.*, 1977; Gressel, 1984).

B. Chemistry of Herbicide Safeners

1. Synthesis

The currently marketed crop safeners for herbicides can be classified chemically in the following groups: (1) naphthopyranone derivatives (e.g., NA); (2) chloroacetamides (e.g., dichlormid, R-29148, etc.); (3) dichloromethyl-dioxolans (e.g., MG-191); (4) oxime ether derivatives (e.g., cyometrinil, oxabetrinil, and CGA-133205); (5) derivatives of 2,4-disubstituted 5-thiazolecarboxylates (e.g., flurazole); and (6) substituted phenyl-pyrimidines (e.g., fenclorim). The chemical structures of these safeners as well as the structures of some of their active or inactive analogs are given in Figs. 3 to 7. The chemical names of all safeners mentioned in the text are given in the Appendix at the end of this book.

The synthetic methods available for the preparation of the currently marketed safeners have been reviewed (Hatzios, 1983) and they will not be repeated here. Additional information on the chemical synthesis of various safeners has been reported by the following investigators: Riden (1976) for NA; Pallos *et al.* (1977) and Stephenson *et al.* (1978) for dichlormid and other chloroacetamide safeners; Martin (1978) for oxime ether safeners; Howe and Lee (1980) and Sacher *et al.*

(1983) for 2,4-disubstituted thiazoles; Dutka and Komives (1987) and Mueller and Seyferth (1969) for dichloromethyl dioxolans; and Quadranti and Ebner (1984) for fenclorim.

2. Chemical, Physical, and Toxicological Properties

The chemical, physical, and toxicological properties of the commercially developed safeners are presented in Table III. Dichlormid and MG-191 are liquids with good solubility in water. All the other safeners are solids with low water solubility and low vapor pressure. The mammalian toxicity of all marketed safeners is quite low.

3. Structure–Activity Relationships

Changes in the chemical structure of a safening agent could result in marked alterations of either the protective or the phytotoxic activity of this chemical. In addition, modifications in the structure of safeners may be needed for enhancing their uptake by the protected plant or to make the active ingredient compatible with formulation additives. Information on structure–activity relationships (SAR) of herbicide safeners is rather scarce. Most of the published reports refer to SAR studies conducted by university or government scientists. Although SAR studies are an integral part of the industrial discovery and development of herbicide safeners, many of these studies remain unpublished to protect proprietary information. Selected information on SAR studies related to marketed safeners, however, is usually included in the patent description of these products.

Structure–activity correlation studies with NA and chemical analogs have been limited. The original patent for NA (Hoffman, 1971) described eight additional compounds with safening activity on corn against injury from the herbicide EPTC. Most of these compounds are structurally related to NA and include naphthalene-1,8-dicarboxylic acid, its lower alkyl esters and its barium and tin salts; *N,N*-diallyl naphthalene-1,8-dicarboxylic acid; *N,N'*-diallyloxamide; *N,N'*-dipropynyloxamide; *N,N,N',N'*-tetrapropynyloxamide; and dipropynylmalonamide (Hoffman, 1971).

Nine structural analogs of NA were evaluated recently as corn protectants against EPTC injury under greenhouse conditions (Hatzios and Zama, 1986). The nine analogs tested included the following compounds designated here by their abbreviated names: ACQ, NH_2NA, NDiA, NHNA, ClNA, DA, PGA, PHA, and PA. The chemical structures and complete chemical names of these analogs are given in Fig. 3. The presence of the dicarboxylic anhydride group and of at least one aromatic ring attached directly to the anhydride appeared to be essential for the exhibition of corn safening activity by these structural analogs of NA against EPTC injury. Thus, ACQ, NH_2NA, NDiA, NHNA, ClNA, DA, and PHA applied as seed dressings at 0.5 or 1.0% (w/w) of corn

Table III Chemical, Physical, and Toxicological Properties of Commercially Developed Herbicide Safeners

Safener	MW	Physical state	Color	Solubility in water at 20°C (ppm)	Melting point (°C)	Vapor pressure (mm Hg)	Acute oral LD_{50} in rats (mg/kg)
Naphthalic anhydride	198.2	Solid	Light tan	<2	270–274	—	12,300
Dichlormid	208.09	Liquid	Colorless	5000	—	6×10^{-3} (25°C)	2,000
LAB-145138	293.2	Solid	White	30	—	—	>5,000
Cyometrinil	185.2	Solid	White	95	55–56	—	2,277
Oxabetrinil	232.24	Solid	White	20	77–79	5.2×10^{-4} (25°C)	>5,000
CGA-133205	271.2	Solid	White	10	—	—	>3,000
Flurazole	321.72	Solid	White	Low	56–58	3.8×10^{-5} (25°C)	7,700
Fenclorim	225.08	Solid	Colorless	2.5	96.9	9.0×10^{-5} (20°C)	>2,500
MG-191	170.1	Liquid	Colorless	Low	—	—	—
CGA-154281	260.12	Solid	Colorless	20	107.6	4.4×10^{-6} (20°C)	5,000

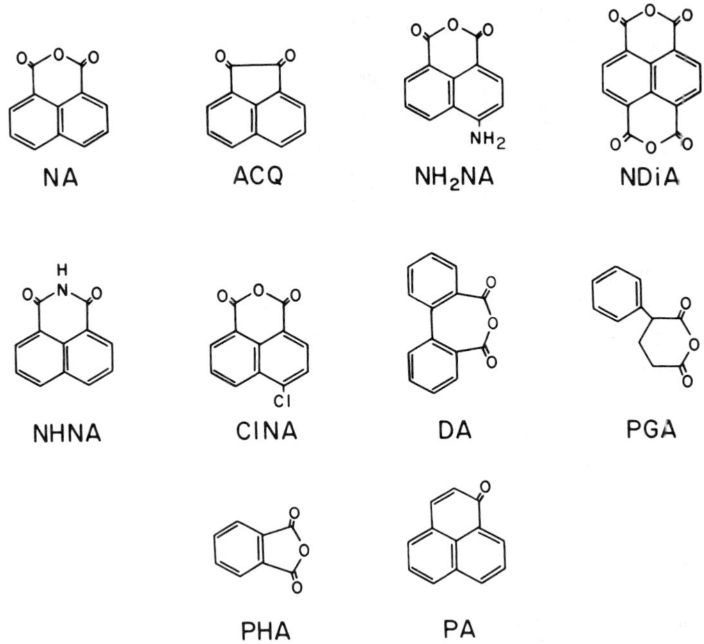

Fig. 3. Chemical structures of naphthopyranone safeners. Abbreviations used: NA = naphthalic anhydride; ACQ = acenaphthenequinone; NH_2NA = 4-aminonaphthalene-1,8-dicarboxylic acid; NDiA = 1,8:4,5-naphthalenetetracarboxylic acid dianhydride; NHNA = naphthalene-1,8-carboximide; ClNA = 4-chloronaphthalene-1,8-dicarboxylic acid; DA = diphenic anhydride; PGA = 2-phenylglutaric anhydride; PHA = phthalic anhydride; and PA = phenale-1-one.

seeds offered satisfactory protection to corn against EPTC doses higher than 6.7 kg/ha (Hatzios and Zama, 1986). The analogs PGA and PA were very toxic and inhibited germination of corn seeds. Chlorination of NA appeared to increase the phytotoxicity of NA since ClNA caused chlorosis and stunting in corn seedlings grown from seeds treated with this analog (Hatzios and Zama, 1986). Narasimhachari and Vining (1963) have described a number of perinaphthenone derivatives produced by bacteria of the genus *Penicillium*. Among them, atrovenetin and atrovenetinone are minor pigments accounting for the color of the mycelium of *Penicillium* cultures. These natural pigments are known to undergo oxidation to their anhydrides and are structurally very similar to NA (Narashimhachari and Vining, 1963). The safening activity of these natural products has not been examined.

Structure–activity relationship studies with several chloroacetamides evaluated as corn safeners against thiocarbamate herbicide injury have been conducted by a number of investigators (Dutka and Komives, 1983; Dutka *et al.*, 1979; Görög *et al.*, 1982; Nagy and Balogh, 1985; Pallos *et al.*, 1975, 1977; Stephenson *et al.*, 1978, 1979; Wuertzer *et al.*, 1983). From such studies it has been concluded that the most active safeners of corn against EPTC are the N,N-disubstituted 2,2-dichloroacetamides. Mono- and trichloroacetamides were less effective as safeners. Variations in the nitrogen substituents resulted also in differential safening activity. In general, dichloroacetamides with two substituents on the nitrogen atom are more effective than those with one substituent (Dutka and Komives, 1983; Pallos *et al.*, 1975 & 1977; Stephenson *et al.*, 1978). Dichloro-substituted *N*-acetyl-1,3-oxazolidines have also been established as effective safeners of corn against EPTC injury (Dutka and Komives, 1983; Görög *et al.*, 1982; Pallos *et al.*, 1975, 1977; Wuertzer *et al.*, 1983). CGA-154281, a substituted dichloroacetyl-1,4-benzoxazine, was recently introduced as a new safener for corn against the herbicide metalachlor (Peek *et al.*, 1988). The chemical structures of selected dichloroacetamides and *N*-acetyl-1, 3-oxazolidines are illustrated in Fig. 4. The chemical names of these safeners are given in the Appendix of this volume.

From all these studies it appeared that the dichloroacetamido moiety (Cl_2CH—CO—N=) is the essential part of all molecules which are effective as corn safeners against thiocarbamate herbicide injury (Dutka *et al.*, 1979). Based on this conclusion, the involvement of a transacylation reaction in the biological activity of these safeners was proposed (Dutka and Komives, 1983). However, recent studies by Dutka and Komives (1987) suggested that the dichloroaceto group (Cl_2CH—CO—) rather than the dichloroacetamido group may be primarily responsible for safening activity. This hypothesis proved to be correct with the discovery and recent introduction of the safener MG-191, which protects corn from thiocarbamate herbicide injury (Dutka and Komives, 1987).

SAR studies with oxime ether safeners have been limited. Chang and Merkle (1982, 1983) showed that the safening activity of these chemicals is affected by the number of nucleophilic sites that are present in their molecules. An oxime with two nucleophilic sites in its molecule was a more effective safener than an oxime possessing only one nucleophilic site. The chemical structures of the commercially developed oxime ether safeners cyometrinil, oxabetrinil, and CGA-133205 as well as the general structure of pyridinealdoxime O-ethers introduced by Chang and Merkle (1983) are shown in Fig. 5. The chemical names of these safeners are given in the Appendix.

SAR studies on the safening activity of selected 2,4-disubstituted 5-thiazolecarboxylates have been reported by Howe and Lee (1980). Out of more than than 60 derivatives tested, the thiazole alkyl esters of the thiazolecarboxylic acids that had a chlorine atom at the 2-position and a trifluoromethyl group at the 4-position of the thiazole ring were found to possess the best safening activity for protecting grain sorghum against chloroacetanilide herbicide injury (Howe and

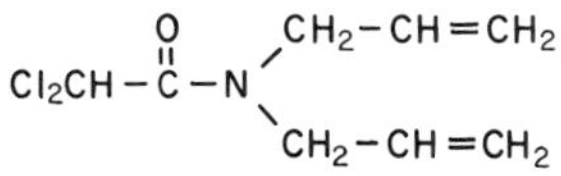

Dichlormid; R-25788; DDCA
(Stauffer, 1971)

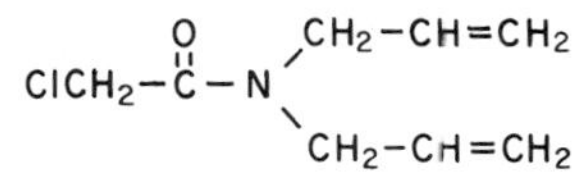

CDAA; Allidochlor (RANDOX®)
(Monsanto, 1960)

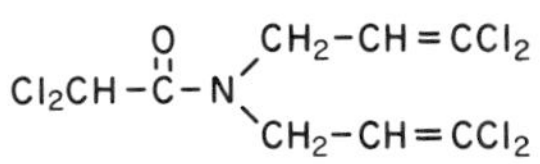

H-31866
(Montedison, 1980)

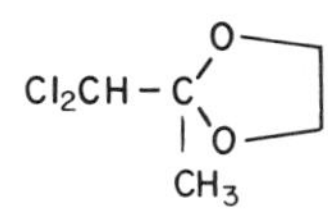

MG-191
(Nitrokemia, 1986)

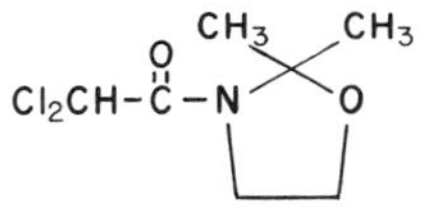

R-28725
(Stauffer, 1973)

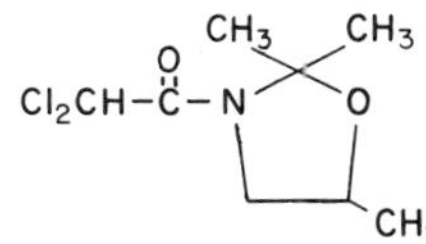

R-29148
(Stauffer, 1973)

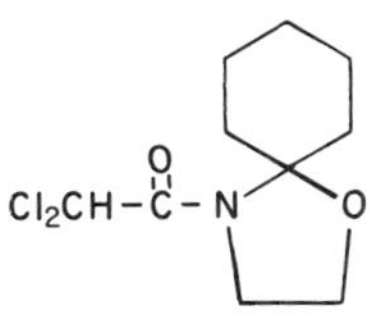

AD-67
(Nitrokemia, 1979)

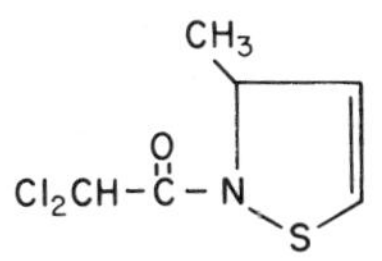

M-32988
(Gulf, 1980)

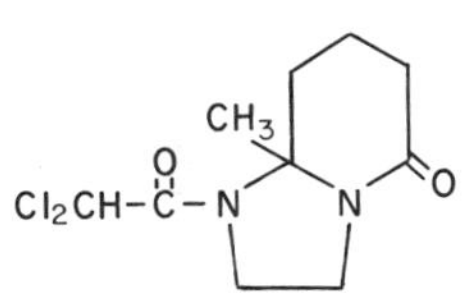

LAB-147886
(BASF, 1984)

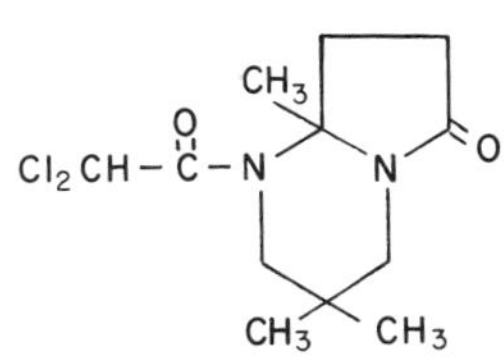

LAB-145138
(BASF, 1984)

Fig. 4. Chemical structures of dichloroacetamide safeners. (The chemical names of coded compounds are given in the Appendix.)

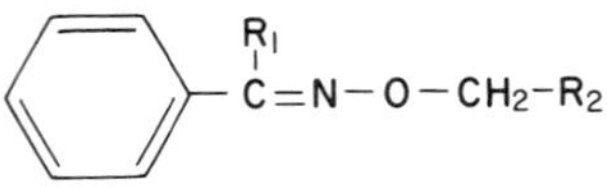

CGA-43089; cyomeтrinil (CONCEP I®) $R_1 = C{\equiv}N$
(CIBA-GEIGY, 1978) $R_2 = C{\equiv}N$

CGA-92194 (CONCEP II®) $R_1 = C{\equiv}N$
(CIBA-GEIGY, 1982) $R_2 =$

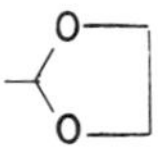

CGA-133205 (CONCEP III®) $R_1 = CF_3$
(CIBA-GEIGY, 1986) $R_2 =$

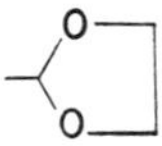

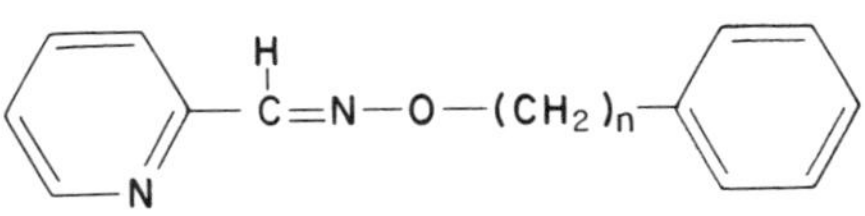

Pyridine-2-aldoxime *O*-ethers
(Chang and Merkle, 1983)
n = 1: benzyl ether n = 2: phenylethyl ether

Fig. 5. Chemical structures of oxime ether safeners. (The chemical names of coded compounds are given in the Appendix.)

Lee, 1980; Sacher *et al.*, 1983). The most effective safener from this class, chosen for commercial development, is flurazole (Fig. 6).

The chemical structure of the safener fenclorim is shown in Fig. 7. This safener has been introduced recently by Ciba-Geigy as an effective protectant of rice against injury from the chloroacetanilide herbicide pretilachlor (Quadranti and Ebner, 1984). Information on SAR studies with this class of safeners is not available.

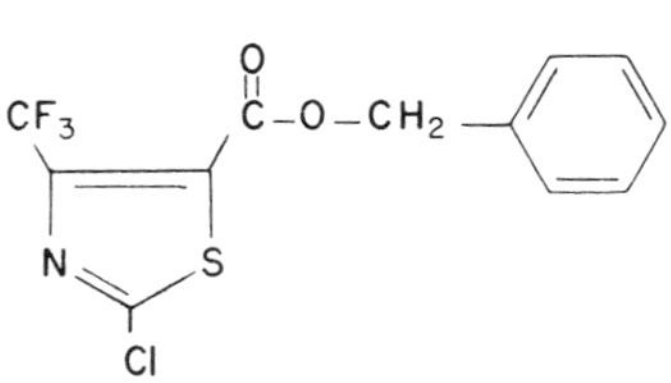

Flurazole; MON-4606
(SCREEN®, Monsanto 1980)

Fig. 6. Chemical structure of the 2,4-disubstituted 5-thiazolecarboxylate safener flurazole.

C. Future Approaches in Safener Development

1. Empirical and Imitative Screening for Future Safeners

Random screening techniques based on the empirical and imitative methods for selecting candidate chemicals will continue to play an important role in the discovery and development of crop safeners for herbicides in the future.

The useful insight provided by the observation and understanding of antagonistic interactions of herbicides with other agrochemicals in plants should be exploited to advantage for the development of new crop safeners for herbicides. The interactions of herbicides with other pesticides in higher plants have been reviewed (Hatzios and Penner, 1985; Putnam and Penner, 1974). Exploitation of all known antagonistic combinations involving herbicides and other agrochemicals that result in less crop injury could provide useful leads for the identification of new chemistries for future crop safeners. In addition, SAR studies of chemicals involved in antagonistic interactions could be used as predictive tools for the synthesis of targeted safeners for specific herbicides.

If the functional groups required for herbicidal activity are clearly defined, structurally similar molecules lacking these functional groups may have a chance of being effective safeners for such herbicides. Stephenson *et al.* (1979) verified this hypothesis by preparing, through targeted synthesis, effective safeners tailored for counteracting the toxicity of several thiocarbamate herbicides to corn. Dichloroacetamide derivatives most similar to selected thiocarbamate herbicides with respect to the N,N-disubstituted alkyl groups were found to be the most active safeners for these herbicides on corn. Thus *N,N*-dipropyl-2, 2-dichloroacetamide was most effective against EPTC and vernolate, whereas *N,N*-diisobutyl-2,2-dichloroacetamide was most active against butylate. This approach was particularly successful in synthesizing an effective safener for cycloate (Stephenson *et al.*, 1979).

Additional evidence supporting the concept that nonphytotoxic chemical analogs of herbicides may be effective safeners for these herbicides comes from recent observations of the safening activity of the fungicide metalaxyl against metolachlor injury to grain sorghum (J. Peek, Ciba-Geigy Corp., personal communication). Metalaxyl is a nonphytotoxic structural analog of the

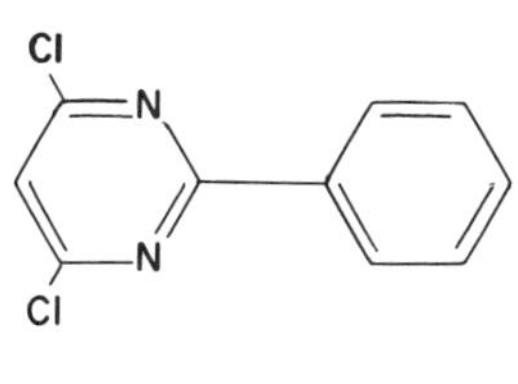

CGA-123407; Fenclorim
(CIBA-GEIGY, 1983)

Fig. 7. Chemical structure of the rice safener fenclorim.

herbicide metolachlor (Hubele *et al.*, 1983). The antagonism of diclofop-methyl effects on oats by the structurally similar 2,4-D and a number of active and inactive analogs of 2,4-D is also supportive of this hypothesis (Taylor and Loader, 1984). Success, however, has been rather limited in attempts to antagonize the effects of paraquat with inactive structural analogs of this herbicide at the chloroplast level (Lewinsohn and Gressel, 1984; Upham and Hatzios, 1987).

More recently, Stephenson and Ezra (1985) evaluated the potential crop safening activity of early pretreatments with subtoxic levels of a particular herbicide against later higher rates of the same or chemically related herbicides. The concept proved particularly successful with the herbicide CDAA. Pretreatments of corn seedlings with sublethal doses of CDAA increased their tolerance to subsequent toxic doses of this herbicide (Ezra *et al.*, 1985). Promising results were obtained also in experiments with the herbicide metribuzin and its effects on tomatoes as well as the herbicide pyrazon and its effects on sugar beets (Stephenson and Ezra, 1985). Enhanced detoxication rates of these herbicides resulting from the induction of selected degradative enzymes were suggested as causes of the increased tolerance of these crops to the aforementioned herbicides (Ezra *et al.*, 1985). Similar studies evaluating the potential safening activity of subtoxic doses of metribuzin on soybean and corn, atrazine on soybean and corn, alachlor and EPTC on corn, and chlorsulfuron on corn and oats did not produce any promising results (Stephenson and Ezra, 1985).

2. Selection of Safeners by the Rational Screening Approach

This method calls for the examination of the potential safening activity of all chemicals known to act as regulators (inhibitors or inducers) of any enzymatic or other physiological systems involved in the expression of herbicidal activity or the detoxication of these toxicants by plants. Although our knowledge and understanding of the exact mechanisms of herbicide action and herbicide detoxication by plants are far from complete, recent advances in these fields offer great opportunities for the use of the rational approach in the development of crop safeners for specific herbicides.

Biosynthesis of amino acids in plants has long been recognized as a desirable target site for herbicides because of the low mammalian toxicity of such inhibitors (LaRossa and Falco, 1984). Following several years of development, a number of potent inhibitors of amino acid biosynthesis have been introduced recently as effective herbicides. These herbicides include: glyphosate, which inhibits the biosynthesis of the aromatic amino acids phenylalanine, tryptophan, and tyrosine by interfering with the activity of the enzyme EPSP synthase (EC 2.5.1.19) (Amhrein *et al.* 1983); glufosinate, which inhibits the biosynthesis of glutamine by interfering with the activity of the enzyme glutamine synthetase (EC 6.3.1.2) (Lea *et al.*, 1984); and sulfonylurea and imidazolinone derivatives, which inhibit the biosynthesis of branched amino acids by interfering with the activity of the enzyme acetolactate synthase (EC 4.1.3.18, also known as acetohydroacid synthase) (Chaleff and Mauvais, 1984; Muhitch *et al.*, 1987). Reversal of the inhibitory action of these herbicides by exogenous applications of the depleted respective amino acids has been reported (Anderson and Hibberd, 1985; Gresshoff, 1979; Jaworski, 1972; Ray, 1984). In addition, the inhibitory action of glyphosate on plants was also reversed with exogenous applications of the amino acids aspartate and glutamate and several organic acids of the tricarboxylic acid cycle such as α-ketoglutarate, succinate, or malate (Killmer *et al.*, 1981). These examples demonstrate that utilization of the rational approach in safener development for such herbicides is possible and will attract more attention in the future. Additional approaches for increasing the selectivity of sulfonylurea and imidazolinone herbicides are discussed by Barrett in Chapter 9, this volume.

Treatment of plants with the herbicide paraquat enhances the production of toxic oxygen species to levels far exceeding those that can be protected by the natural defense systems of plants. Superoxide radicals (O_2^-), hydrogen peroxide (H_2O_2), and hydroxyl radicals (OH·) are all known to be formed in plants treated with paraquat (Halliwell, 1984). Plants are normally protected by the harmful effects of these toxic oxygen species through the action of several enzymes and antioxidants, including superoxide dismutase (SOD, EC 1.15.1.1), ascorbate peroxidase (EC 1.11.1.7), catalase (EC 1.11.1.6), and α-tocopherol (Badger, 1985; Hossain *et al.*, 1984). The tolerance of perennial ryegrass and of other plants to paraquat has been attributed to increased levels and/or activity of the endogenous protectants of toxic oxygen species such as the enzymes SOD, catalase, and peroxidase (Gressel, 1986; Harvey and Harper, 1982). These observations have stimulated considerable interest in the chemical regulation of plant tolerance to paraquat and several approaches have been attempted.

D-Penicillamine, a copper chelate with superoxide dismutating activity, offered partial protection to flax (*Linum ussitatisimum* L.) against paraquat (Youngman *et al.*, 1979). Preconditioning of plants to sublethal doses of stresses which increase oxygen toxicity, such as ferrous sulfate (Hoffman, 1978) and sulfite (Rabinowitch and Fridovitch, 1985), causes an increase of the levels of the endogenous mechanisms protecting plants toward toxic oxygen species and partially protected plants from paraquat injury. Ferrous sulfate, applied as a spray with paraquat, could protect wheat and oats from paraquat injury but the safener to herbicide ratio (100:1) for this combination was extremely uneconomical (Hoffman, 1978).

Current work on the primary electron acceptor of photosystem I of plant chloroplasts, however, appears promising and may provide an alternative site for the chemical regulation of plant tolerance to paraquat. Upham *et al.* (1987) reported recently on the existence of an unidentified inhibitor from hemolyzed rabbit sera which is capable of inhibiting paraquat- or ferredoxin-mediated O_2 evolution without inhibiting $NADP^+$-dependent O_2 evolution. These results coupled with the observations of others (see review by Rutherford and Heathcotte, 1985) support the notion that the pathway of O_2 reduction by photosystem I is branched. This offers a potential target site for regulating paraquat toxicity since electron flow to paraquat could be inhibited without interfering with the reduction of $NADP^+$. Preliminary studies with hemoglobin showed that this approach may be viable. Hemoglobin provided protection to pea (*Pisum sativum* L.) chloroplasts by inhibiting only the paraquat-mediated O_2 reduction and not the reduction of O_2 mediated by ferredoxin or $NADP^+$ (Upham and Hatzios, 1987). Additional research is needed, however, to test this approach at the whole plant level under field conditions.

The potential use of the rational approach in the development of crop safeners for herbicides was demonstrated also by Hilton and Pillai (1986, 1987), who introduced the chemicals L-2-oxathiazolidine-4-carboxylic acid (OTC) and L-thiazolidine-4-carboxylic acid (TC) as moderately effective protectants of corn and grain sorghum against injury from the herbicide tridiphane. Both of these chemicals are metabolized by plants to cysteine, thus providing increased levels of a key amino acid needed for the biosynthesis of the tripeptide glutathione (GSH). GSH plays a very important role in the detoxication of tridiphane and of many other herbicides in plants (Hatzios and Penner, 1982).

Established antagonistic interactions of drugs with endogenous substrates or protective molecules which are well known in the fields of pharmacology and medicine should also be considered as possible leads for the employment of the rational approach in safener research. A well-documented example is given here to illustrate this point. The antagonism of the bacteriocide sulfanilamide by *p*-aminobenzoic acid, a precursor of the vitamin folic acid, has been documented in the medical literature for more than 40 years (Woods, 1940). It was only recently, however, that Killmer *et al.* (1980) and Veerasekaran *et al.* (1981a,b) reported on the reversal of the inhibitory action of the herbicide asulam on wheat, wild oats, and carrots by exogenous applications of folic acid and *p*-aminobenzoic acid. Asulam is a derivative of sulfanilamide which acts as an inhibitor of the biosynthesis of the vitamin folic acid (Veerasekaran *et al.*, 1981a).

Conversely, information generated from studies on the mechanisms of action of herbicide safeners and their effects on plant biochemistry could be used for the rational design of safeners protecting other biological systems from herbicide or other xenobiotic injury. A notable example illustrating this point is the development of a safener protecting the fish Japanese carp (*Cyprinus carpio* var. Yamato Koi) from the thiocarbamate herbicide molinate. Based on the well-established stimulatory effects of dichloroacetamide safeners on glutathione and glutathione *S*-transferases (Lay and Casida, 1978), the compound *N*-ethyl-*N*-benzyldichloroacetamide was introduced as an effective safener reducing significantly the mortality of Japanese carp caused by molinate (Lay and Menn, 1987).

3. Prosafeners

The term "prosafeners" is used to describe safeners which must undergo biotransformation prior to exhibiting their safening effect. The concept is rather new. It was introduced recently by Rubin *et al.* (1985), who showed that *N*-phenylmaleimides and *N*-isomaleimides behaved as prosafeners being hydrolyzed rapidly to *N*-phenylmaleamic acids, which are the actual safeners of sorghum against injury from the herbicide alachlor.

Theoretically, the aforementioned safeners OTC and TC, which are converted in plants to the amino acid cysteine, could also be classified as prosafeners since they have to be metabolized to exhibit their safening action (Hilton and Pillai, 1986, 1987).

The concept of prosafeners is quite interesting because it provides an alternative way for increasing the selectivity of chemical safeners. Differences in the ability of crop and weed plants in biotransforming prosafeners to actual safeners could be exploited to advantage and enhance the practical usefulness of this concept. The potential of this approach is supported by the successful use of "proherbicides" such as 2,4-DB and isomethiozin. 2,4-DB is oxidized to 2,4-D, whereas isomethiozin is hydrolyzed to metribuzin in susceptible plants, thus facilitating the selective use of these herbicides (Fedtke, 1982; Wain and Smith, 1976). Prosafeners are discussed further in Chapter 14, this volume, by Rubin and Kirino.

4. Microbial Safeners

Recent advances in recombinant DNA technology offer great promise for the use of genetically engineered microbes as crop safeners for herbicides in the future. Genes coding for degradative enzymes which break down certain herbicides could be incorporated into appropriate plasmids and transferred to bacteria (e.g., *Pseudomonas fluorescens*) which colonize the roots or seeds of susceptible crop plants. Such microbial safeners could protect susceptible plants from selected soil-applied and root-absorbed herbicides.

Although the practical utilization of this concept seems remote at present, recent advances in the utilization of a microbial insecticide demonstrate the feasibility of this approach (Kaufman, 1986). This model system utilized the cloning of the gene coding for the active endotoxin of *Baccillus thurigiensis kurstani* into isolates of *Pseudomonas fluorescens* thus constructing a microbial insecticide which controls effectively root-feeding insects of corn (Kaufman, 1986). In addition, earlier research conducted in Poland showed that injury caused by the soil-applied herbicide EPTC to sugar beets could be reduced significantly with the use of selected bacterial fertilizers such as azotobacterin (Geller and Nikolaenko, 1972; Wegrzyn, 1975).

Theoretical aspects of the molecular biology and cloning of microbial genes into plasmids and the future potential of microbial crop safeners for herbicides are discussed by Karns in Chapter 15, this volume.

IV. APPLICATION OF HERBICIDE SAFENERS IN THE FIELD

The major agricultural uses, formulations, and methods of application of herbicide safeners which are currently marketed or undergoing development for commercialization in the United States or other parts of the world are summarized in Table IV.

Under practical conditions, the application of herbicide safeners in the field does not involve any extra operation, because the respective companies that manufacture herbicide safeners market them as either prepackaged tank mixtures with the herbicide or treated crop seeds (seed safeners). A basic reason for the application of safeners as a prepackaged tank mixture with the herbicide or as seed treatments is the selectivity of the respective safener–herbicide combinations on crops and weeds. However, other reasons such as stability of the safener during storage, suitability of active ingredient for formulation, etc., are also important.

Statistics on the sale volume of the currently marketed safeners are not readily available. The volume of oxabetrinil sold worldwide for the protection of grain sorghum against metolachlor exceeded 20 metric tons in 1985 (J. Peek, Ciba-Geigy, personal communication). Statistics for other safeners could not be obtained. From a commercial point of view, however, the larger volume of herbicide sales that results from the use of auxiliary chemicals such as safeners is more important than the volume of safeners sold by themselves. The current status of chemical safening of crops against herbicides and the characteristics of safeners applied as prepackaged tank mixtures with herbicides or as seed dressings are outlined in this section.

A. Current Uses of Crop Safeners for Herbicides

1. Safeners for Corn

The commercial use of NA for the protection of corn against injury from thiocarbamate herbicides was very brief. As pointed out earlier, the subsequent introduction and commercialization of the safener dichlormid diminished the marketing competitiveness of NA. Today, NA is not marketed as a corn safener for thiocarbamate herbicides. However, because of its versatility, NA has provided scientists with a valuable tool for identifying promising crop–herbicide–safener combinations. Extensive lists of crop–herbicide combinations that are safened by NA have been published by several investigators (Hatzios, 1983; Parker, 1983; Stephenson and Ezra, 1983) and they will not be repeated here. Apart from its effectiveness in safening corn from thiocarbamate herbicide injury, NA could protect corn from a number of chemically unrelated herbicides. Such herbicides include phenylcarbamates, dithiocarbamates, chloroacetanilides, sulfonylureas, imidazolinones, cyclohexenones, and arylophenoxyalkanoic acids (Hatzios,

Table IV Agricultural Uses, Formulations, and Application Methods of Commercial Herbicide Safeners

Safener	Crop protected	Herbicide counteracted	Formulation	Trademark	Application method	Recommended rate	Storage stability of treated seed	Manufacturer
Naphthalic anhydride	Corn	EPTC, butylate, vernolate	95% seed protectant powder	Protect	Seed dressing	0.5% by weight	More than 1 year	Gulf Oil
Dichlormid	Corn	EPTC, butylate, vernolate	Tank mixture, with herbicide	Eradicane, Sutan, Vernam	Preplant incorporation	0.56 kg/ha 0.25 kg/ha 0.56 kg/ha	—	Stauffer
	Wheat	Triallate	20% seed protectant powder	—	Seed dressing	0.5% by seed weight	More than 1 year	Stauffer
LAB-145138	Corn	Metazachlor	Tank mixture with herbicide	—	Preemergence	0.5 kg/ha	—	BASF
MG-191	Corn	EPTC	Tank mixture with herbicide	—	Preemergence	0.56 kg/ha	—	Nitrokemia
CGA-154281	Corn	Metolachlor	Tank mixture with herbicide	—	Preemergence preplant incorporated	0.067 kg/ha	3 years	Ciba-Geigy
Cyometrinil	Grain sorghum	Metolachlor	2.09 S	Concep I	Seed dressing	0.125% by seed weight	More than 1 year	Ciba-Geigy
Oxabetrinil	Grain sorghum	Metolachlor	70 WP or 50 SD	Concep II	Seed dressing	0.05–0.3% by seed weight	More than 1 year	Ciba-Geigy
CGA-133205	Grain sorghum	Metolachlor	70 WP	Concep III	Seed dressing	0.05–0.3% by seed weight	More than 1 year	Ciba-Geigy
Flurazole	Grain sorghum	Alachlor	80 WP	Screen	Seed dressing	0.06–0.25% by seed weight	More than 1 year	Monsanto
			1% granular		In-furrow application	0.07–0.25 kg/ha	—	Monsanto
Fenclorim	Rice	Pretilachlor	Tank mixture with herbicide	Sofit	Preemergence	0.56 kg/ha	—	Ciba-Geigy

1983; Parker, 1983). In addition, NA is the only safener capable of providing protection to grass crops such as corn and oats against postemergence applications of selected herbicides (e.g., chlorsulfuron, diclofop-methyl, etc.) (Parker, 1983).

In contrast to the diversity of NA, dichlormid exhibits a remarkable degree of chemical and botanical specificity by being highly effective only as a safener of corn against injury caused primarily by the thiocarbamate herbicides (Pallos *et al.*, 1977; Stephenson and Chang, 1978). Dichlormid affords only marginal protection to other grass crops and weeds against thiocarbamate herbicide injury and because of this selectivity it is used primarily as a prepackaged tank mixture with the respective thiocarbamate herbicides. These formulations are marketed under the trade names Eradicane (EPTC + dichlormid), Sutan^{+} (butylate + dichlormid), and Vernam^{+} (vernolate + dichlormid). Apart from the thiocarbamates, dichlormid safens corn against chloroacetanilide herbicides (Leavitt and Penner, 1978). It also provides some degree of protection to corn against the herbicide sethoxydim (Hatzios, 1984a) and to several new herbicides with bleaching activity such as clomazone (FMC-57020) and SC-0774 (Devlin and Koszanski, 1987a,b). These results illustrate that dichlormid may be chemically more diverse than previously believed.

In addition to dichlormid, a number of substituted *N*-dichloroacetyl-1,3-oxazolidines, including the experimental chemicals R-29148, R-28725, AD-67, and MG-191, have shown promising potential as safeners of corn against thiocarbamate herbicides (Dutka and Komives, 1987; Görög *et al.*, 1982; Hatzios, 1983).

More recently, the experimental chemical LAB-145138 has shown excellent activity in safening corn against injury caused from the chloroacetanilide herbicide metazachlor and it is under commercialization by the BASF company (Wuertzer *et al.*, 1983). LAB-145138 is a selective safener and it will be formlated as a prepackaged tank mixture with metazachlor in a 1:6 safener-to-herbicide ratio. CGA-154281, a corn safener for the chloroacetanilide herbicide metolachlor, is also under development by Ciba-Geigy Corporation. This safener is formulated as a tank mixture with metolachlor in a 1:30 safener-to-herbicide ratio (Peek *et al.*, 1988).

2. Safeners for Sorghum

Apart from its effectiveness in protecting corn, NA is also capable of safening grain sorghum against injury caused from several herbicides (Hatzios, 1983; Parker, 1983). None of these uses of NA, however, has been exploited commercially.

The commercialization of the oxime ether safeners (Concep I, II, and III) by Ciba-Geigy allowed the safe use of the grass herbicide metolachlor and other chloroacetanilide herbicides in grain sorghum. A number of other oxime ether derivatives with good safening activity against chloroacetanilide herbicide injury to sorghum have been reported by Chang and Merkle (1982, 1983).

The oxime ether safeners of the Concep type exhibit good chemical and botanical specificity by being highly selective to grain sorghum against chloroacetanilide herbicides. They are applied, however, as seed dressings because they provide good protection to several weeds of the genus *Sorghum* (e.g., johnsongrass, sudangrass, etc.) (Ellis *et al.* 1980; Muller and Nyffeler, 1981). The performance of cyometrinil (Concep I) as a sorghum safener for metolachlor was inconsistent under field conditions. Adverse effects on seed germination and stand establishment were frequently caused by this safener (Turner *et al.*, 1982). As a result, oxabetrinil, a chemical analog of cyometrinil, was developed and is currently marketed by Ciba-Geigy as Concep II. Oxabetrinil is not phytotoxic to grain sorghum or sweet sorghum hybrids. However, undesirable interactions of this safener with the downy mildew of sorghum have been observed (J. Peek, Ciba-Geigy, personal communication). To eliminate these undesirable interactions, Ciba-Geigy introduced CGA-133205, a chemical analog of oxabetrinil, which is expected to replace Concep II in the sorghum market (Dill, 1986; Helseth and Dill, 1986).

Flurazole, a sorghum safener developed by Monsanto under the trade name Screen, has allowed the safe use of the chloroacetanilide herbicides alachlor and acetochlor in grain sorghum (Schafer *et al.*, 1981). Flurazole is used primarily as a seed safener. It could also be effective when applied in-furrow, but because it affords protection to weed species of the genus *Sorghum,* flurazole is marketed as treated sorghum seeds (Schafer *et al.*, 1981). Flurazole, applied as a broadcast treatment, was not very effective in protecting corn against thiocarbamate and chloroacetanilide herbicides (Hatzios, 1986). R-30228, a safener whose chemistry has not been released, is currently under development as a protectant of sorghum against thiocarbamate herbicide injury (Ketchersid and Merkle, 1985).

3. Safeners for Rice

Fenclorim is the first commercially developed safener which protects rice against injury from the chloroacetanilide herbicide pretilachlor (Pyon, 1986; Quadranti and Ebner, 1983). It is formulated as a prepackaged tank mixture with pretilachlor and marketed primarily in Japan under the trade name Sofit.

NA has been reported to ameliorate the adverse effects of the thiocarbamate herbicide molinate on rice (Smith, 1971), but this successful combination was never exploited commercially.

The herbicide methoxyphenone is used commercially in Japan to prevent the formation of a dechlorinated derivative of the thiocarbamate herbicide benthiocarb, which injures paddy rice (Wakabayashi and Matsunaka, 1982). Additional approaches for the safening of rice against herbicide injury are discussed by Matsunaka and Wakabayashi in Chapter 2, this volume.

4. Safeners for Other Crops

The product label of the safener dichlormid also lists wheat as a crop which could be safened by this chemical against injury from the herbicide diallate. However, in this case, dichlormid is applied as a seed dressing rather than as a broadcast treatment with the herbicide (Mullison, 1979). The practical importance of this use of dichlormid is obscure at the present time, since a number of new herbicides with greater activity and selectivity than diallate have been introduced recently for use in wheat and other winter cereals (Sanders, 1981).

NA has been reported to provide satisfactory protection to several grass crops such as oats, wheat, barley, and ryegrass against injury caused from several herbicides (Hatzios, 1983; Parker, 1983). Again, however, none of these successful crop–herbicide–safener combinations has been exploited commercially. Recently, a safener of undisclosed chemistry has been developed by Hoechst-Roussel Company as a protectant of wheat against injury from the herbicide fenoxaprop-ethyl. The safener is formulated as a prepackaged mixture with the herbicide and is marketed under the trade name Puma in Europe and other wheat-producing countries.

The promising potential of octamethylene diamine for protecting wheat against injury from a herbicidal urea derivative has been reported by Japanese researchers (Okii *et al.*, 1979). Research is also under way to evaluate the efficacy of the insecticide phorate as a seed safener of wheat against injury from the herbicide clomazone (Dr. H. P. Wilson, Eastern Shore Agricultural Experiment Station, Painter, VA).

Development or commercialization of chemical safeners protecting broadleaf crops against herbicide injury has not been successful. Chemical adsorbents such as activated charcoal and lignin by-products could be used for such crops and this approach will be discussed later in this volume (Hoagland, Chapter 11; Riggle and Penner, Chapter 12).

B. Prepackaged Mixtures of Safeners with Herbicides

To be effective, a safener formulated as a prepackaged mixture with a given herbicide should have several common features with this herbicide. The most important of them are listed in Table V.

The comparative solubility and biodegradability of a safener and a herbicide packaged together in a mixture are of particular importance (Prochnow, 1978). If the safener is more soluble than the herbicide then leaching of the safener from

Table V Characteristics of Prepackaged Herbicide/Safener Tank Mixtures

Safener should remain in the protection zone for as long as the crop is susceptible to the herbicide
Safener should not be more water-soluble than the herbicide
Safener should not be more biodegradable than the herbicide
Safener should not interfere with the effectiveness of the herbicide on target weeds
Safener should be compatible with other pesticides that may be added to the spray tank or applied to the soil when the herbicide/safener mixture is present

the zone of protection may result after heavy rain or irrigation. Similarly, if the safener is more amenable to biodegradation than the herbicide then its effectiveness could be diminished since it may fail to provide adequate protection to the desirable crop. The ratio of herbicide-to-safener doses is also very important for soil-applied prepackaged mixtures of herbicides and safeners. This ratio reflects the effectiveness of a specific safener in counteracting the effects of herbicides on crops and it is a deciding factor for the commercialization of a particular safener.

Dichlormid was the first safener to be formulated as a prepackaged mixture with thiocarbamate herbicides for soil application in corn production. The safener-to-herbicide ratio in these formulations ranges from 1:12 in Eradicane and Vernam^{+} to 1:24 in Sutan^{+} (Mullison, 1979). Other safeners which are formulated as prepackaged mixtures with respective herbicides include the fenclorim–pretilachlor combination for rice and the LAB-145138–metazachlor combination for corn. CGA-154281, a corn safener for metolachlor being developed currently by Ciba-Geigy, is also formulated as a prepackaged mixture with this herbicide.

Prepackaged herbicide–safener mixtures appear to have a distinct advantage over seed safeners because this pattern of application combines technical and marketing concepts into a single package. From an industrial point of view, this approach is very desirable because a single company controls all components of the formulation. From the user's point of view, this approach is also attractive because the farmer purchases and uses a single and reliable product which allows a wider selection of crop cultivars.

On the negative side, problems that may be encountered with prepackaged herbicide/safener formulations include the potential shifting of weed populations toward species tolerant to this formulation following repeated applications of the mixture and the reduced selectivity of these formulations under certain environmental conditions (Stephenson and Ezra, 1983).

C. Seed Safeners

The seed represents a critical stage in the life cycle of all plants. Unsown seeds of crops are separated from weeds and their dressing with safeners provides the ultimate in safener selectivity (Hoffman, 1962, 1978). Seed dressing with a chemical is always a challenge for seed technologists. Some operational mechanics of the dressing of seeds with safeners have been reported by Hoffman (1978). Traditionally, the dressing of crop seeds with safeners could be accomplished through two methods (Muller and Nyffeler, 1981): (1) as a sequential treatment after the seed has received the usual fungicide/insecticide protection, or (2) as a "one-shot treatment" together with the needed fungicide/insecticide dressing. It is obvious that, in each case, a seed safener should be compatible with the fungicides and insecticides which are also applied as seed dressings. Additional factors that should be considered during the coating of crop seeds with any chemical have been discussed by Leaver and Roberts (1984). The most important of them are outlined in Table VI.

Seed safeners should provide protection for as long as the herbicide is active and their stability in seed storage should be evaluated. Tests with most seed safeners currently in use have shown that storing for more than one year did not result in any loss of their safening activity (Muller and Nyffeler, 1981; Schafer *et al.*, 1981). It is recommended, however, that in any new growing season, high quality crop seeds should be dressed with safeners some time prior to planting. Dressing with oxime ether safeners works best when performed 4 weeks prior to planting (Muller and Nyffeler, 1981). When seed dressing is considered as the preferable method for applying a safener, the potential short- and long-term effects of seed safeners on seed quality (e.g., germination) and stand establishment should be determined to establish the margin of its crop safety. As mentioned earlier, the narrow margin of crop safety provided by the sorghum safener cyometrinil was responsible for its replacement by oxabetrinil, a less phytotoxic analog. The potential interactions of seed safeners with other biological systems such as phytopathogenic agents causing crop seed or seedling diseases should also be evaluated. The undesirable interaction of the seed safener oxabetrinil with the downy mildew of sorghum prompted the replacement of this safener with another oxime ether analog coded CGA-133205 (J. Peek, Ciba-Geigy, personal communication).

During the early stages of the development of the safener concept, the necessity of seed treatment for the application of safeners was viewed as a considerable disadvantage which could limit the potential of these chemicals in the market. This proved to be true in the case of the NA whose performance in the corn market was a disappointment (Gray *et al.*, 1982). However, new strategies employed for the marketing of currently available seed safeners have eliminated the problems encountered with NA. Basically, the major change was to have the seed treatment with safeners being done at the level of the chemical manufacturer or seed distributor rather than at the user's level. Thus, chemical companies developing seed safeners are marketing them through either affiliated or cooperating seed companies (e.g. Ciba-Geigy/Funks Seed affiliation). Apart from the facilitation of the sales of seed safeners, the cooperation of chemical and seed companies is desirable because it promotes the market sales of

Table VI Characteristics of Safeners Applied as Seed Dressings

Safener should be present in equal amounts per seed (uniformity)
Safener should be stable during processing and subsequent seed storage
Safener should be held sufficiently strongly to the surface of the treated seed so that it is not removed in handling and seed storage
Safener and the process of seed coating should not affect the subsequent performance of the seed
The size and particle-flow properties of the coated seed must be consistent so that they do not interfere with the mechanical sowing of treated seeds

proprietary seed hybrids. However, this marketing approach has been criticized for putting some pressure on the farmer since these safeners are sold only in the form of treated seeds (Parker, 1983).

V. FACTORS AFFECTING THE FIELD PERFORMANCE OF HERBICIDE SAFENERS

Several environmental factors are known to affect the performance and selectivity of herbicides under field conditions (Hammerton, 1967; Hatzios and Penner, 1982; Muzik, 1976). Therefore, it is expected that the performance of crop safeners for herbicides will also be affected by selected environmental conditions. Indeed, temperature, light, soil moisture and soil type have all been reported to influence the efficacy of herbicide safeners under field conditions (Hatzios, 1983). In addition, varietal differences and interactions of safeners with each other are also important.

Güneyli (1971) showed that the efficacy of NA as a seed safener of corn against the herbicide EPTC was greater at high rather than low temperatures. Similarly, Leek and Penner (1981) demonstrated a good correlation of the safening activity of cyometrinil against metolachlor effects on grain sorghum and high temperatures. Information on the effects of temperature on the efficacy of other safeners is not available.

The herbicidal activity of thiocarbamate and chloroacetanilide herbicides is known to require light (Wilkinson, 1978). However, the role of light in the efficacy of the currently marketed safeners for these herbicides has not been studied thoroughly. A significant correlation of light and the activity of the safener cyometrinil has been reported by Ebert (1982). She reported that the distinct reduction of metolachlor uptake into sorghum seedlings caused by the safener cyometrinil was very pronounced under light conditions but absent under dark.

Soil moisture appears to be very important for the safening activity of cyometrinil and other oxime safeners against chloroacetanilide herbicide injury in the field. Under extremely wet conditions, cyometrinil was not effective in protecting grain sorghum against metolachlor injury (Ketchersid *et al.*, 1981). Variations in the efficacy of cyometrinil as a seed safener for sorghum as a result of changes in soil moisture have also been reported by Simkins *et al.* (1979). Muller and Nyffeler (1981) suggested that the observed influences of temperature and soil moisture on the efficacy of oxime ether safeners for sorghum are indirect and result from the direct effect of these environmental parameters on the response of grain sorghum to the herbicide metolachlor. Chloroacetanilide herbicide injury to grass weeds and crops is greater under wet conditions (Nyffeler *et al.*, 1980) and higher doses of these safeners may be needed for acceptable protection of grain sorghum. The efficacy of NA as a seed safener of corn against injury from thiocarbamate herbicides appeared to be independent of soil moisture (Hahn, 1974). Similar results have been reported for the safeners dichlormid (Burt and Akinsorotan, 1976) and flurazole (Gingerisch *et al.*, 1981). However, a difference in the mobility of the safener dichlormid as compared to that of the herbicide EPTC has been observed in studies evaluating the performance of Eradicane (EPTC + dichlormid) under wet field conditions (Burt and Buzio, 1978; Buzio and Burt, 1977). This separation of the herbicide EPTC and the safener dichlormid has been proposed as a possible reason for the observed sporadic toxicity of Eradicane on corn under wet field conditions (Burt and Buzio, 1979).

The efficacy of herbicide safeners may be also affected by soil type. Thiessen *et al.* (1980) reported that the efficacy of NA as a safener of oats against injury from the herbicide barban was much greater in light sandy soils than in silty clay soils. In contrast, dichlormid was very ineffective as a safener of corn against thiocarbamate herbicides in light soils (Crook, 1975). The activity of flurazole as a seed safener of sorghum against alachlor injury was independent of soil types. This safener was equally effective on soils with textures ranging from silt loam to muck (Schafer *et al.*, 1981). Information on the effects of soil fertility on the efficacy of herbicide safeners is limited. Soil fertility did not influence the safening activity of NA (Thiessen *et al.*, 1980).

The potential interactive effects of two or more safeners used for protecting the same crop are poorly understood. In the only study of this kind, Parker (1982) reported that NA and dichlormid did not interact synergistically in protecting corn from herbicide injury. The crop-protecting activity of selected safeners (e.g. NA) may be synergized by exogenously applied plant hormones such as GA (Hahn, 1974; Stephenson and Chang, 1978).

Intraspecific differential responses of several inbred lines or varieties of hybrid crops to combinations of herbicides and safeners have been reported (Lay and Niland, 1985; Poneleit, 1974). The inbred corn line 'Oh 551' and the DeKalb hybrids 'XL-25A,' 'XL-45A,' 'XL-67,' 'XL-55A,' and 'XL-379' are not protected by the safener dichlormid against injury from the herbicide EPTC (Lay and Niland, 1985; Sagaral and Foy, 1982). Other cultivars of corn such as 'TXS114' are highly tolerant to the herbicide EPTC even in the absence of the safener dichlormid (Sagaral and Foy, 1982). Similarly, Hatzios and Zama (1986) showed that NA and several of its analogs did not protect the thiocarbamate-susceptible 'XL-67' corn hybrid from EPTC injury. NA, dichlormid, and the oxime ether safeners cyometrinil and oxabetrinil were more effective in protecting the 'XL-72AA' corn hybrid (thiocarbamate-tolerant) against injury from the herbicides chlorsulfuron, fluazifop-butyl and sethoxydim rather than the hybrid 'XL-67' (Hatzios, 1984a). NA was also ineffective in protecting selected cultivars of oats against injury caused by the herbicide barban (Thiessen *et al.*, 1980).

Research with the sorghum seed safeners has shown that, in general, most of the commercially used hybrids of this crop are responsive to safener treatments and they are protected from chloroacetanilide herbicide injury (Davidson *et al.*, 1978; Schafer *et al.*, 1980). However, the sorghum hybrid 'G623GBR' is not protected by flurazole against alachlor injury (Schafer *et al.*, 1980).

In general, the crop tolerance offered by safeners against selected herbicides in the early stages of crop growth persists throughout the life of the crop. Work by Wright *et al.* (1974), however, showed that the protection offered by dichlormid to many corn genotypes against EPTC injury early in the season was greatly reduced, in many of the genotypes, and symptoms of EPTC injury appeared later in the season.

VI. RESIDUES AND ADVERSE EFFECTS OF HERBICIDE SAFENERS

A. Residues of Safeners in Plants and Soils

Information reported on the analysis of safener residues in plants, soils, or the environment has been rather limited. Several analytical methods, however, are available for such analysis and they have been reviewed briefly by Hatzios (1983). Additional information could be found in reports by other investigators as follows: Riden (1976) for NA; Miaulis *et al.* (1978) and Stephenson *et al.* (1978) for dichlormid and other dichloroacetamide safeners; Anonymous (1981a) and Egli (1982) for oxime ether safeners; Breaux *et al.* (Chapter 7, this volume) for flurazole; and Anonymous (1981b) for fenclorim.

Metabolism studies with radiolabeled NA showed that this safener is metabolized to naphthalic acid by corn plants (Frear *et al.*, 1987; Riden and Asbell, 1975). No residues of NA or naphthalic acid were detectable in mature corn plants beyond the fifth or sixth week after emergence (Riden and Asbell, 1975). Naphthalic acid appears to possess safening activity comparable to that of NA against herbicide injury (Frear *et al.*, 1987).

Corn seedlings grown in soil treated with radiolabeled dichlormid metabolized this safener into several metabolites (Miaulis *et al.*, 1978). Dichlormid was also metabolized rapidly in soils and by rats (Miaulis *et al.*, 1978). The safener MG-191 was also found to be metabolized by corn seedlings (Jablonkai *et al.*, 1986). Information on the metabolism of oxime ether safeners by sorghum or other plant species is very limited. Recently, Brooks *et al.* (1987) reported that sorghum coleoptiles were capable of metabolizing oxime ether safeners as well as flurazole through oxidation reactions. The exact metabolites of these safeners, however, were not identified. Corn seedlings are capable of rapidly conjugating flurazole to glutathione (see Breaux *et al.*, Chapter 7, this volume). Two other metabolites of flurazole were also detected in these studies, but they were minor compared to the GS–flurazole conjugate.

Undoubtedly, our understanding of the metabolism and residual activity of herbicide safeners in plants, soils, and the environment is far from being a closed book.

B. Phytotoxicity of Herbicide Safeners

As pointed out earlier in this chapter, low potential for phytotoxicity and reliability under field conditions are two very important considerations for the selection and development of candidate chemicals as crop safeners against herbicide injury (Table II). Therefore, it is expected that, when applied at recommended rates, the currently available safeners would not cause any adverse phytotoxic effects to safened crops. However, since the molecules of the currently marketed herbicide safeners contain functional chemical groups similar to those found in many herbicides, safeners have the potential of causing adverse phytotoxic effects under certain circumstances. The remarkable structural similarity of the herbicide CDAA to the safener dichlormid (Fig. 6) illustrates this point.

Early and recent reports have documented the potential of NA to interfere with the growth of corn, sorghum, oats, rice, and navy-beans (*Phaseolus* sp.) even at the recommended dose of 0.5% by seed weight (Blair, 1979; Eastin, 1972; Guneyli, 1971; Hatzios, 1984a,b; Thiessen *et al.*, 1980). Chlorosis and stunting are typical symptoms induced by NA on safened plants. The phytotoxic effects of NA to oats appeared to be more pronounced following long periods of storage (Thiessen *et al.*, 1980). Recently, Hatzios and Zama (1986) showed that changes in the chemical structure of NA could reduce the phytotoxic potential of this safener without any significant loss of its protective activity against the thiocarbamate herbicide EPTC. However, chlorination of NA increased the phytotoxicity of this seed safener (Hatzios and Zama, 1986).

Adverse effects caused by the safener dichlormid on the growth of corn and navy beans have been reported by Ezra *et al.* (1982) and Blair (1979), respectively. Reduction of the viability of sweet and yellow-endosperm sorghum seeds treated with the safeners cyometrinil and oxabetrinil has also been reported (Davidson *et al.*, 1978; Rufener *et al.*, 1982). In addition, cyometrinil at rates higher than 1.88 g/kg of seed caused significant phytotoxic effects on grain sorghum (Davidson *et al.*, 1978; Warmund *et al.*, 1981). Seed viability and growth of grain sorghum did not appear to be adversely affected by the safener flurazole (Schafer *et al.*, 1980). However, measurable effects of flurazole, cyometrinil, and oxabetrinil on the growth and respiration of grain sorghum seedlings have been observed by Ketchersid and Merkle (1983) during early stages of germination of sorghum seeds. The potential of flurazole to inhibit respiration of germinating grain sorghum seeds has also been confirmed by Warmund *et al.* (1985).

Cyometrinil, oxabetrinil, and flurazole, applied at 100 μ*M* were found to inhibit several metabolic processes of isolated leaf cells of soybean, including photosynthesis, protein synthesis, nucleic acid synthesis, and lipid synthesis (Zama and Hatzios, 1986). However, at lower concentrations (0.1 and 1 μ*M*) all three safeners were stimulatory rather than inhibitory of these processes. Similar results were also observed in comparable studies conducted with the safener oxabetrinil and leaf protoplasts from grain sorghum (Zama and Hatzios, 1985). Interference of flurazole with the incorporation of [^{14}C]acetate into triglyceride lipids of grain sorghum seeds has been reported also by Warmund *et al.* (1985). Finally, the shikimic and phenylpropanoid acid pathways have been proposed as sites for the potential interference of flurazole with plant metabolism. Support for such a postulation comes from a recent report by Mollin *et al.* (1986) demonstrating the inhibitory effects of flurazole on anthocyanin and lignin biosyntheses of grain sorghum even at the extremely small concentration of 10 n*M*.

VII. CONCLUDING REMARKS

Born out of keen observations of antagonistic herbicide interactions, the concept of enhancing crop tolerance to herbicides with chemical safeners has evolved into a recognized agricultural practice. Successful safeners protect crop plants from herbicide injury without reducing weed control. The benefits offered by the safener concept are many. Herbicide safeners may allow: (1) the selective chemical control of weeds in botanically related crops; (2) the use of nonselective herbicides for selective weed control; (3) the counteraction of residual activity of soil-applied persistent herbicides in crop rotation systems; (4) an increase in the spectrum of herbicides available for weed control in "minor" crops; (5) an expansion and extension of the uses and marketability of "out-of-patent" herbicides; (6) the use of cheaper herbicides and reduction in the cost of weed control, and (7) the elucidation of sites and mechanisms of herbicide action by serving as useful biochemical probes. Because of these benefits, herbicide safeners are expected to remain competitive in the future even if new effective and selective herbicides used on the same crop markets become available.

Practical success in crop safening against herbicide injury has been achieved primarily with grass crops such as corn, grain sorghum and rice and soil-applied herbicides such as the thiocarbamates and chloroacetanilides. Other grass crops such as oats, wheat, barley, and ryegrass can also be safened against injury from soil-applied herbicides. Dichloroacetamide derivatives have been the most successful commercialized safeners and they are very effective in protecting corn against injury from thiocarbamate and chloroacetanilide herbicides. These safeners are applied primarily as prepackaged tank mixtures with the respective herbicides. Oxime ether derivatives and 2,4-disubstituted thiazoles are currently marketed as successful safeners of grain sorghum against chloroacetanilide herbicide injury. These safeners are applied primarily as seed dressings. Fenchlorim is the first commercially developed safener for rice against injury from the chloroacetanilide herbicide pretilachlor. NA, the first commercially developed safener, is not marketed today, but it has been quite useful in the identification of potential crop–herbicide–safener combinations.

The performance of herbicide safeners in the field is affected by several environmental factors, including temperature, light, soil moisture, and soil type. In addition, differential intraspecific tolerance of crop cultivars to combinations of herbicides plus safeners needs to be established for optimum effectiveness under field conditions.

Although the commercial development of safeners for herbicides is primarily an industrial activity, the cooperation between industry and academia has been very productive in the development and proper use of these chemicals in modern agriculture. Future approaches in safener development that may result from such a cooperation between industry and academia include: (1) a more extensive use of the rational approach in discovering and developing new safeners for herbicides; (2) "preconditioning" of crops to subtoxic levels of selected herbicides, which may induce the tolerance of these crops to subsequent higher levels; (3) exploitation of the concept of "prosafeners" which could be biotransformed differentially in tolerant crops and susceptible weeds; and (4) the development of genetically engineered microbial safeners which could be used selectively to protect crops from herbicide injury.

Acknowledgments

I express my sincere appreciation to Drs. D.E. Schafer of Monsanto Chemical Company and J.W. Peek of Ciba-Geigy Corporation for providing useful information on general aspects of the industrial development of herbicide safeners.

References

Adler, E. F., Wright, W. L., and Klingman, G. C. (1977). Development of the American pesticide industry. *ACS Symp. Ser.*, **37**, 39–55.

Amhrein, N, Johanning, D., Schab, J., and Schultz, A. (1983). Biochemical basis for glyphosate tolerance in a bacterium and a plant tissue culture. *FEBS Lett.* **157**, 191–196.

Anderson, P. C., and Hibberd, K. A. (1985). Evidence for the interaction of an imidazolinone herbicide with leucine, valine, and isoleucine metabolism. *Weed Sci.* **33**, 479–483.

Anonymous (1981a). Status Report of CGA-92194. Ciba-Geigy Ltd., Agricultural Division, Basle, Switzerland.

Anonymous (1981b). Status Report of CGA-123407. Ciba-Geigy Ltd., Agricultural Division, Basle, Switzerland.

Ashton, F. M., and Harvey, W. A. (1971). Selective chemical weed control. *Circ. Calif. Agric. Exp. Stn.* **558**.

Ashton, F. M., DeVilliers, O. T., Glenn, R. K., and Duke, W. B. (1977). Localization of metabolic sites of action of herbicides. *Pestic. Biochem. Physiol.* **7**, 122–141.

Badger, M. R. (1985). Photosynthetic oxygen exchange. *Annu. Rev. Plant Physiol.* **36**, 27–53.

Blair, A. M. (1979). The interaction of protectants with EPTC on field bean and triallate on wheat. *Ann. Appl. Biol.* **92**, 105–111.

Blair, A. M., Parker, C., and Kasasian, L. (1976). Herbicide protectants and antidotes—A review. *Pestic. Artic. News Summ.*. **22**, 65–74.

Brooks, R. L., Ketchersid, M. L., and Merkle, M. G. (1987). Effects of the sorghum seed safeners on sorghum mixed-function oxidases. *Proc., South. Weed Sci. Soc.* **40**, 328.

Burt, G. W., and Akinsorotan, A. O. (1976). Factors affecting thiocarbamate injury to corn. I. Temperature and soil moisture. *Weed Sci.* **24**, 319–321.

Burt, G. W., and Buzio, C. A. (1978). EPTC + R-25788 injury to corn: The influence of soil movement and stage of growth. *Proc. Annu. Meet. Northeast. Weed Sci. Conf.* **32**, 62.

Burt, G. W., and Buzio, C. A. (1979). EPTC plus R-25788 injury to corn (*Zea mays*) as affected by plant age at treatment. *Weed Sci.* **27**, 460–463.

Buzio, C. A., and Burt, G. W. (1977). EPTC + R-25788 movement in Maryland soils. *Proc. Annu. Meet. Northeast. Weed Sci. Conf.* **31**, 115.

Chaleff, R. S., and Mauvais, C. J. (1984). Acetolactate synthase is the site of action of two sulfonylurea herbicides in higher plants. *Science* **224**, 1443–1445.

Chang, T. S., and Merkle, M. G. (1982). Oximes as seed safeners for grain sorghum (*Sorghum bicolor*) to herbicides. *Weed Sci.* **30**, 70–73.

Chang, T. S., and Merkle, M. G. (1983). Structure–activity relationships of oximes as seed safeners. *Proc., South. Weed Sci. Soc.* **36**, 383.

Comai, L., and Stalker, D. M. (1984). Impact of genetic engineering on crop protection. *Crop Prot.* **3**, 399–408.

Crook, J. J. (1975). A status report of thiocarbamate injury to corn on the Maryland eastern shore. *Proc. Annu. Meet. Northeast. Weed Sci. Conf.* **29**, 49–54.

Dale, J. E. (1979). Application equipment for Roundup: The ropewick applicator. *Proc. Beltwide Cotton Prod. Res. Conf.—Cotton Weed Sci. Res. Conf.* **3**, 138–141.

Davidson, W. E., Gagnon, S. A., Chrustensen, M. D., and Dorr, J. E. (1978). A new herbicide safener which permits effective grass control in sorghum. *Proc. West. Soc. Weed Sci.* **31**, 123–129.

Devlin, R. M., and Koszanski, Z. K. (1987a). R-25788 as an antidote for SC-0774 in corn. *Proc. Annu. Meet. Northeast. Weed Sci Conf.* **41**, 74–77.

Devlin, R. M., and Koszanski, Z. K. (1987b). Effect of FMC-57020 on plant growth and pigment synthesis. *Proc. Annu. Meet. Northeast. Weed Sci. Conf.* **41**, 95–99.

Dill, T. R. (1986). CGA-133205—A new seed applied safener for sorghum. *Proc., South. Weed Sci. Soc.* **39**, 521.

Donald, W. W., and Fawcett, R. S. (1975) Prevention of EPTC injury to hybrid corn with the antidote R-29148 and exogenous GA_3. *Proc.—North Cent. Weed Control Conf.* **20**, 28–29.

Donald, W. W., and Fawcett, R. S. (1976). Prevention of EPTC injury to corn: A developmental and anatomical study. *Abstr. Weed Sci. Soc. Am.* **16**, 70.

Dutka, F., and Komives, T. (1983). On the mode of action of EPTC and its antidotes. *In* "Pesticide Chemistry: Human Welfare and the Environment" (J. Miyamoto and P. C. Kearney, eds.), Vol 3, pp. 213–218. Pergamon, Oxford.

Dutka, F., and Komives, T. (1987). MG-191: A new selective herbicide antidote *In* "Pesticide Science and Biotechnology" (R. Greenhalgh and T. R. Roberts, eds.), pp. 201–204. Blackwell, Oxford.

Dutka, F., Komives, T., Marton, A. F., Hulesch, A., Foder-Csobra, K., and Kappati, K. (1979). Structure–activity relationships of herbicide antidotes. *Proc. Hung. Annu. Meet. Biochem.* **19**, 1–4.

Eastin, E. F. (1972). Evaluation of a sorghum seed treatment to prevent injury from acetanilide herbicides. *Agron. J.* **64**, 556–557.

Ebert, E. (1982). The role of waxes in the uptake of metolachlor into sorghum in relation to the protectant CGA-43089. *Weed Res.* **22**, 305–311.

Egli, H. (1982). Storage stability of pesticide residues. *J. Agric. Food Chem.* **30**, 861–866.

Ellis, F. J., Peek, J. W., Boehle, J. Jr., and Muller, G. (1980). Effectiveness of a new safener for protecting sorghum (*Sorghum bicolor*) from metolachlor injury. *Weed Sci.* **28**, 1–5.

Ezra, G., Krochmal, E., and Gressel, J. (1982). Competition between a thiocarbamate herbicide and herbicide protectants at the level of uptake into maize cells in culture. *Pestic. Biochem. Physiol.* **18**, 107–112.

Ezra, G., Rusness, D. G., Lamoureux, G. L., and Stephenson, G. R. (1985). The effect of CDAA (*N*,*N*-diallyl-2-chloroacetamide) pretreatments on subsequent CDAA injury to corn (*Zea mays* L.). *Pestic. Biochem. Physiol.* **23**, 108–115.

Fedtke, C. (1982). "Biochemistry and Physiology of Herbicide Action." Springer-Verlag, Berlin and New York.

Frear, D. S., Swanson, H. R., and Mansager, E. R. (1987). 1,8-Naphthalic anhydride/auxin protection against chlorsulfuron inhibition of corn seedling growth. *In* "Pesticide Science and Biotechnology" (R. Greenhalgh and T. R. Roberts, eds.), pp. 499–503. Blackwell, Oxford.

Geller, I. A., and Nikolaenko, I. Zh. (1972). The effects of microbiological preparations on the phytotoxicity of herbicides. *Sakh. Svekla* **17**, 25–26.

Gingerich, L. L., Schneider, R. P., Schultz, G. E., and Rieck, W. L. (1981). Efficacy of MON-4606 against acetanilide injury to grain sorghum. *Abstr. Weed Sci. Soc. Am.* **21**, 26.

Glass, E. E., ed. (1986). "Pesticide Resistance: Strategies and Tactics for Management." National Academy Press, Washington, D.C.

Görög, K., Muschinek, G. Y., Mustardy, L. A., and Faludi-Daniel, A. (1982). Comparative studies of safeners for the prevention of EPTC injury in maize. *Weed Res.* **22**, 27–33.

Gray, R. A., Green, L. L., Hoch, P. E., and Pallos, F. M. (1982). The evolution of practical crop safeners. *Proc. Br. Crop Prot. Conf.—Weeds* **2**, 431–437.

Gressel, J. (1984). Plant tissue culture systems for screening of plant growth regulators: Hormones, herbicides and natural phytotoxins. *Adv. Cell Cult.* **3**, 93–181.

Gressel, J. (1986). Modes and genetics of herbicide resistance in plants. *In* "Pesticide Resistance: Strategies and Tactics for Management" (E. H. Glass, ed.), pp. 54–73. National Academy Press, Washington, D.C.

Gresshoff, P. M. (1979). Growth inhibition by glyphosate and reversal of its action by phenylalanine and tyrosine. *Aust. J. Plant Physiol.* **6**, 177–185.

Güneyli, E. (1971). Factors affecting the action of 1,8-naphthalic anhydride in corn treated with *S*-ethyl dipropylthiocarbamate (EPTC). *Diss. Abstr. Int. B* **32**, 1957–1958.

Hahn, R. R. (1974). Use of 1,8-naphthalic anhydride to improve grain sorghum tolerance to alachlor. *Diss. Abstr. Int. B* **35**, 1483.

Halliwell, B. (1984). "Chloroplast Metabolism. The Structure and Function of Chloroplasts in Green Leaves." Oxford Univ. Press, London and New York.

Hammerton, J. L. (1967). Environmental factors and susceptibility to herbicides. *Weeds* **15**, 330–336.

Harvey, B. M. R., and Harper, D. B. (1982). Tolerance to bipyridylium herbicides. *In* "Herbicide Resistance in Plants" (H. M. LeBaron and J. Gressel, eds.), pp. 215–234. Wiley, New York.

Hatzios, K. K. (1983). Herbicide antidotes: Development, chemistry, and mode of action. *Adv. Agron.* **36**, 265–316.

Hatzios, K. K. (1984a). Interactions between selected herbicides and protectants on corn (*Zea mays*). *Weed Sci.* **32**, 51–58.

Hatzios, K. K. (1984b). Potential safeners for protecting sorghum (*Sorghum bicolor* (L.) Moench) against chlorsulfuron, fluazifop-butyl and sethoxydim. *Weed Res.* **24**, 249–254.

Hatzios, K. K. (1986). Interactions of the safener flurazole with chloroacetanilide and thiocarbamate herbicides on maize (*Zea mays* L.). *Can. J. Plant Sci.* **66**, 353–359.

Hatzios, K. K. (1987). Biotechnology applications in weed management: Now and in the future. *Adv. Agron.* **41**, 325–375.

Hatzios, K. K., and Penner, D. (1982). "Metabolism of Herbicides in Higher Plants." Burgess, Minneapolis, Minnesota.

Hatzios, K. K., and Penner, D. (1985). Interactions of herbicides with other agrochemicals on higher plants. *Rev. Weed Sci.* **1**, 1–63.

Hatzios, K. K., and Zama, P. (1986). Physiological interactions between the herbicide EPTC and selected analogues of the antidote naphthalic anhydride on two hybrids of maize. *Pestic. Sci.* **17**, 25–32.

Helseth, N. T., and Dill, T. R. (1986). Greenhouse results with CGA-133205 on sorghum. *Proc., South. Weed Sci Soc.* **39**, 35–40.

Hilton, J. L., and Pillai, P. (1986). L-2-Oxathiazolidine-4-carboxylic acid protection against tridiphane toxicity. *Weed Sci.* **34**, 669–675.

Hilton, J. L., and Pillai, P. (1987). L-Thiazolidine-4-carboxylic acid, a new herbicide protectant. *Abstr. Weed Sci. Soc. Am.* **27**, 67.

Hoffman, O. L. (1962). Chemical seed treatments as herbicide antidotes. *Weeds* **10**, 322–323.

Hoffman, O. L. (1969). Chemical antidotes for EPTC on corn. *Abstr. Weed Sci. Soc. Am.* **9**, 12.

Hoffman, O. L. (1971). U.S. Patent 3,564,768.

Hoffman, O. L. (1978). Herbicide antidotes: From concept to practice. *In* "Chemistry and Action of Herbicide Antidotes" (F. M. Pallos and J. E. Casida, eds.), pp. 1–13. Academic Press, New York.

Hoffman, O. L., Gull, P. W., Zeisig, H. C., and Epperly, J. R. (1960). Factors influencing wild oat control with barban. *Proc.—North Cent. Weed Control Conf.* **17**, 20.

Hossain, M. A., Nakano, Y., and Asada, K. (1984). Monodehydroascorbate reductase in spinach chloroplasts and its participation in regeneration of ascorbate for scavenging hydrogen peroxide. *Plant Cell Physiol.* **25**, 385–395.

Howe, R. K., and Lee, L. F. (1980). U. S. Patent 4,199,506.

Hubele, A., Kunz, W., Eckhardt, W., and Sturm, E. (1983). The fungicidal activity of acylanilines. *In* "Pesticide Chemistry: Human Welfare and Environment" (J. Miyamoto and P. C. Kearney, eds.), pp. 233–242. Pergamon, Oxford.

Jablonkai, I., Elker, Zs., Komives, T., and Dutka, F. (1986). Uptake, translocation and biodegradation of the herbicide antidote MG-191. *IUPAC Int. Congr. Pectic. Chem., 6th,* Abstr. No. 3D-11.

Jaworski, E. G. (1972). Mode of action of *N*-phosphonomethyl-glycine: Inhibition of aromatic amino acid biosynthesis. *J. Agric. Food Chem.* **20**, 1195–1198.

Kaufman, R. J. (1986). The development and risk assessment of a genetically engineered microbial insecticide. *IUPAC Int. Congr. Pestic. Chem., 6th,* Abstr. No. 2F-01.

Ketchersid, M. L., and Merkle, M. G. (1983). Effects of protectants on sorghum seedling respiration. *Proc., South. Weed Sci. Soc.* **36**, 384.

Ketchersid, M. L., and Merkle, M. G. (1985). R-30228 as a safener for sorghum against thiocarbamate herbicides. *Proc., South. Weed Sci. Soc.* **38**, 439.

Ketchersid, M. L., Norton, K., and Merkle, M. G. (1981). Influence of soil moisture on the safening effect of CGA-43089 in grain sorghum (*Sorghum bicolor*). *Weed Sci.* **29**, 281–287.

Killmer, J. L. , Widholm, J. M., and Slife, F. W. (1980). Antagonistic effect of *p*-aminobenzoate or folate on asulam (methyl-(4-amino benzenesulphonyl carbamate)) inhibition of carrot suspension cultures. *Plant Sci. Lett.* **19**, 203–208.

Killmer, J. L., Widholm, J. M., and Slife, F. W. (1981). Reversal of glyphosate inhibition of carrot cell culture growth by glycolytic intermediates and organic and amino acids. *Plant Physiol.* **68**, 1299–1302.

Kirby, C. (1980). "The Hormone Weedkillers: A short History of Their Discovery and Development." Br. Crop Prot. Counc. Publ., Groydon.

Kydonieus, A. F. (1980). "Controlled Release Technologies: Methods, Theories, and Applications." CRC Press, Boca Raton, Florida.

LaRossa, R. A., and Falco, S. C. (1984). Amino acid biosynthetic enzymes as targets of herbicide action. *Trends Biotechnol.* **2**, 158–161.

Lay, M.-M., and Casida, J. E. (1978). Influence of glutathione and glutathione *S*-transferase in the action of dichloroacetamide antidotes for thiocarbamate herbicides. *In* "Chemistry and Action of Herbicide Antidotes" (F. M. Pallos and J. E. Casida, eds.), pp. 151–160. Academic Press, New York.

Lay, M.-M., and Menn, J. J. (1987). Amelioration of toxicity to molinate to Japanese carp (*Cyprinus carpio*) with selected dichloroacetamides. *Pestic. Biochem. Physiol.* **28**, 149–154.

Lay, M.-M., and Niland, A. M. (1985). Biochemical response of inbred and hybrid corn (*Zea mays* L.) to R-25788 and its distribution with EPTC in corn seedlings. *Pestic. Biochem. Physiol.* **23**, 131–140.

Lea, P. J., Joy, K. W., Ramos, J. L., and Guerrero, M. G. (1984). The action of 2-amino-4-(methylphosphinyl)-butanoic acid (phosphinothricin) and its 2-oxo-derivative on the metabolism of cyanobacteria and higher plants. *Phytochemistry* **23**, 1–6.

Leaver, J. P., and Roberts, E. H. (1984). Peroxides in seed coatings. *Outlook Agric.* **13**, 147–153.

Leavitt, J. R. C., and Penner, D. (1978). Protection of corn (*Zea mays*) from acetanilide herbicidal injury with the antidote R-25788. *Weed Sci.* **27**, 47–50.

LeBaron, H. M. (1986). Resistance of weeds to herbicides. Presentation to the Illinois Custom Spray Training School.

LeBaron, H. M., and Gressel, J., eds. (1982). "Herbicide Resistance in Plants." Wiley, New York.

Leek, G. L., and Penner, D. (1981). Environmental conditions affecting efficacy of CGA-43089 as an antidote for sorghum. *Abstr. Weed Sci. Soc. Am.* **21**, 101.

Lewinsohn, E., and Gressel, J. (1984). Benzyl viologen-mediated counteraction of diquat and paraquat phytotoxicities. *Plant Physiol.* **76**, 125–130.

Martin, H. (1978). U. S. Patent 4,070,389.

Mersie, W., and Parker, C. (1983). Selective control of grass weeds in teff with and without use of a safener. *Trop. Pest Manage.* **29**, 333–338.

Miaulis, J. B., Thomas, V. M., Gray, R. A., Murphy, J. J., and Hollingworth, R. M. (1978). Metabolism of R-25788 (*N,N*-diallyl-2,2-di-chloroacetamide) in corn plants, rats, and soil. *In* "Chemistry and Action of Herbicide Antidotes" (F. M. Pallos and J. E. Casida, eds.), pp. 109–131. Academic Press, New York.

Molin, W. T., Anderson, E. J., and Porter, C. A. (1986). Effects of alachlor on anthocyanin and lignin synthesis in etiolated sorghum (*Sorghum bicolor* (L.) Moench) mesocotyls. *Pestic. Biochem. Physiol.* **25**, 105–111.

Mueller, D. C., and Seyferth, R. (1969). Halomethyl–metal compounds. XXIV. Rearrangements of mercurial-derived (dialkoxymethyl)chlorocarbenes. *J. Am. Chem. Soc.* **91**, 1754–1758.

Muhitch, M. J., Shaner, D. L., and Stidham, M. A. (1987). Imidazolinones and acetohydroacid synthase from higher plants. Properties of the enzyme from maize suspension culture cells and evidence for the binding of imazapyr to acetohydroacid synthase *in vivo*. *Plant Physiol.* **83**, 451–456.

Muller, G., and Nyffeler, A. (1981). CONCEP (CGA-43089), a safening agent protecting sorghum (*Sorghum bicolor*) from metolachlor injury. *Acta Phytopathol. Acad. Sci. Hung.* **16**, 245–248.

Mullison, W. R., ed. (1979). "Herbicide Handbook," 4th ed. Weed Sci. Soc. Am., Champaign, Illinois.

Muzik, T. J. (1976). Influence of environmental factors on toxicity to plants. *In* "Herbicides: Physiology, Biochemistry, Ecology" (L. J. Audus, ed.), 2nd ed., Vol. 2, pp. 203–247. Academic Press, London.

Nagy, J., and Balogh, K. (1985). A new safener for EPTC in corn. *Proc. Br. Crop Prot. Conf.—Weeds* **1**, 107–111.

Narasimhachari, N., and Vining, L. C. (1963). Studies on the pigments of *Penicillium herquei*. *Can. J. Chem.* **41**, 641–648.

Netzer, W. J. (1984). Engineering herbicide tolerance: When is it worthwhile? *Bio/Technology* **2**, 939–944.

Nyffeler, A., Gerber, H. R., and Hensley, J. R. (1980). Laboratory studies on the behavior of the herbicide safener CGA-43089. *Weed Sci.* **28**, 6–10.

Okii, M., Teranishi, M., Konnai, M., and Takematsu, T. (1979). Studies on the antidote octamethylenediamine (OM) as foliar application type. I. Antidotal effect of amines against the herbicidal compound (*N*-3,4-dichlorophenylcarbamoyl)-*N*-methyl glycine monohydrate (CD). *Weed Res. (Tokyo)*. **24**, 96–100.

Pallos, F. M., and Casida, J. E., eds. (1978). "Chemistry and Action of Herbicide Antidotes." Academic Press, New York.

Pallos, F. M., Gray, R. A., Arneklev, D. R., and Brokke, M. E. (1975). Antidotes protect corn from thiocarbamate herbicide injury. *J. Agric. Food Chem.* **23**, 821–822.

Pallos, F. M., Brokke, M. E., and Arneklev, D. R. (1977). U. S. Patent 4,021,224.

Parker, C. (1982). Further studies with herbicide safeners on rice and maize. *Proc. Br. Crop Prot. Conf.—Weeds* **2**, 475–482.

Parker, C. (1983). Herbicide antidotes—A review. *Pestic. Sci.* **14**, 40–48.

Parker, C., Richardson, W. G., and West, T. M. (1980). Potential for extending the selctivity of DPX-4189 by use of herbicide safeners. *Proc. Br. Crop Prot. Conf.—Weeds* **1**, 15–21.

Paul, L. E., and Knake, E. L. (1984). Control of volunteer corn in soybean. *Abstr. Weed Sci. Soc. Am.* **24**, 3.

Peek, J. W., Collins, H. A., Porpiglia, P. J., and Ellis, J. F. (1988). CGA-154281—a new safener for improved herbicide tolerance in corn. *Abstr. Weed Sci. Soc. Am.* **28**, 13–14.

Peek, J. W., Dill, T. R., and Turner, W. E. (1981). Use of safeners to protect sorghum from herbicide injury. *Proc. Annu. Corn Sorghum Res. Conf.* **36**, 31–47.

Poneleit, C. G. (1974). Review of thiocarbamate herbicide research and genetic resistance studies. *Proc. Annu. Corn Sorghum Res. Conf.* **29**, 142–152.

Prochnow, C. L. (1978). Chemical safeners for herbicides to increase crop tolerance. *Proc. West. Soc. Weed Sci.* **31**, 62–63.

Putnam, A. R., and Penner, D. (1974). Pesticide interactions in higher plants. *Residue Rev.* **50**, 73–110.

Pyon, J. Y. (1986). Effect of 1,8-naphthalic anhydride and CGA-123407 on reducing injury of rice in subirrigated seedbed. *Abstr. Weed Sci. Soc. Am.* **26**, 4.

Quadranti, M., and Ebner, L. (1984). Sofit, a new herbicide for use in direct-seeded rice (wet-sown rice). *Proc. Asian-Pac. Weed Sci. Soc. Conf., 9th, 1983,* pp. 405–412.

Rabinowitch, H. D., and Fridovitch, I. (1985). Growth of *Chorella sorokiniana* in the presence of sulfite elevates cell content of superoxide dismutase and imparts resistance towards paraquat. *Planta* **164**, 524–528.

Ray, T. B. (1984). Site of action of chlorsulfuron: Inhibition of valine and isoleucine biosynthesis in plants. *Plant Physiol.* **75**, 827–831.

Riden, J. (1976). PROTECT. *Anal. Methods Pestic. Plant Growth Regul.* **8**, 483–489.

Riden, J., and Asbell, W. J. (1975). The fate of protect (1,8-naphthalic anhydride) in corn plants. *Environ. Qual. Saf., Suppl.*, **3**, 175–179.

Rubin, B., Kirino, O., and Casida, J. E. (1985). Chemistry and action of *N*-phenylmaleamic acids and their progenitors as selective herbicide antidotes. *J. Agric. Food Chem.* **33**, 489–494.

Rufener, J., Nyffeler, A., and Peek, J. W. (1982). CGA-92194, a new safener to protect sorghum from injurious effects of metolachlor. *Proc. Br. Crop Prot. Conf.—Weeds 2*, 461–467.

Rutherford, A. W., and Heathcote, P. (1985). Primary photochemistry in photosystem-1. *Photosynth. Res.* **6**, 295–317.

Sacher, R. M., Lee, L. F., Schafer, D. E., and Howe, R. K. (1983). Synthesis and application of novel thiazoles as herbicide antidotes. *In* "Pesticide Chemistry: Human Welfare and the Environment" (J. Miyamoto and P. C. Kearney, eds.), Vol. 1, pp. 165–168. Pergamon, Oxford.

Sagaral, E. G., and Foy, C. L. (1982). Response of several corn (*Zea mays*) cultivars and weed species to EPTC with and without the antidote R-25788. *Weed Sci.* **30**, 64–69.

Saggers, D. T. (1976). The search for new herbicides. *In* "Herbicides: Physiology, Biochemistry, Ecology" (L. J. Audus, ed.), 2nd ed., Vol. 2, pp. 447–473. Academic Press, London.

Sanders, H. J. (1981). Herbicides: Special report. *Chem. Eng. News* **59**, 20–35.

Schafer, D. E. (1985). Herbicide antidotes: An industrial perspective. Presented at the 12th Annual Meeting of the Plant Growth Regulator Society of America, Boulder, Colorado.

Schafer, D. E., Brinker, R. J., and Radke, R. O. (1980). Introduction of a new safening agent for alachlor on grain sorghum. *Proc.—North Cent. Weed Control Conf.* **35**, 67–71.

Schafer, D. E., Brinker, R. J., and Radke, R. O. (1981). Evaluation of the behavior and performance of MON-4606 for alachlor in grain sorghum. *Abstr. Weed Sci. Soc. Am.* **21**, 26.

Simkins, G. S., Moshier, L. J., and Russ, O. G. (1979). Effect of temperature and moisture on phytotoxicity of acetanilide herbicides and CGA-43089 protection in grain sorghum. *Proc.—North Cent. Weed Control Conf.* **34**, 45.

Smith, R. J. (1971). Red rice: A problem weed in rice. *Weeds Today* **2**(2), 12–13.

Stephenson, G. R., and Chang, F. Y. (1978). Comparative activity and selectivity of herbicide antidotes. *In* "Chemistry and Action of Herbicide Antidotes" (F. M. Pallos and J. E. Casida, eds.), pp. 35–61. Academic Press, New York.

Stephenson, G. R., and Ezra, G. (1983). Herbicide antidotes—A new era in selective chemical weed control. *In* "Plant Growth Regulating Chemicals" (L. G. Nickell, ed.), Vol. 2, pp. 193–212. CRC Press, Boca Raton, Florida.

Stephenson, G. R., and Ezra, G. (1985). Use of subtoxic pretreatments to improve crop tolerance to herbicides. *ACS Symp. Ser.* **276**, 69–84.

Stephenson, G. R., and Pallos, F. M. (1983). Discovery and development of antidotes to improve herbicide selectivity. *In* "Pesticide Chemistry: Human Welfare and the Environment" (J. Miyamoto and P. C. Kearney, eds.), Vol. 3, pp. 209–211. Pergamon, Oxford.

Stephenson, G. R., Bunce, J. J., Makowski, R. I., and Curry, J. C. (1978). Structure–activity relationships for *S*-ethyl *N,N*-dipropylthiocarbamate (EPTC) antidotes in corn. *J. Agric. Food Chem.* **26**, 137–140.

Stephenson, G. R., Bunce, J. J., Makowski, R. I., Bergsma, M. D., and Curry, J. C. (1979). Structure–activity relationships for antidotes to thiocarbamate herbicides in corn. *J. Agric. Food Chem.* **27**, 543–547.

Sweetser, P. B. (1985). Safening of sulfonylurea herbicides to cereal crops: Mode of herbicide antidote action. *Proc. Br. Crop Prot. Conf.—Weeds*, **3**, 1147–1154.

Taylor, H. F., and Loader, M. P. C. (1984). Research on the control of wild oats and broad-leaved weeds by herbicide mixtures. *Outlook Agric.* **13**, 58–68.

Thiessen, E. P., Stephenson, G. R., and Anderson, G. W. (1980). Factors influencing 1,8-naphthalic anhydide activity as antidote for barban in oats. *Can. J. Plant Sci.* **60**, 1005–1013.

Turner, W. E., Clark, D. R., Helseth, N., Dill, T. R., King, J., and Seifried, E. B. (1982). CGA-92194—A new safener to protect sorghum from metolachlor injury. *Proc., South. Weed Sci. Soc.* **35**, 388.

Upham, B. L., and Hatzios, K. K. (1987). Counteraction of paraquat toxicity at the chloroplast level. *Z. Naturforsch., C: Biosci..* **42C**, 824–828.

Upham, B. L., Hatzios, K. K., and Jahnke, L. S. (1987). Inhibition of the photosynthetic electron transport of isolated thylakoids by hemolyzed rabbit sera. Evidence for the potential involvement of parallel electron transport in photosystem I Mehler reactions. *Photosynth. Res.* **12**, 63–72.

Veerasekaran, P., Kirkwood, R. C., and Parnell, E. W. (1981a). Studies on the mechanism of action of asulam in plants. Part I.Antagonistic interaction of asulam and 4-amino-benzoic acid. *Pestic. Sci.* **12**, 325–329.

Veerasekaran, P., Kirkwood, R. C., and Parnell, E. W. (1981b). Studies on the mechanism of action of asulam in plants. Part II.Effect of asulam on the biosynthesis of folic acid. *Pestic. Sci.* **12**, 330–338.

Wain, R. L., and Smith, M. S. (1976). Selectivity in relation to metabolism. *In* "Herbicides: Physiology, Biochemistry, Ecology" (L. J. Audus, ed.), 2nd ed., Vol.1, pp. 279–302. Academic Press, London.

Wakabayashi, K., and Matsunaka, S. (1982). Crop safening in Japan. *Proc. Br. Crop Prot. Conf.—Weeds* **2**, 439–450.

Warmund, M. R., England, D. C., and Kerr, H. D. (1981). Anatomical site of action of MON-4606 in milo, *Sorghum bicolor*. *Proc.—North Cent. Weed Control Conf.* **36**, 123–124.

Warmund, M. R., Kerr, H. D., and Peters, E. J. (1985). Lipid metabolism in grain sorghum (*Sorghum bicolor*) treated with alachlor plus flurazole. *Weed Sci.* **33**, 25–28.

Wegrzyn, T. (1975). Some aspects of the relationship between *Azotobacter chroococcum* and herbicides. *Rocz. Glebozn.* **26**, 79–89.

Wiese, A. F. (1985). "Weed Control in Limited-Tillage Systems." Weed Sci. Soc. Am., Champaign, Illinois.

Wilkinson, R. E. (1978). Physiological response of lipid components to thiocarbamates and antidotes. *In* "Chemistry and Action of Herbicide Antidotes" (F. M. Pallos and J. E. Casida, eds.), pp. 85–108. Academic Press.

Woods, D. D. (1940). The relation of *p*-amino-benzoic acid to the mechanism of action of sulfanilamide. *Br. J. Exp. Pathol.* **21**, 74–90.

Wright, T. H., Rieck, C. E., and Poneleit, C. G. (1974). Effect of R-25788 on EPTC injury to corn genotypes. *Abstr. Weed Sci. Soc. Am.* **14**, 1.

Wuertzer, B., Nuyken, W., and Rohr, W. (1983). Reliability of safeners for metazachlor under varying precipitation levels. *Aspects Appl. Biol.* **4**, pp. 227–234.

Youngman, R. J., Dodge, A. D., Lengfelder, E., and Elstner, E. F. (1979). Inhibition of paraquat phytotoxicity by a novel copper chelate with superoxide dismutating activity. *Experientia* **35**, 1295–1296.

Zaitlin, M., Day, P., and Hollaender, A. (1986). "Biotechnology in Plant Science: Relevance to Agriculture in the Eighties." Academic Press, Orlando, Florida.

Zama, P., and Hatzios, K. K. (1985). Interactive effects of metolachlor and the safener CGA-92194 on the metabolism of sorghum protoplasts. *Proc. South. Weed Sci. Soc.* **38**, 441.

Zama, P., and Hatzios, K. K. (1986). Comparative effects of CGA-92194, cyometrinil, and flurazole on selected metabolic processes of isolated soybean leaf cells. *J. Plant Growth Regul.* **5**, 59–72.

Chapter 2

Crop Safening against Herbicides in Japan

Shooichi Matsunaka
Faculty of Agriculture
Kobe University
Nada-ku, Kobe 657, Japan

Ko Wakabayashi
Mitsubishi Kasei Corporation
Research Center, Kamoshida-cho,
Midori-ku, Yokohama 227, Japan

I. INTRODUCTION

After World War II, rice (*Oryza sativa* L.) production in Japan was very important in supplying food to the increasing population, including the repatriated peoples from outside Japan. Chemical fertilizers, fungicides, and insecticides contributed much to this purpose. Later the labor shortage in agriculture became a very severe problem because expanding industries absorbed manual laborers from rural areas. The development of herbicides by Japanese chemical industries, the use of the granular formulations in flooded and transplanted paddy fields, and the government subsidy for rice production increased the utilization of herbicides. Statistics show, for instance, that since 1973, applications of herbicides in paddy rice fields have been used more than two times per year in Japan.

During the history of herbicide utilization in Japan, several problems and issues have been raised and most of them have been solved (Matsunaka, 1983). Crop injury by herbicides was among them. Injury to crops may be solved by several countermeasures, including improvements of formulations, structural modifications of herbicides, and use of safener/herbicide combinations. These measures will be outlined and discussed in this chapter.

II. IMPROVEMENTS OF HERBICIDE FORMULATIONS FOR CROP SAFENING

A. Granular Formulations of Herbicides for Rice

In the past 35 years, many herbicides have been introduced for crops produced in Japan, especially for rice culture. The most remarkable success in this field is the utilization of granular formulations of herbicides for rice weed control. Initially, 2,4-D and MCPA were structurally modified to ester forms and their granules were applied in the paddy field to avoid injury to neighboring crop plants, especially mulberry (*Morus* spp.) trees for silkworm culture.

Then, pentachlorophenol (PCP) granules followed the utilization of phenoxy-acetic acid herbicides. When PCP–Na solutions were applied to transplanted rice plants, the leaves suffered severe injury. However, PCP granules did not adhere to the foliage of plants and did not injure rice plants. PCP granules sank, swelled, and bound weakly to the mud of paddy rice fields, forming a uniform herbicide layer on the soil surface. PCP granules proved reasonably effecctive in killing weeds, mainly barnyard grass [*Echinochloa crus-galli* (L.) Beauv.] whose growing parts are always concentrated in the region of the herbicide-treated layer, and prevented rice plant injury because the growing points or roots are located under the soils.

Currently almost all herbicides used in paddy rice fields are formulated as granules, even though there are a few disadvantages with such formulations, such as higher cost of formulation, storage, and transportation. These are compensated by the advantages of granules, such as easy application by hand or simple machines, and also the above-mentioned crop safening effect to rice plants.

Many granular formulations of herbicides have been markedly successful in flooded rice culture in Japan. When sales of herbicides in Japan are classified from the standpoint of formulations, the weight percentages of 1985 are: granules 84.9%, emulsifiable concentrates 10.9%, wettable powders 3.3%, and others 0.8%. Among the ten most successful rice herbicides used in Japan, oxadiazon and its combination with butachlor are formulated as emulsifiable concentrates. All others are formulated as granules.

B. Emulsifiable Concentrate Formulations of Herbicides for Rice

1. Oxadiazon Emulsifiable Concentrate: Formulation and Application Timing

An unusual herbicidal formulation available for paddy rice culture is the emulsifiable concentrate of oxadiazon. The concentrate is applied by hand directly, without any dilution, from special holes in a handy container onto the muddy water surface at the time of puddling. The floating soil particles absorb all the herbicide and settle to the bottom, forming a firm soil–herbicide layer. This provides good weed control and does not cause injury to the transplanted rice seedlings.

When the oxadiazon concentrate is applied after transplanting, it causes crop damage and its selectivity is not maintained. Granules of oxadiazon form a thin herbicide layer similar to PCP granules, but some of the herbicide dissolves in water, is absorbed by the stems, and causes crop injury, mainly brown spots on the leaf sheath or leaf-floating damage. Thus the emulsifiabale concentrate formulation of this herbicide can be used safely only in muddy water conditions (Kawamura, 1973).

Emulsifiable concentrate formulations of rice herbicides other than oxadiazon are currently commercially available and are based on the same principle.

2. Propanil Emulsifiable Concentrate: Enhancement of Differential Herbicide Absorption between Crops and Weeds

Propanil shows a considerable selectivity between rice and weeds. Its selectivity results from the presence or absence in plants of an aryl acylamidase (EC 3.5.1.13), which can detoxify the parent herbicide to 3,4-dichloroaniline. An isolated mutant of rice which lacks this enzyme and is very susceptible to propanil has been reported (Matsunaka, 1982). Crossing experiments with the propanil-susceptible mutant and the original tolerant rice plants showed that the biosynthesis of the propanil-detoxifying enzyme is controlled by a single dominant gene (Matsunaka, 1982).

In addition to this biochemical basis, Konnai and associates (1967) investigated the improvement of formulations of this herbicide and reported that the selectivity of propanil was intimately connected with the hyprophilic–lipophilic balance of each formulation and differences in fine leaf structure of these plants. Further physiological studies showed that the surface of rice leaves was relatively more lipophilic than that of barnyard grass. Much empirical work was necessary to optimize the propanil formulation using surfactants, solvents, and other substances to achieve maximum selectivity. As a result, 2-(1-cyclohexenyl)cyclohexanone (CHCH), a more hydrophilic solvent, was found to increase the selectivity of propanil. The typical formulation of propanil used currently in paddy rice culture in Japan consists of propanil (35%), isophorone (25%), CHCH (7%), xylene (25%), and Solpol (surfactant)(8%).

C. Use of Controlled Release Herbicide Formulations

Controlled release pesticide technology promises to improve the efficacy of some existing pesticides and solve the environmental problems associated with others. Several parameters, including volatility, photodecomposition, leaching, microbial metabolism, binding, and accumulation, determine the persistence, distribution, and indirectly the efficacy of the pesticide. A major benefit that controlled release technology can bring to weed control is to expand herbicide selectivity to additional crops.

1. Microencapsulation of Herbicides

Microcapsules are 1- to 200-μm particles of a solid or liquid core surrounded by a wall. The wall is usually polymeric in nature and protects the core material (pesticide) in storage, but is designed to release the core in a controlled fashion when the microcapsules are used. The core material can be released by crushing the wall, breaking the wall by pressure from within, dissolving the wall, hydrolyzing the wall, or by diffusion through the wall.

Microencapsulation of pesticides using many types of wall materials is currently being investigated in Japan. Yamamoto and Katsuda (1980) have indicated that the pyrethroid insecticides become more stable when included in β-cyclodextrin, particulary because photodecomposition is reduced. I. Yamamoto (personal communication, 1982) has recently formulated thiocarbamate herbicides such as molinate using β-cyclodextrin and starch-xanthate. He has demonstrated great improvements by reducing the volatility of the herbicide.

2. Control of Herbicide Movement in Soil: Modification of Herbicide Leaching

Modification of the leaching properties of soil-applied herbicides has great significance in selective weed control. Reduced leaching of a relatively mobile herbicide could lead to better selectivity if the selectivity depends largely on the depth at which herbicides are located. Moreover, keeping the herbicide in the target area could also reduce the dosage required, thus decreasing phytotoxicity. The use of certain surfactants or latex emulsions as adjuvants is currently being investigated in Japan and is resulting in additional benefits, such as a decrease of soil residue problems.

Saito and Hayabe (1962) reported that the leaching of simazine was greatly reduced when used with the surfactant oxyethylene docosanol (OED). For example, the growth of the test plant, radish (*Raphanus sativus* L. var. *radicula* Pers.), seeded on the soil taken at a depth of 0.51–1.50 cm in a simazine-treated plot under precipitation of 30 mm, was inhibited by 40%, whereas that of the combination of simazine with OED showed no inhibition.

Takematsu *et al.* (1978) studied the effects of latex emulsions on leaching and adsorption of many types of soil-applied herbicides and found that Sanpex CE-15, a urethane type of latex emulsion, was the most promising. Simazine is normally leached to a depth of 3–4 cm, but, with the addition of Sanpex CE-15, it forms a firm herbicide–soil layer at a depth of 0–2 cm.

The same latex prevents leaching of many other herbicides such as diuron, atrazine, trifluralin, and chlorophthalim, and reduces their phytotoxicity as shown in Table I (H. Watanabe, unpublished, 1982).

III. STRUCTURAL MODIFICATION OF HERBICIDES FOR CROP SAFENING

Structural modification of a parent herbicide may influence the mobility, absorption, and selective properties of this herbicide, although in some cases the herbicidal spectrum of the original compound may be slightly altered.

2,4-D and MCPA are much more active against dicotyledonous than monocotyledonous plants, including rice. However, the vapors of these chemicals caused phytotoxicity to nearby mulberry trees and sometimes even to rice plants. Esters normally exhibit greater herbicidal activity than the parent acids and appear to widen the difference between the susceptibility of weeds and the tolerance of useful plants (Takematsu, 1968). The ethyl and allyl esters of 2,4-D or MCPA may be cited as examples. Application of granular formulations of these ester herbicides to rice seedlings 25 to 30 days after transplanting catches most of the weeds at their more susceptible stage, but does not injure rice plants or mulberry trees.

Table I Crop Safening Effect of Sanpex CE-15[a]

Herbicide dosage (100 g/ha)	Sanpex CE-15 dosage (kg/ha)	Phytotoxicity[b]	
		Wheat	Radish
Diuron	0	1	2
	2.5	0	1
	5.0	0	0
Atrazine	0	1	1
	2.5	0	0
	5.0	0	0
Trifluralin	0	2.5	0
	2.5	1	0
	5.0	0	0
Chlorophthalim	0	2	1[c]
	2.5	2	0.5[c]
	5.0	1	0[c]

Note. Four hundred liters per hectare of the mixture of Sanpex and herbicide was sprayed in upland conditions. Toxicity effect was evaluated 20 days after application. Crabgrass (*Digitaria adscendens* Henr.) and livid amaranth (*Amaranthus lividus* Loisel.) were completely killed in all application plots, showing a score of 5.

[a] From Watanabe (unpublished, 1982).

[b] Toxicity scale: 0 = no injury and 5 = dead.

[c] Test plant is soybean.

An additional example is the structural modification of alachlor to form the herbicide butachlor. Alachlor is a selective preemergence herbicide for maize (*Zea mays* L.), soybean [*Glycine max* (L.) Merr.], cotton (*Gossypium hirsutum* L.), sugar cane (*Saccharum officinarum* L.), peanuts (*Arachis hypogaea* L.), radish, rapeseed (*Brassica napus* L.), and other brassicas in upland conditions, but causes severe injury in paddy rice culture. Butachlor results from a slight change in the methoxy group of alachlor, which is replaced by a butoxy radical, and has been used for rice culture in Japan and Korea. Recently, bensulfuron-methyl, containing a benzyl structure in place of the phenyl one in chlorsulfuron, was found to be selective to rice (Takeda *et al.*, 1985).

IV. USE OF CROP SAFENER/HERBICIDE COMBINATIONS IN JAPAN

A. Historical Background of Crop Safeners for Herbicides in Japan

Twenty-five years ago, Matsunaka (1963), in his review "Development of Antidotes for Herbicides," classified antidotes (crop safeners) into five groups: (1) binding type, (2) degradation type, (3) competition type, (4) compensation type, and (5) exciting type.

A typical example of *binding-type* safeners is the use of activated charcoal to inactivate herbicidal residues, such as diuron (Bayer, 1967) and simazine or linuron (Anderson, 1968) in soil. In Japan, Yamane *et al.* (1962) reported the utilization of charcoal for dichlobenil in wheat (*Triticum aestivum* L.) cultivation. They reported that, when sown 6 days after dichlobenil application, 90% of wheat seeds treated with charcoal could emerge, whereas no emergence was observed in the absence of charcoal. The use of activated charcoal as a safener for herbicides is reviewed in detail by Hoagland, Chapter 11, this volume.

Inside the maize plant, simazine and atrazine may be transformed to their inactive hydroxy derivatives by means of a chemical reaction mediated by an endogenous catalyst, 2,4-dihydroxy-7-methoxy-1,4-benzoxazin-3-one. Other biochemical detoxication mechanisms, including conjugation to glutathione (GSH) and N-dealkylation reactions, are also known to inactivate triazine herbicides in maize (Shimabukuro *et al.*, 1978). In soils, pyridine, hydroxylamine, or calcium polysulfide can enhance the degradation of simazine (Castelfranco and Brown, 1962). These compounds are referred to as *degradation-type* safeners. Riboflavin accelerates the photodegradation of amitrole (Castelfranco *et al.*, 1963). Matsunaka (1963) confirmed that pretreatment of rice seedlings with riboflavin solution before transplanting enhanced rice tolerance to amitrole applied to the soil. If the glutathione and glutathione-*S*-transferase theory for the action of the safener dichlormid which protects maize from thiocarbamate herbicides (Lay and Casida, 1976) is true, this compound can be classified as a degradation-type safener.

Examples of *competition-type* safeners are the many safeners for barban injury on oat (*Avena sativa* L.), which were described by Hoffman (1962).

The herbicidal symptoms of the Hill reaction inhibitors, phenylureas and *s*-triazines, can be relieved by the addition of sugars (Gentner and Hilton, 1960; Moreland *et al.*, 1959). In this case sugars may be referred to as *compensation-type* safeners, but this has not found practical uses. Purine derivatives, such as adenine or guanine for amitrole (Sund *et al.*, 1960) and pantothenic acid for dalapon (Hilton *et al.*, 1959), may also be considered as safeners of this type.

Exciting-type safeners excite or promote the total activities of crop plants for detoxication or recovery from the injury of herbicides, but could not be found in the literature.

In Japan, Ichizen (1980) has also reviewed "Development of Antidotes for Herbicides," in which he compiled and discussed recent advances in chemical and physical safening for herbicides.

The following combinations of safeners and herbicides would be cited as successful case histories of crop safening in Japan.

B. Reduction of Phytotoxicity of Thiobencarb

1. Methoxyphenone

Thiobencarb, also known as benthiocarb, is sufficiently nonphytotoxic to rice when used in paddy fields. Nevertheless, a dwarfing injury to rice plants has occurred in paddy fields under special conditions and this has been attributed to the use of this herbicide (Shigematsu *et al.*, 1980). A dechlorinated metabolite of thiobencarb, *S*-benzyl-*N*,*N*-diethylthiocarbamate, has been shown to be the cause of dwarfing (Ishikawa *et al.*, 1980; Tatsuyama *et al.*, 1981). After extensive biological and biochemical studies using rice seedlings, it was found that dechlorinated thiobencarb is formed only in the soil under strong reductive conditions, but not in sterilized soil even in the presence of organic matter. This suggested that the dechlorination of thiobencarb is mediated by soil microorganisms (Abe and Kuwatsuka, 1983; Kuwatsuka, 1983). If the activities of the soil microorganisms responsible for the dechlorination of thiobencarb could be suppressed, this herbicide could be used safely without any risk of adverse effects to rice.

Researchers of Kumiai Chemical Industries, the producer of thiobencarb, evaluated the effects of many practical herbicides on the dwarfing injury of rice plants caused by thiobencarb. They found a decrease of such injury when thiobencarb was used in combination with methoxyphenone (Fig. 1). Methoxyphenone has been shown to act as an inhibitor of the dechlorination of thiobencarb.

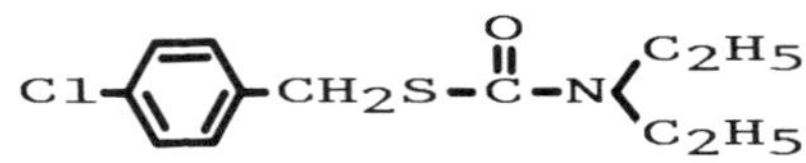

Thiobencarb
(Benthiocarb)

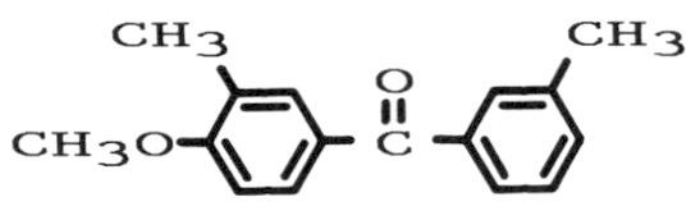

Methoxyphenone

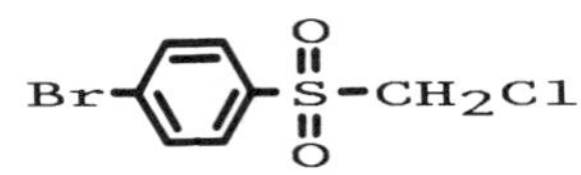

4-Bromophenyl chloromethyl sulfone (BCS)

Fig. 1. Thiobencarb herbicide and safeners.

Shigematsu *et al.* (1981) have observed safening activity of methoxyphenone in their model experiments. In rice plants grown hydroponically, dechlorinated thiobencarb was not detected by residue analysis, although it caused plant injury when added to the hydroponic solution. Under these conditions no safening effect of methoxyphenone for either thiobencarb or dechlorinated thiobencarb was found. The remarkable safening effect of methoxyphenone is observed only when it is applied with thiobencarb in paddy soil culture conditions. It was concluded that this safener may inhibit the microbial dechlorination of thiobencarb.

Chisaka (1981) studied the behavior of methoxyphenone in combination with thiobencarb and observed that methoxyphenone was rapidly reduced in soil to form its alcohol derivative. This fact may suggest that methoxyphenone serves as an alternative substrate, like methylendioxyphenyl compounds in the synergism of pyrethroid insecticides (Yamamoto, 1973). The discovery of such a biochemical effect of methoxyphenone has given new impetus to develop more effective and cheaper compounds having such safening property.

2. 4-Bromophenyl Chloromethyl Sulfone (BCS)

4-Bromophenyl chloromethyl sulfone (BCS) (Fig.1) has been developed recently by Kumiai Chemical Industries as a safener of rice against injury caused by the herbicide thiobencarb (Hirai *et al.*, 1986; Mitsumori *et al.*, 1985; Tauchi *et al.*, 1986). Small amounts of this chemical incorporated into the formulated product of thiobencarb are effective in reducing the phytotoxicity of rice and BCS has no biological activity on any crops and weeds (Mitsumori *et al.*, 1985).

The consumption of thiobencarb herbicide and its combinations with simetryn, chlornitrofen, or simetryn plus MCPB increased steadily until 1979. In 1982, methoxyphenone was added to the thiobencarb combined formulations for the purpose of safening. However, in 1983, the formulations of thiobencarb combined with other herbicides excluding methoxyphenone increased to 26,805 tons from 17,419 tons in 1982, as shown in Table II. From these statistics we can deduce that starting in 1983, a new safener was added to the formulations of thiobencarb herbicide replacing methoxyphenone. There is no publication indicating that the new safener is BCS. However, one of the above-mentioned toxicological studies of BCS (Hirai *et al.*, 1986) was conducted by researchers of Kumiai Chemical Industries and such a deduction concerning the new safener, BCS, seems to be quite reasonable. If so, BCS may be considered as one of the most successful practical safeners for herbicides all over the world.

3. Others

Ichizen (1976) has investigated the phytotoxicity of thiobencarb itself to improve its selectivity between rice plants and barnyard grass and to find effective safeners. 1-Naphthoxyacetic acid, 2,6-dichlorophenoxyacetic acid, 2,4,6-trichlorophenoxyacetic acid, and 3-chloro-4-methoxyaniline were all active as safeners of rice against thiobencarb. On the other hand, naphthalic anhydride (NA) and dichlormid were less active.

C. 1,8-Octamethylene Diamine: A Safener for Specific Urea Herbicides

N-(3,4-Dichlorophenylcarbamoyl)-*N*-methylglycine monohydrate and *N*-(3,4-dichlorophenylcarbamoyl)-*N*-methyl-β-alanine methylester are two new urea herbicides effective against broad-leaved weeds. Okii *et al.* (1979) studied their selectivity between wheat and weeds and found that 1,8-octamethylene diamine acts as a wheat safener for both compounds. This chemical can be used as a foliar application and its safening effects may be attributed to the prevention of the foliar uptake of these herbicides. This chemical also reduces the phytotoxicity of fenac to rice and wheat and that of chloramben to soybeans.

D. Safeners for Chlorophthalim

Chlorophthalim is one of the most potent and cheapest compounds in the imide series of herbicides. K. Wakabayashi and H. Watanabe (unpublished, 1982)

Table II Sales of Thiobencarb-Combined Herbicides in Japan

	Sales of thiobencarb herbicides in tons		
Year	Containing methoxyphenone	Others	Total
1982	9,218	17,419	26,637
1983	527	26,805	27,332
Difference	−8,691	+9,406	+695

screened many chemicals in searching for a suitable safener for chlorophthalim and NA was found to be the most active. NA improved the selectivity of chlorophthalim in maize plants as shown in Table III.

E. Thiocarbamate Herbicides as Safeners for the Rice Herbicide Bensulfuron-methyl

Bensulfuron-methyl (BSM, DPX-F5384) is a member of the sulfonylurea group of herbicides, which possess extremely high herbicidal activity and very low mammalian toxicity.

Although *Indica*-type rice varieties have a high tolerance to bensulfuron-methyl, *Japonica* rice varieties are slightly injured by higher dosages of this herbicide. The degree of injury caused by 100 g a.i./ha of bensulfuron-methyl was rated as a 31% reduction in dry weight of transplanted rice plants and a crop injury rating of 4 (rating 0 = no injury and 10 = plant killed) (Yuyama *et al.* 1986).

The following facts on safening effects on bensulfuron-methyl were reported by Yuyama *et al.* (1986): Table IV summarizes the safening effect of several thiocarbamate herbicides combined with bensulfuron-methyl. Thiobencarb, dimepiperate (MY-93), isopropylate (CH-83), and SC-2957 (Fig. 2) significantly reduced bensulfuron-methyl injury to the paddy rice plants, whereas molinate, although being similar to these thiocarbamates in structure and herbicidal activity, provided marginal or no safening.

Japonica-type rice is more prone to herbicide injury when it is directly seeded (water-sown) than when transplanted. Most thiocarbamate herbicides and bensulfuron-methyl cause marginal to moderate injury on direct-seeded rice at the first leaf growth stage. The herbicide MY-93, however, exhibits reasonable

Table III Safening Activity of 1,8-Naphthalic Anhydride for Chlorophthalim[a]

Plot	Chlorophthalim dosage (kg a.i./ha)	Injury[b]: Crops: Maize	Injury[b]: Crops: Soybean	Injury[b]: Weeds: Crabgrass	Injury[b]: Weeds: Lambsquarter[c]	Injury[b]: Weeds: Velvetleaf[d]
Chlorophthalim (WP) alone	1	3	1	10	10	7
	2	5	3	10	10	9
Chlorophthalim (WP) with NA[e]	1	0	1	10	10	7
	2	0	3	10	10	9

[a] From K. Wakabayashi and Watanabe (unpublished, 1980).
[b] Toxicity scale: 0 = no injury, 10 = plant dead.
[c] *Chenopodium album* L.
[d] *Abutilon theophrasti* Medik.
[e] NA: 0.5% seed dressing.

Table IV Safening of Bensulfuron-methyl (BSM) with Thiocarbamate Herbicides in Transplanted *Japonica* Rice[a]

		Crop injury[b]		Shoot dry weight (% of untreated check)	
Additives to BSM	Dosage (kg a.i./ha)	Combination	Additive only	Combination	Additive only
BSM only[c]	—	3.4	—	77	—
Thiobencarb[d]	2.1	1.7	0	97	103
MY-93	3.0	1.3	0	100	105
CH-83	3.0	1.5	0	95	95
SC-2957	2.8	1.5	0.5	96	98
Molinate	2.4	2.5	0	88	98
Untreated check (no BSM)	—	0	—	100	—

[a] Modified from Yuyama *et al.* (1986).
[b] Toxicity scale: 0 = no injury, 10 = plant dead; estimated at 3 weeks after application.
[c] BSM dosage = 100g a.i./ha.
[d] Thiobencarb = benthiocarb.

safety as shown in Fig. 3. It was concluded that mixtures of MY-93 at 2.0–3.0 kg a.i./ha with bensulfuron-methyl at 40–50 g a.i./ha would be safe for use on direclty water-seeded rice plants.

The combination of bensulfuron-methyl with thiobencarb or MY-93 was effective in controlling barnyard grass and Japanese bulrush (*Scirpus juncoides* Roxb. var. *hotarui* Ohwi) when chemicals were applied at the 1-, 2-, and 3-leaf

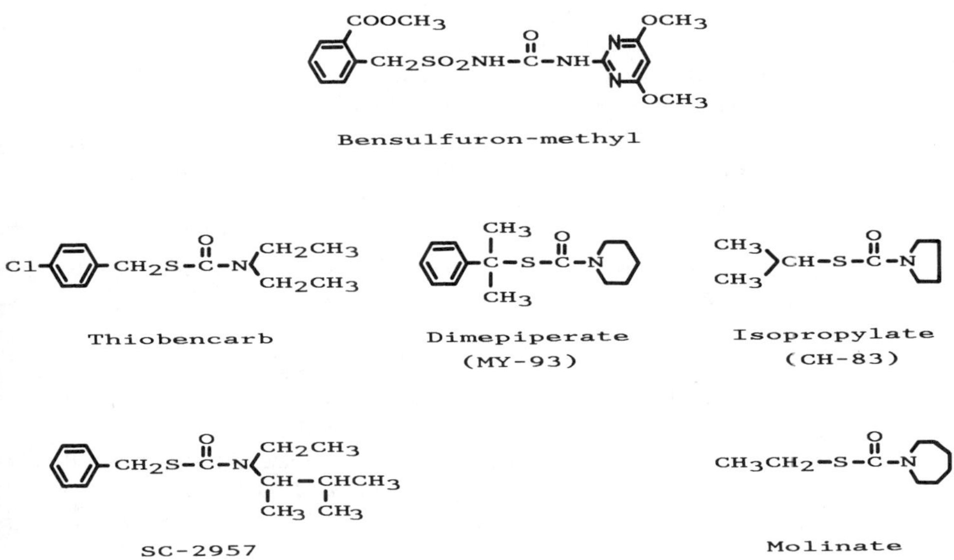

Fig. 2. Bensulfuron-methyl herbicide and safeners.

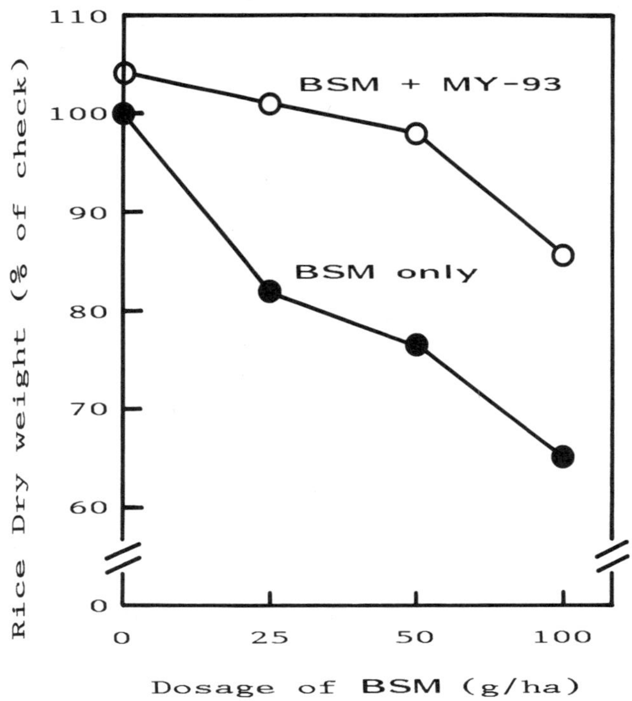

Fig. 3. Rice safening against bensulfuron-methyl (BSM) with dimepiperate (MY-93). (From Yuyama *et al.*, 1986.)

stages, although bensulfuron-methyl and thiobencarb or MY-93 alone did not provide sufficient control of barnyard grass and Japanese bulrush, respectively.

The metabolism of bensulfuron-methyl by rice appears to play an important role in the plant tolerance to this herbicide and involves mainly the change of the methoxy compound into an inactive hydroxy metabolite (Takeda *et al.*, 1986). Foliar treatment of rice (24 hr prior to bensulfuron-methyl application) with either thiobencarb or MY-93 gave a three- to fourfold increase in the rate of the metabolism of this herbicide by the excised rice leaves (Table V). For example, following treatments with MY-93 or thiobencarb, the half-life ($T_{1/2}$) of bensulfuron-methyl decreased from 8.4 hr to 2.6 and 1.9 hr, respectively. Molinate did not significantly increase the rate of bensulfuron-methyl metabolism, and also did not appreciably reduce the bensulfuron-methyl injury to rice plants (Table IV).

Treatment of rice with thiobencarb significantly increased the metabolism of bensulfuron-methyl even under conditions favoring rapid metabolism. Thus, in young rice leaves, bensulfuron-methyl had a $T_{1/2}$ of 4.6 hr and a foliar spray with thiobencarb decreased the $T_{1/2}$ to 2.9 hr. The addition of thiobencarb granules (2.5 kg a.i./ha) to the water of a paddy rice field almost doubled the rate of bensulfuron-methyl metabolism.

Table V Effect of Thiocarbamate Herbicides on the Metabolism of Bensulfuron-methyl by Excised Rice Leaves[a]

Foliar spray[b]	Half-life of bensulfuron-methyl[c] in excised leaves (hr)
Solvent control[d]	8.4
Thiobencarb (2 kg a.i./ha)	1.9
MY-93 (2 kg a.i./ha)	2.6
Molinate (2 kg a.i./ha)	7.9

[a] From Yuyama *et al.* (1986).
[b] Foliar spray of rice plants was made 24 hr prior to excised leaf uptake of bensulfuron-methyl.
[c] Bensulfuron-methyl was taken through cut ends of excised leaves (1-hr uptake from 2 ppm solution of the herbicide).
[d] Solvent: 20% acetone plus 0.1% Tween 20.

These studies indicate that the safening action of thiobencarb and MY-93 is primarily due to their ability to significantly increase the rate of bensulfuron-methyl metabolism by rice plants. Since the bensulfuron-methyl metabolism in common weeds is very slow, the acceleration of the metabolism of bensulfuron-methyl in weeds is negligible, resulting in no decrease of herbicidal activity.

It is very interesting to note that thiobencarb is a unique herbicide which can be both safened on rice by methoxyphenone or BCS (see Sections IV,B,1 and IV,B,2 of this chapter) and also serve as a safener of rice against injury from the herbicide bensulfuron-methyl.

F. Brassinolide as a Safener against Herbicides

Brassinolide is a new plant growth regulator which was isolated from pollen of rapeseed (*Brassica napus* L.), and its structure was determined by Grove and associates (1979). In Japan, brassinolide itself and other brassinosteroids are chemically synthesized and are evaluated for agricultural uses. Fujita's review (1985) shows that brassinolide has almost no activity on crop plants which are growing normally but does affect them when they grow under various stresses, such as low temperature, herbicide treatment, disease infection, physical cutting, or salt accumulation. The review also shows that this growth regulator accelerates the growth of special recessive plant parts, i.e., weak tillers or spikelets of rice or wheat (Luo *et al.*, 1986), and that it has a somewhat delayed action after application but extremely high activities, being effective at 0.01–100 ppb.

Uesono and others (1985) tried to utilize brassinolide as a safener against herbicides and found that this growth regulator was effective for safening rice against simetryn, butachlor, and pretilachlor and wheat against simazine. Crop safening by brassinolide was observed with seed or root treatment, but not with foliar applications of this growth regulator. The herbicides applied 5 to 7 days after brassinolide treatment were more effectively detoxified than those applied just after the treatment. They also found that the absorption of simetryn by rice plant seedlings was inhibited 44–73% by treatment at 1–10 ppt of the safener. The safening effect of brassinolide against phytotoxicity by simetryn, prometryn, atrazine, and diuron in rice plants was confirmed also in our laboratory (S. Matsunaka *et al.*, unpublished, 1987).

V. CONCLUDING REMARKS

In Japan, rice production is very important, and for effective and safe weed control in paddy fields many granular herbicide formulations are being utilized. The relative higher cost of such formulations is compensated by the ease of application and safety to the transplanted rice plants. In the case of special herbicides, such as oxadiazon, emulsifiable concentrates are being applied to flooded muddy fields directly from small herbicide bottles without dilution during the puddling process, forming herbicide-bound soil particles which form a thin herbicide layer on the soils which prevents rice injury and kills weeds.

Although there have been some attempts to use controlled release herbicides, such formulations are not currently practical in Japan.

The endeavors for structural modification of herbicides to get higher selectivity continue and utilize fundamental information of biochemical and genetic mechanisms of herbicide selectivity.

The use of crop safener/herbicide combinations has been studied extensively in Japan for the past 25 years. Practical success is exemplified by the development of safeners for the protection of rice against dechlorinated thiobencarb and the use of thiocarbamate safeners of rice against bensulfuron-methyl.

An alternative idea for crop safening against herbicides is the breeding of herbicide-resistant crop varieties (LeBaron and Gressel, 1982). For this purpose, an atrazine-tolerant tobacco (*Nicotiana tabacum* L.) plant was created by Furusawa *et al.* (1984) through a selection method of protoplast population. The Mitsubishi Plant-Tech Research Institute has begun a search for imide-herbicide-tolerant crops through a cell fusion method, and research in the laboratory of one of the authors (Hirata and Matsunaka, 1985; Kawaguchi and Matsunaka, 1987) is directed toward the selection of paraquat-tolerant cell-fusant from protoplasts of carrot (*Daucus carota* var. *sativus* L.) and paraquat-resistant Canadian fleabane (*Erigeron canadensis* L.). The Meiji Seika Kaisha is investigating the selection of tolerant crops to bialaphos, the first practical antibiotic herbicide, through a gene-manipulation procedure.

With these new technologies available for crop safening, herbicides will be more widely and safely utilized and will continue to contribute to the modernization of agriculture in Japan and in other countries.

References

Abe, H., and Kuwatsuka, S. (1983). Enrichment of microorganisms capable of decomposing benzyl thiocarbamate herbicides in the soil. *Int. Congr. Pestic. Chem., 5th, 1982,* Abstr. No. Vf-11.

Anderson, A.H. (1968). The inactivation of simazine and linuron in soil by charcoal. *Weed Res.* **8**, 58–60.

Bayer, D.E. (1967). Effect of surfactants on leaching of substituted urea herbicides in soil. *Weeds* **15**, 249–252.

Castelfranco, P., and Brown, W.S. (1962). Purification and properties of simazine resistance factor of *Zea mays*. *Weeds* **10**, 131–136.

Castelfranco, P., Oppenheim, A., and Yamaguchi, S. (1963). Riboflavin-mediated photodecomposition of amitrole in relation to chlorosis. *Weeds* **11**, 111–115.

Chisaka, H. (1981). Formulation of dechlorinated thiobencarb in soils and its inhibition by methoxyphenone. *Weed Res. (Tokyo)* **26**, Suppl., 163–164 (in Japanese).

Fujita, F. (1985). Expectation to brassinolide for agricultural utilization. *Kagaku to Seibutsu* **23**, 717–725 (in Japanese).

Furusawa, I., Mizuguchi, A., Hasiguchi, K., Tanaka, K., and Asada, K. (1984). Herbicide tolerant calli from protoplasts and their regenerated plants. *Weed Res. (Tokyo)* **29**, Suppl., 157–158 (in Japanese).

Gentner, W. A., and Hilton, J.L. (1960). Effect of sucrose on the toxicity of several phenylurea herbicides to barley. *Weeds* **8**, 413–417.

Grove, M. D., Spencer, G. F., Rohwedder, W. K., Mandava, N., Worley, J. F., Warthen, J. D., Jr., Steffens, G. L., Flippen-Anderson, J. L., and Cook, J. C., Jr., (1979). Brassinolide, a plant growth-promoting steroid isolated from *Brassica napus* pollen. *Nature (London)* **281**, 216–217.

Hilton, J. L., Ard, J. S., Jansen, L. L., and Gentner, W.A. (1959). The pantothenate synthesizing enzyme, a metabolic site in the herbicidal action of chlorinated aliphatic acids. *Weeds* **7**, 381–396.

Hirai, A., Mizuhashi, F., Murata, T., and Kitagaki, T. (1986). Toxicology studies on the safety of 4-bromophenyl chloromethyl sulfone. *Kiso to Rinsho* **20**, 117–125 (in Japanese).

Hirata, T., and Matsunaka, S. (1985). Paraquat tolerant mechanism of *Erigeron canadensis*. *Weed Res. (Tokyo)* **30**, Suppl., 127–128 (in Japanese).

Hoffman, O.L. (1962). Chemical seed treatments as herbicidal antidotes. *Weeds* **10**, 322–323.

Ichizen, N. (1976). Studies on herbicidal activities and synergistic effects of benthiocarb. *Chem. Regul. Plants* **11**, 56–63 (in Japanese).

Ichizen, N. (1980). Development of antidotes for herbicides. *Weed Res. (Tokyo)* **25**, 245–252 (in Japanese).

Ishikawa, K., Shinohara, R., Yagi, A., Shigematsu, S., and Kimura, I. (1980). Identification of *S*-benzyl *N,N*-diethylthiocarbamate in paddy field soil applied with benthiocarb herbicide. *J. Pestic. Sci* **5**, 107–109.

Kawaguchi, S., and Matsunaka, S. (1987). Cell fusion between paraquat tolerant *Erigeron canadensis* and carrot. *Weed Res. (Tokyo)* **32**, Suppl., 100–101 (in Japanese).

Kawamura, Y. (1973). A simplified application method for weed control in paddy fields with RP-17623 (G-315). *Proc. Asian-Pac. Weed Sci. Soc. Conf., 3rd,* 1971, pp. 176–184.

Konnai, M., Ichizen, N., and Takahashi, Y. (1967). Synergistic action of cyclohexenylcyclohexanone on the herbicidal activity of propanil. *Weed Res. (Tokyo)* **6**, 74–79 (in Japanese with English summary).

Kuwatsuka, S. (1983). Fate of herbicides in flooded paddy soils. *In* "Pesticide Chemistry: Human Welfare and the Environment" (J. Miyamoto and P.C. Kearney, eds.), Vol. 2, pp., 345–349. Pergamon, Oxford.

Lay, M.M., and Casida, J.E. (1976). Dichloroacetamide antidotes enhance thiocarbamate sulfoxide detoxication by elevating corn root glutathione content and glutathione-*S*-transferase activity. *Pestic. Biochem. Physiol.* **6**, 442–456.

LeBaron, H. M., and Gressel, J., eds. (1982). "Herbicide Resistance in Plants." Wiley, New York.

Luo, B.-S., Kumura, A., Ishii, R., and Wada, D. (1986). The effects of brassinolide treatments on growth and development processes in wheat plants. *Jpn. J. Crop Sci.* **55**, 291–298 (in Japanese).

Matsunaka, S. (1963). Development of antidotes for herbicides. *Weed Res. (Tokyo)* **2**, 5–11 (in Japanese).

Matsunaka, S. (1982). Evolution of rice weed control practices and research: World perspective. *In* "Weed Control in Rice," pp. 5–17. Int. Rice Res. Inst., Manila, Philippines.

Matsunaka, S. (1983). Essential characteristics of herbicides for rice culture. *In* "Pesticide Chemistry: Human Welfare and the Environment" (J. Miyamoto and P.C. Kearney, eds.), Vol. 2, pp. 325–330. Pergamon, Oxford.

Mitsumori, K., Maita, K., Kosaka, T., Miyaoka, T., and Shirasu, Y. (1985). Thirteen-week subacute toxicity study of 4-bromophenyl chloromethyl sulfone (BCS) in rats. *J. Toxicol. Sci.* **10**, 295–308.

Moreland, D. E., Gentner, W. A., Hilton, J. L., and Hill, K.L. (1959). Studies on the mechanism of herbicidal action of 2-chloro-4,6-bis(ethylamino)-*s*-triazine. *Plant Physiol.* **34**, 432–435.

Okii, S., Nishimuta, T., Konnai, M., and Takematsu, T. (1979). Studies on the antidote, octamethylenediamine (OM), as foliar application type. V. The efficacy of OM under the solution culture condition and speculative mechanisms of antidotal action by OM. *Weed Res. (Tokyo)* **24**, 226–232 (in Japanese).

Saito, H., and Hayabe, K. (1962). Studies on the movability of herbicide in the soil by mixing OED. *Weed Res (Tokyo)* **1**, 78–80 (in Japanese).

Shigematsu, S., Yamada, Y., Yamauchi, S., and Kimura, I. (1980). Studies on rice injury by benthiocarb herbicides. I. Culture conditions causing the injury and the safening measures. *Weed Res. (Tokyo)* **25**, Suppl., 25–26 (in Japanese).

Shigematsu, S., Yamada, Y., Fukai, Y., Abe, H., Kimura, I., Ozawa, K., and Nakano, T. (1981). Studies on rice injury by benthiocarb herbicides. VI. Activities of NK-049 as a safener. *Weed Res (Tokyo)* **26**, Suppl., 163–164 (in Japanese).

Sund, K. A., Putala, E. C., and Little, H.N. (1960). Reduction of 3-amino-1,2,4-triazole phytotoxicity in tomato plants. *J. Agric. Food Chem.* **8**, 210–212.

Takeda, S., Yuyama, T., Ackerson, R. C., and Weigel, R.C. (1985). Selection of rice herbicides from several sulfonylurea compounds. *Weed Res. (Tokyo)* **30**, 278–283.

Takeda, S., Erbes, D. L., Sweetser, P. B., Hay, J. V., and Yuyama, T. (1986). Mode of herbicidal and selective action of DPX-F5384 between rice and weeds. *Weed Res. (Tokyo)* **31**, 157–163.

Takematsu, T. (1968). "Chemical Weed Control in Paddy Field." Hakuyu-sha Press, Tokyo (in Japanese).

Takematsu, T., Konnai, M., and Takeuchi, Y. (1978) Studies on utilization of simazine with latex emulsion in turf. *J. Jpn. Turfgrass Res. Assoc.* **7**, 129–132 (in Japanese).

Tatsuyama, K., Yamamoto, H., and Egawa, H. (1981). Bioassay of dechlorination of thiobencarb herbicide in flooded soil using germinated grains of rice plants. *J. Pestic Sci.* **6**, 193–199 (in Japanese with English summary).

Tauchi, K., Kawanishi, H., and Takeshima, T. (1986). Teratogenicity study on 4-bromophenyl chloromethyl sulfone (BCS) in the rat. *J. Toxicol. Sci.* **11**, 125–134.

Uesono, T., Nishi, S., Fujiwara, S., and Hamada, K. (1985). On the safening effect of brassinolide against herbicide injury. *10th Annu. Meet. Pestic. Sci. Soc. Jpn.,* Abstr. No. A108 (in Japanese).

Yamamoto, I. (1973). Mode of action of synergists in enhancing the insecticidal activity of pyrethrum and pyrethroids. *In* "Pyrethrum: The Natural Insecticide" (J.E. Casida, ed.), pp. 195–210. Academic Press, New York.

Yamamoto, I., and Katsuda, Y. (1980). β-Cyclodextrin inclusion complexes of pyrethroids. *Pestic. Sci.* **11**, 134–140.

Yamane, K., Ishida, T., and Otani, R. (1962). Injury of 2,6-dichlorobenzonitrile to wheat seed and protection by coating with active charcoal. *Rep. Hyogo Prefect. Agric. Exp. St.* **10**, 53–54 (in Japanese).

Yuyama, T., Sweetser, P. B., Ackerson, R. C., and Takeda, S. (1986). Safening of DPX-F5384 on rice by combination with thiocarbamate herbicides. *Weed Res. (Tokyo)* **31**, 164–170.

Part Two

Physiological and Biochemical Mechanisms of Action of Herbicide Safeners

Chapter 3

Mechanisms of Action of Herbicide Safeners: An Overview

Kriton K. Hatzios
Department of Plant Pathology, Physiology and Weed Science
Virginia Polytechnic Institute and State University
Blacksburg, Virginia 24061-0330

I. INTRODUCTION

Investigations into the possible mechanisms of action of herbicide safeners were initiated in the early 1970s, soon after the introduction of naphthalic anhydride (NA) and dichlormid as protectants of corn against thiocarbamate herbicide injury, and have continued since that time. Much of this work has been reviewed previously by a number of investigators in varying levels of detail (Blair *et al.*, 1976; Fedtke, 1985; Gressel *et al.*, 1982; Hatzios, 1983c, 1984c; Pallos and Casida 1978; Parker, 1983; Stephenson and Ezra, 1982, 1983, 1985).

Although the effects of safeners as crop protectants against herbicide injury are not difficult to demonstrate under field, greenhouse, and laboratory conditions, the exact physiological and biochemical mechanisms of the protective action of safeners have not been elucidated. Most, if not all, possible interactions between safeners and herbicides have been tested and several theories have been proposed. Of the mechanisms proposed, two have received the most attention, and suggest either a safener-induced enhancement of herbicide detoxication (Lay and Casida, 1976) or a competitive antagonism of herbicides and safeners at a common target site of action (Stephenson and Chang, 1978). Early investigations on the physiological activity of safeners postulated that safeners may act by a single mechanism which was assumed to be common in all crop–herbicide–safener combinations. The evidence, however, for the exclusive involvement of one of these mechanisms in the action of safeners is not always conclusive and does not preclude the possibility of an additional contribution by other mechanisms. In fact, more recent investigations (Ezra *et al.*, 1983a) suggest that the action of the currently available safeners is most likely the result of a series of multilevel interactions between safeners and herbicides which contribute to the overall protection of grass crops from herbicide injury.

The purpose of this chapter is to present an overview of the fundamental concepts underlying the physiological, biochemical, and molecular aspects of the potential mechanisms of action of the currently marketed safeners which protect grass crops against thiocarbamate and chloroacetanilide herbicides. Schematic diagrams and models illustrating the principles involved in each mechanism of safener action are presented and representative examples supporting each mechanism are discussed. Additional information and data demonstrating the involvement of these mechanisms in specific crop–herbicide–safener interactions are presented in subsequent chapters of this book.

II. MECHANISMS OF SAFENER ACTION: FUNDAMENTALS

A. Specificity of Crop–Herbicide–Safener Combinations

At present, successful crop safening against herbicide injury is exemplified by only a few crop–herbicide–safener combinations, which were reviewed in Chapter 1, this volume. A critical examination of the existing literature on the physiological aspects of these successful crop–herbicide combinations leads to the recognition of at least four important characteristics underlying the protective action of the currently marketed herbicide safeners. These characteristics are summarized in Table I.

All the presently available safeners exhibit a high degree of botanical and chemical specificity and protect only certain crops against injury from selected groups of herbicides. Crops that can be protected include primarily large-seeded grasses such as corn (*Zea mays* L.), grain sorghum [*Sorghum bicolor* (L.) Moench], rice (*Oryza sativa* L.), and oats (*Avena sativa* L.). Herbicides antagonized by chemical safeners include primarily the soil-applied thiocarbamates and chloroacetanilides and to a lesser extent the phenylcarbamates, sulfonylureas, imidazolinones, and arylophenoxyalkanoic acids. Marketed safeners are derivatives of the following chemical classes: dichloroacetamides (e.g., dichlormid), oxime ethers (e.g., cyometrinil), thiazole carboxylates (e.g.,

Table I Characteristics of the Protective Action of Currently-Marketed Safeners

1. Chemical specificity
   Successful safening against soil-applied herbicides (thiocarbamates, chloroacetanilides)
2. Botanical specificity
   Large-seeded grass crops (corn, grain sorghum, rice, oats)
   Protected grass crops are moderately tolerant to antagonized herbicides
3. Safeners are most effective when applied prior to or simultaneously with the herbicides whose injury they prevent
4. Emerging shoot (coleoptile) of protected grass seedlings appears to be the site of uptake and action of both safeners and antagonized herbicides

flurazole), naphthopyranones (e.g., naphthalic anhydride, NA), and substituted phenyl-pyrimidines (e.g., fenclorim). Success in safening broadleaf crops against herbicide injury or in protecting grass crops against other groups of herbicides has been limited (Hatzios, 1983c; Parker, 1983).

The remarkable botanical specificity of the commercialized safeners coupled with the observation that the protected grass crops are moderately tolerant to the antagonized herbicides indicates that the protective action of the safeners may be related to some physiological, biochemical, or molecular function which is either unique or highly efficient in these grass crops. Thus, the biological system(s) affected by the thiocarbamate and chloroacetanilide herbicides or altered by chemical safeners in the protected grass crops is (are) either not present in other plants (e.g., broadleaf crops or weeds) or they may not be affected to the same extent by herbicides and safeners in unprotected herbicide-sensitive plants. The herbicidal effects of thiocarbamate and chloroacetanilide herbicides are complex and variable and our understanding of the exact biochemical mechanisms of action of these herbicides is far from complete (Fedtke, 1982). The general chemical structures of thiocarbamate and chloroacetanilide herbicides are shown in Fig. 1, whereas the main herbicidal characteristics shared by these two classes of herbicides are summarized in Table II.

The chemical specificity exhibited by safeners such as the dichloroacetamides has been attributed, at least partially, to their structural similarity to thiocarbamate and chloroacetanilide herbicides (Leavitt and Penner, 1978b; Stephenson and Chang, 1978; Stephenson *et al.*, 1978). In fact, the α-chloroacetamide derivative allidochlor or CDAA, which is chemically almost identical to the

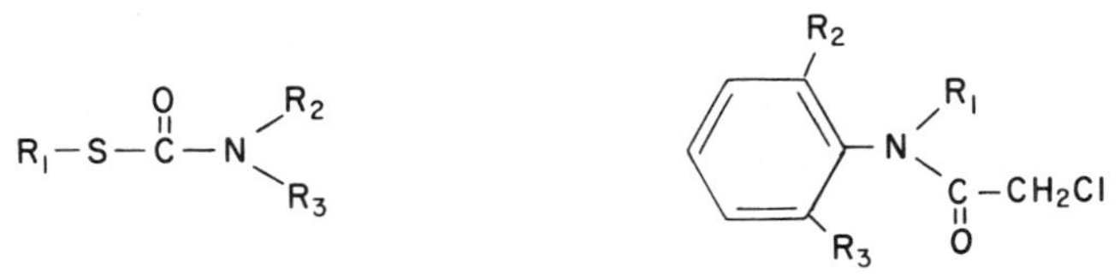

Fig. 1. General chemical structures of thiocarbamate and chloroacetanilide herbicides.

Table II Herbicidal Characteristics Shared by Thiocarbamate and Chloracetanilide Herbicides

Efficacy:	More toxic to grasses than broadleaved plants.
Site of uptake:	Shoot (coleoptile) uptake more effective than root uptake.
Site of action:	Most likely at or near the coleoptilar node of grasses.
Mechanism of action:	Generally classified as inhibitors of shoot growth of germinating seedlings. Chloroacetanilides affect root growth too. Specific mechanism of action is unknown.
Pattern of toxicity:	Toxicity is chronic rather than acute even on grasses.
Symptomology:	Shoot growth of treated grasses slows and eventually stops. Treated seedlings are stunted and malformed with dark green leaves which may also be twisted, bent, tightly rolled, and brittle.
Detoxication mechanisms:	Rapid conjugation with glutathione (GSH). GSH-conjugation of thiocarbamates is preceded by sulfoxidation. The GS-conjugates of both groups are catabolized to cysteine conjugates which are frequently N-acylated with malonic acid or deaminated to thiolactic acid conjugates.

safener dichlormid, is known to act both as a safener and as a herbicide (Chang *et al.*, 1973). Alternatively, it has been proposed that similar, but less phytotoxic molecules may act as safeners by elevating the same substrates and the activity of the same enzymes required for the detoxication of their more phytotoxic analogs in moderately tolerant plant species (Stephenson and Ezra, 1985; Stephenson *et al.*, 1983). However, other mechanisms are also possible and they are outlined in the next sections of this chapter.

B. Potential Mechanisms of Action of Herbicide Safeners

Any attempts to understand the possible mechanisms through which safeners confer crop tolerance on herbicides with marginal selectivity should start with a thorough evaluation of the bases of herbicide selectivity. The phenomenon of herbicide selectivity, whereby plants exhibit differential sensitivity toward a herbicide, is well known and has been a determining factor for the successful and extensive use of organic herbicides in modern crop production (Fedtke, 1982; Hatzios and Penner, 1982).

In a broad sense, tolerant plants are able to survive and grow at agriculturally recommended rates of a herbicide either by excluding the herbicide from reaching its site of action in an active form or by modifying the sensitivity of a target site (commonly a protein or enzyme) to this herbicide. Extensive biochemical investigations during the past two decades have identified several physiological and biochemical mechanisms as key determinants of plant tolerance to herbicides (Duessing, 1984; Wakabayashi and Matsunaka, 1982). These mechanisms include: (1) reduced uptake and/or transport of a herbicide from its site of application to the site of action inside a tolerant plant, (2) extensive metabolic detoxication of the herbicide in tolerant plants, (3) modification of the sensitivity of a target site of the herbicide in tolerant plants, (4) increased synthesis of target enzymes in tolerant plants, and (5) increased synthesis of substrates able to reverse the herbicide-induced inhibition of growth in tolerant plants.

These determinants of herbicide selectivity are under genetic control and, at a molecular level, plant tolerance or resistance to specific herbicides may result from any of the following qualitative or quantitative genetic changes: (1) alteration of structural genes, (2) gene regulation and (3) gene amplification (Hardy, 1986). Attempts to manipulate the determinants of plant tolerance to herbicides by means of recombinant DNA technology or traditional plant breeding procedures are currently the subject of intensive investigation (Gressel, 1986; Hatzios, 1987; LeBaron and Gressel, 1982; Zaitlin *et al.*, 1986). In addition, the patterns of translocation, penetration, and biotransformation of certain herbicides are known to be influenced by the prior or simultaneous applications of other pesticides giving rise to significant agrochemical interactions (Hatzios and Penner, 1985).

Therefore, it is not surprising that the major theories put forward for explaining the protective action of herbicide safeners propose the following mechanisms: (1) a safener-induced reduction of herbicide uptake and/or translocation in protected plants, (2) a safener-induced enhancement of herbicide detoxication mechanisms in the protected plants, (3) competition of safeners and herbicides for binding to or reacting with a common target site, and (4) combinations of the above mechanisms.

According to these mechanisms, safeners may act either as "bioregulators," influencing the amount of a herbicide that reaches its target site in an active form, or as "antagonists" of herbicidal effects at a common site of action. A safener-induced reduction in the amount of a herbicide reaching its site of action can be accomplished in two ways. The first involves a direct reaction of the safener with the herbicide to form an inactive chemical complex and has been described in the literature as "chemical antagonism" (Hatzios and Penner, 1985). The second mechanism is known as "biochemical antagonism" and results from a safener-induced alteration of the patterns of penetration, translocation, and biotransformation of a herbicide in the protected plants. When it is not possible to exclude the herbicide from reaching its site of action, safeners could elicit immunity to herbicide injury on certain crops by competing with herbicides at a common target site ("competitive antagonism") or by inducing alterations of the conformation or function of the physiologically important target site of a herbicide ("physiological antagonism"). The principles of these mechanisms and their potential involvement in the protective action of the currently available safeners are discussed next.

III. MECHANISMS OF SAFENER ACTION: PHYSIOLOGICAL AND BIOCHEMICAL ASPECTS

A. Exclusion of Herbicides from Reaching Their Site of Action

1. Chemical Antagonism

In this type of interaction a safener reacts chemically with the respective herbicide to form an inactive chemical complex, thus reducing the amount of herbicide that reaches its site of action. A schematic representation of this mechanism is given in Fig. 2.

Safeners may react chemically with herbicides in the soil, on plant surfaces, or inside the protected plant. A direct chemical reaction of safeners with herbicides in the soil appears to be the basis of the protective action of "herbicide adsorbents" such as activated carbon (charcoal), which has been used as a crop protectant against injury from soil-applied herbicides for many years (Blair *et al.*, 1976; Gupta, 1976). Activated carbon exhibits a great adsorptive capacity and acts as a physical barrier to herbicide uptake, excluding the herbicide from reaching its site of action. Other polymeric substances such as PC-671 (a formulated lignin by-product) and humic acids are also effective as crop protectants acting externally by adsorbing phytotoxic chemicals. The biological activity and other aspects of activated charcoal and lignin by-products are discussed by Hoagland (Chapter 11, this volume) and Riggle and Penner (Chapter 12, this volume).

Chemical antagonism may also be important as a mechanism of action of safeners which work inside the protected plant and its possible involvement in crop safening should not be overlooked. However, this mechanism has not been studied as extensively as other proposed mechanisms for the action of herbicide safeners. Chemical antagonism appears promising for the development of

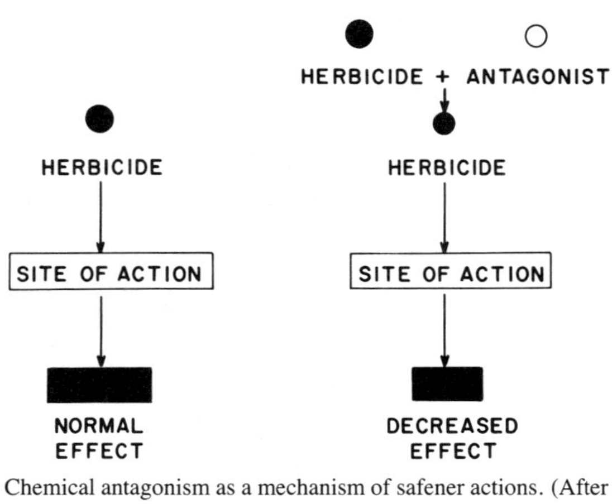

Fig. 2. Chemical antagonism as a mechanism of safener actions. (After Hatzios and Penner, 1985.)

safeners against herbicides which contain specific functional groups such as aliphatic or aromatic amino (R–NH_2) groups, carboxylates (COO–), phosphonate groups, etc., in their chemical structures. Such functional groups are capable of serving as effective electron donors in the binding of metal ions and metal chelation of organic herbicides bearing these groups is quite possible (Martell, 1971). Candidate herbicides that could be antagonized on crop plants by metal chelation include glyphosate, amitrole, picloram, and paraquat. Preliminary studies by some investigators have demonstrated the potential of this mechanism in minimizing the herbicidal effects of glyphosate on selected plant species (Turner and Loader, 1978).

Apart from the possible contribution of inactive herbicide–safener complexes in the reduction of the amount of active herbicide reaching its site of action, chemical complexes of herbicides with safeners may also be responsible for the removal of herbicides from their target site following the binding of such herbicides to receptor proteins. Such a mechanism has been postulated recently by Chang (1983) for the action of oxime ether safeners which protect grain sorghum against chloroacetanilide herbicide injury. Taking into consideration the electrophilic character of chloroacetanilide herbicides and the nucleophilic character of oxime ether safeners, he proposed a model which postulates the formation of a chemical complex between oxime safeners and chloroacetanilides ("acetamidized safener"). This complex results in the removal of these herbicides from their site of action, freeing a hypothetical receptor site which is essential for normal plant metabolism. Additional work is needed, however, to substantiate the validity and importance of this mechanism.

2. Biochemical Antagonism

Biochemical antagonism occurs when a safener reduces or eliminates completely the amount of a given herbicide that would otherwise be available at the site of action in the absence of the safener. This mechanism is illustrated in Fig. 3 and could result from either a safener-induced reduction in the rate of herbicide penetration and/or translocation or a safener-induced enhancement of the detoxication of a given herbicide in the protected plant.

a. Safener-Induced Reduction of Herbicide Uptake and/or Translocation. Studies on the physiological behavior of the safeners NA, dichlormid,cyometrinil, flurazole, and oxabetrinil (CGA-92194) have shown that the coleoptile of grass seedlings appears to be the main site of entry of these safeners into grass plants (Donald and Fawcett, 1976; Gray and Joo, 1978; Jackson *et al.*, 1986; Jordan and Jolliffe, 1971; Ketchersid *et al.*, 1982; Nyffeler *et al.*, 1980). The coleoptile of grass plants is also the main site of entry and action of the thiocarbamate and chloroacetanilide herbicides (Table II). Therefore, the possibility that safeners may protect grass crops from thiocarbamate or chloroacetanilide herbicide injury by preventing the uptake and/or translocation of these

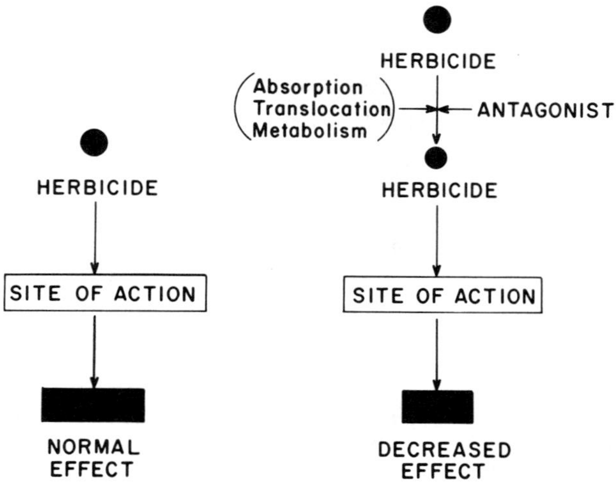

Fig. 3. Biochemical antagonism as a mechanism of safener action. (After Hatzios and Penner, 1985.)

herbicides in protected plants appears promising and has been tested (reviewed by Hatzios, 1983c; Stephenson and Ezra, 1983). The findings of such studies have been variable and often contradictory.

A significant reduction in the uptake and/or distribution of antagonized herbicides by selected safeners has been reported for the following herbicide–safener–crop combinations: metolachlor–NA–grain sorghum (Ahrens and Davis, 1978); metolachlor–cyometrinil–grain sorghum (Ketchersid *et al.*, 1982); barban–NA–oats (Thiessen, 1978); and EPTC–dichlormid–corn (Ezra *et al.*, 1982).

In contrast, a significant stimulation of herbicide uptake and/or distribution has been observed in the following herbicide–safener–crop combinations: EPTC–NA–corn (Ezra *et al.*, 1982; Güneyli, 1971); EPTC–dichlormid–corn (Carringer *et al.*, 1978b; Sagaral, 1978); EPTC–cyometrinil–corn (Ezra *et al.*, 1982); metolachlor–oxabetrinil–grain sorghum (Fuerst and Gronwald, 1986; Zama and Hatzios, 1986; 1987); and alachlor–flurazole–grain sorghum (Rubin *et al.*, 1985).

Finally, a number of investigators have failed to detect any significant alteration (reduction or stimulation) of chloroacetanilide or thiocarbamate herbicide uptake and/or translocation by safeners in protected grass crops (Chang *et al.*, 1974; Christ, 1981, 1985; Holm and Szabo, 1974; Jackson *et al.*, 1986; Marton *et al.*, 1978; Murphy, 1972; Winkle *et al.*, 1980).

The effects of safeners on herbicide uptake and/or translocation could be direct or indirect and appear to be dependent on the concentration of the herbicide and safener used as well as on the plant tissue examined (Ezra *et al.*, 1982; Zama and Hatzios, 1987). A safener-induced alteration in the patterns of uptake and/or translocation of a herbicide is direct when it results from an antagonistic interaction of the safener and the herbicide on membrane permeability or a competition for "active sites" of uptake (Ezra *et al.*, 1982). Safener-induced changes in the patterns of uptake and/or distribution of thiocarbamate and chloroacetanilide herbicides in protected plants could also be indirect. The possible contribution of an indirect safener effect most likely results from the prevention by safeners of the effect of thiocarbamate and chloroacetanilide herbicides on epicuticular wax formation and cuticular integrity of corn and grain sorghum plants (Ebert, 1982; Görög *et al.*, 1982; Leavitt and Penner, 1978a). This effect of safeners causes a decrease in the transpiration rate of the protected plants and therefore a reduced penetration and/or translocation of these soil-applied herbicides into the safener-treated plants (Ebert and Ramsteiner, 1984).

On the basis of current evidence, it is unlikely that the protective action of the presently available safeners against thiocarbamate and chloroacetanilide injury on corn or grain sorghum results totally from a safener-induced alteration of herbicide uptake and/or translocation. Such a mechanism may be a minor step in the series of possible interactions between herbicides and safeners and could contribute only partially to the action of the currently used safeners.

b. Safener-Induced Enhancement of Herbicide Detoxication. Differential herbicide metabolism in tolerant and susceptible plant species has long been recognized as a very important basis for the selectivity of many of the currently available herbicides (Hatzios and Penner, 1982). Metabolic detoxication of most herbicides in higher plants includes degradative and synthetic reactions which are catalyzed by specialized enzymes (Hatzios and Penner, 1982; Lamoureux and Frear, 1979). Alterations in the pathways or the rate of metabolic reactions involved in the detoxication of certain herbicides are known to be brought about by the prior or simultaneous application of other agrochemicals (Hatzios and Penner, 1985) and such a mechanism has attracted considerable attention as a likely explanation for the protective action of the currently marketed safeners. An enhancement of herbicide detoxication in safener-treated plants may result from a safener-induced increase in the synthesis or activity of specialized enzymes which detoxify herbicides in plants or of specific cofactors and natural substrates which facilitate the function of such enzymes.

Our understanding of the major metabolic pathways and reactions involved in the detoxication of thiocarbamate and chloroacetanilide herbicides in higher plants is not complete but has advanced considerably in recent years. A number of *in vivo* and *in vitro* studies have demonstrated the importance of glutathione (GSH) conjugation as the key metabolic pathway contributing to the selectivity of thiocarbamate and chloroacetanilide herbicides (Blattmann *et al.*, 1986; Breaux, 1986; Carringer *et al.*, 1978a; Horváth and Pulay, 1980; Hubbell and Casida, 1977; Lamoureux and Frear, 1987; Lamoureux and Rusness, 1986). The conjugation of thiocarbamate herbicides with GSH is preceded by their oxidation to sulfoxides or sulfones, which are the true carbomoylating agents for glutathione or other thiols (Dutka and Komives, 1983; Horváth and Pulay, 1980; Lay *et al.*, 1975). Chloroacetanilide herbicides conjugate directly with GSH and they do not require any activation step predisposing them for such a reaction (Lamoureux and Rusness, 1986; Leavitt and Penner, 1979). The conjugation of thiocarbamate sulfoxides and chloroacetanilide herbicides with GSH has been shown to be catalyzed by substrate-specific enzymes known as glutathione-*S*-transferases (GSTs; EC. 2.5.1.18) (Ezra *et al.*, 1985; Mozer *et al.*, 1983). Nonenzymatic conjugation of these herbicides with GSH is also possible, but occurs at a very slow rate under physiological pH (Gronwald *et al.*, 1987; Leavitt and Penner, 1979).

The GS-conjugates of chloroacetanilide and sulfoxidized thiocarbamate herbicides are catabolized rapidly to cysteine conjugates, which are subsequently deaminated to thiolactic acid conjugates or N-acylated with malonic acid (Blattmann *et al.*, 1986; Lamoureaux and Frear, 1987). In tolerant plants such as corn, thiolactic acid conjugates are considered as terminal metabolites of these herbicides. Other metabolic pathways which could contribute to the detoxication of these herbicides in plants include hydrolysis or N-dealkylation of thiocarbamates and alkyl or aryl hydroxylation for chloroacetanilides (Hatzios and Penner, 1982; Krause *et al.*, 1985). The occurrence of such biotransformations, however, in the metabolic detoxication of these herbicides in higher plants has not been demonstrated and their potential significance is poorly understood. Additional discussion of the metabolism of thiocarbamates and chloroacetanilide herbicides in higher plants is given by Gronwald in Chapter 4, this volume.

*Elevation of Glutathione and Glutathione-*S*-Transferases by Safeners.* The possibility that safeners may protect grass crops from herbicide injury by enhancing herbicide detoxication was first proposed by Güneyli (1971). He postulated that NA may act by activating the enzyme system(s) responsible for the biotransformation of EPTC in corn seedlings. A marked enhancement of the metabolism of the herbicide cisanilide in NA-treated corn seedlings was also observed by Holm and Szabo (1974). Shortly after these reports with NA, Wright *et al.* (1973) and Chang *et al.* (1974) showed that the metabolism of the thiocarbamate herbicides EPTC and butylate was enhanced significantly in corn safened with dichlormid. Later, Lay and Casida (1976, 1978) documented the occurrence of the conjugation of sulfoxidized thiocarbamates with GSH and proposed that dichlormid and other dichloroacetamide safeners protect corn by stimulating the metabolism of these herbicides by GSH conjugation. A significant elevation of GSH levels and of the activity of the respective GST enzymes catalyzing the conjugation of sulfoxidized EPTC with GSH was demonstrated by Lay and Casida (1976) in roots of corn seedlings grown in the presence of the safener dichlormid. A safener-induced elevation of GSH levels in protected corn and grain sorghum has since been confirmed by a number of investigators. Flurazole, dichlormid, and other chloroacetamide safeners (CDAA, R-29148,

MG-191, and LAB-145138) appear to be highly effective in increasing by two- to threefold the GSH content of corn and grain sorghum and of tobacco (*Nicotiana tabacum* L.) suspension cultures (Adams *et al.*, 1983; Dutka and Komives, 1987; Ezra and Stephenson, 1986; Ezra *et al.*, 1985; Gronwald *et al.*, 1987, T. Komives *et al.*, 1985b; Rennenberg *et al.*, 1982). NA and oxime ether safeners such as cyometrinil and oxabetrinil, however, are rather ineffective as inducers of GSH content in treated plants (Fedtke, 1981; Gronwald *et al.*, 1987; T. Komives *et al.*, 1985b).

Another safener which protects plants against herbicide injury by increasing GSH levels is OTC (L-2-oxothiazolidine-4-carboxylic acid) (Hilton and Pillai, 1986). OTC and its analog thioproline or TC (L-thiazolidine-4-carboxylic acid) are enzymatically converted to *S*-carboxy-L-cysteine which is catabolized through decarboxylation to L-cysteine. The cysteine is then rapidly incorporated into glutathione. The enzyme that converts OTC and TC to cysteine is found in both plants and animals (Mazelis and Creveling, 1978; Meister, 1984). In growth chamber experiments, OTC and TC provided partial protection to corn, grain sorghum and wheat (*Triticum aestivum* L.) against injury from the herbicide tridiphane (Hilton and Pillai, 1986, 1987). However, OTC was ineffective as a safener against injury from the herbicide atrazine, which also conjugates with GSH (Hilton and Pillai, 1986).

Plant cells contain millimolar concentrations of reduced glutathione (GSH) which vary significantly, depending on the species, and range from 0.1 to 0.7 m*M* in the cytosol and from 1.0 to 3.5 m*M* in the chloroplast (Rennenberg, 1982). Apart from its role as a reductant in herbicide and xenobiotic detoxication, GSH is used as a storage form of reduced sulfur in plant cells or serves as the main long-distance transport form of reduced sulfur translocated from mature leaves to roots of higher plants (Rennenberg, 1982). In addition, GSH is very important for protecting chloroplast membranes from oxidative damage caused by H_2O_2 or other oxidants (Halliwell, 1984). When GSH reacts with oxygen it is oxidized to glutathione disulfide (GSSG). The subsequent re-reduction of GSSG to GSH is catalyzed by the enzyme glutathione reductase (EC 1.6.4.2) in an NADPH-dependent reaction. This enzyme plays an essential role in maintaining a high GSH/GSSG ratio, which is necessary not only for the function of GSH as a reductant but also for an optimal rate of protein synthesis in plant and animal cells (Rennenberg, 1982).

It is evident, therefore, that safeners may elevate GSH levels in protected plants either directly, by inducing or activating the enzymes involved in the biosynthesis of GSH in plants cells, or indirectly, by inducing the activity of glutathione reductase. An induction of glutathione reductase by safeners will maintain a high GSH/GSSG ratio in the cells of protected grass crops. In turn, high levels of GSH will compensate for the reduction of GSH content caused by the formation of the GS-conjugates of chloroacetanilide and sulfoxidized thiocarbamate herbicides in protected crops such as corn and grain sorghum.

Thus, in safener-treated plants, sufficient GSH is available for the normal function of key metabolic processes such as protein synthesis, protection of chloroplast membrane from peroxidative damage, and detoxication of the antagonized herbicides. This theory is supported by recent reports by T. Komives and his co-workers (1985a,b), who showed an induction of glutathione reductase in corn treated with the safeners dichlormid and MG-191. Additional discussion on this topic is given by Komives and Dutka in Chapter 5, this volume.

A direct effect of safeners on GSH biosynthesis is also possible. A direct activation of GSH synthetase (EC 6.3.2.3) in corn by dichlormid was postulated by Carringer *et al.* (1978b) as being involved in the protective action of this safener. Renneberg *et al.* (1982), however, proposed that the stimulation of GSH synthesis by dichlormid in heterotrophically grown tobacco cell suspensions was not the result of a direct activation of preexisting enzymes but an enhancement of the amount of enzymes involved in this process. Adams *et al.* (1983) reported that the safeners dichlormid and R-29148 enhance GSH content in safened corn by acting at an early stage of the *de novo* synthesis of GSH from sulfate. The activity of ATP sulfurylase (EC 2.7.7.4), the first enzyme in sulfate assimilation, and the subsequent metabolism of sulfate to "bound sulfide," cysteine, and GSH were greatly increased in roots of corn seedlings grown in the presence of the safeners dichlormid and R-29148 (Adams *et al.*, 1983). These safeners elicited similar responses in grain sorghum and redroot pigweed (*Amaranthus retroflexus* L.) but not in other grass or broadleaf plants (Adams *et al.*, 1983).

The biosynthesis of GSH in animal and plant cells is believed to be a feed back-regulated reaction (Kondo *et al.*, 1984). GS-conjugates have been shown to interfere with the feed back-regulating mechanism of GSH synthesis in mammalian cells and can cause an increase in the synthesis of GSH (Kondo *et al.*, 1984). Therefore, it is likely that safeners such as flurazole, CDAA, and dichlormid, which are known to be metabolized by GSH conjugation in crop plants or mammalian tissues (Breaux *et al.*, Chapter 7, this volume; Ezra *et al.*, 1985; Miaullis *et al.*, 1978), could enhance GSH levels by deregulating the feed back control of GSH synthesis. Such a mechanism has been demonstrated by Kondo *et al.* (1984) in mammalian cells and has been postulated as a potential mechanism for the protective action of the safener flurazole (see Breaux, *et al.*, Chapter 7, this volume).

The potential contribution, however, of a safener-induced increase in GSH levels of protected plants to the mode of action of safeners remains obscure. As mentioned earlier, NA and oxime ether safeners are totally ineffective in eliciting this response, but they are still very effective protectants of corn and grain sorghum against thiocarbamate and chloroacetanilide herbicide injury. In addition, a number of other herbicides and growth regulators have been shown to elevate GSH levels in corn or grain sorghum (Ezra and Gressel, 1982; Fedtke, 1981; Rubin *et al.*, 1985; Stephenson *et al.*, 1983), but they failed to protect corn against injury from herbicides known to be detoxified by GSH conjugation.

Recently, Lay and Niland (1985) showed that dichlormid increased the GSH content of several EPTC-susceptible corn lines to the same extent as it did on EPTC-tolerant lines but failed to protect them from injury. The discrepancy between the safening action of dichlormid and its enhancement of GSH content seems to support the view that multiple requirements are needed for the protective action of herbicide safeners (Adams *et al.*, 1983; Lay and Niland, 1985).

Enhanced metabolism of chloroacetanilide and thiocarbamate herbicides by GSH conjugation could result also from a safener-induced increase of the activity of the respective glutathione *S*-transferase enzymes (GSTs, EC 2.5.1.18), which catalyze the conjugation of these herbicides with GSH in protected grass crops. Plants contain a number of GSTs which exhibit a rather high degree of substrate specificity (Diesperger and Sandermann, 1979; Frear and Swanson, 1970; Guddewar and Dauterman, 1979; Mozer *et al.*, 1983). Recent work by Mikami *et al.* (1986) shows that even the properties of plant GSTs which exhibit high activity on the same xenobiotic substrate could differ. The existence of GST isozymes in higher plants has been documented by Mozer *et al.* (1983).

Lay and Casida (1976) were first to demonstrate that the safener dichlormid increases the activity of GSTs in corn roots by two- to threefold. In addition to dichlormid, a significant elevation of cytosolic and microsomal GST activity of corn tissues has been documented with several other chloroacetamide safeners such as CDAA, MG-191, and LAB-145138 (Ezra and Stephenson, 1986; Ezra *et al.*, 1985; T. Komives *et al.*, 1985a). Mozer *et al.*, (1983) showed that the safener flurazole not only enhanced the activity of corn GSTs that are constitutively present, but it also induced a novel GST isozyme with greater activity in conjugating chloroacetanilide herbicides with GSH. NA and oxime ether safeners such as cyometrinil and oxabetrinil have also been reported to significantly increase the activity of GSTs in corn and grain sorghum (Frear *et al.*, 1987; Gronwald *et al.*, 1987; T. Komives *et al.*, 1985a; Mozer *et al.*, 1983).

It is of interest to note, however, that in studies by Lay and Niland (1985), dichlormid failed to significantly increase GST activity in shoot tissues of 'Dekalb XL72B' corn, which is protected by this safener against injury from herbicide EPTC. In addition, a number of herbicides and growth regulators were shown to increase GST activity in corn plants, but they failed to act as safeners against injury from herbicides that conjugate to GSH (A. V. Komives *et al.*, 1985b; Stephenson and Ezra, 1985).

In most of the reported cases, the increase of GST activity by safeners has been demonstrated with the use of chlorodinitrobenzene (CDNB), which is a general substrate for plant or animal GSTs. CDNB, however, is not a very effective substrate for differentiating the enhanced activity induced by different safeners. In a recent study, Gronwald *et al.* (1987) demonstrated the importance of using the actual herbicidal substrate for differentiation of increases in GST activity induced by several safeners. A strong correlation between the efficacy of a safener in protecting grain sorghum from chloroacetanilide injury and its ability to increase GST activity in grain sorghum shoots has been demonstrated (Fuerst and Gronwald, 1986; Gronwald *et al.*, 1987). Flurazole was the most effective sorghum safener, eliciting a 30-fold increase in GST activity. Oxabetrinil and NA were also effective safeners of sorghum against metolachlor injury, causing a 20-fold and 17-fold increase of GST (metolachlor) activity, whereas dichlormid was the least effective safener of sorghum, causing only a 5-fold increase in GST activity (Gronwald *et al.*, 1987).

The exact mechanism of the safener-induced enhancement of GST activity is not known. It appears that safeners do not act directly on GSTs since in *in vitro* studies dichlormid and oxabetrinil did not alter the activity of GST enzymes (Gronwald *et al.*, 1987; Lay and Casida, 1976). It is more likely that safeners act by an enzyme induction process rather than an enzyme activation.

Enhancement of Oxidative Herbicide Metabolism by Safeners. Contrary to earlier reports (Lay and Casida, 1976; Lay *et al.*, 1975) the sulfoxidation of thiocarbamate herbicides is presently viewed as being of equal importance to GSH conjugation in the metabolic detoxication of these herbicides in corn. The possibility that dichlormid may protect corn from EPTC injury by stimulating the rate of EPTC sulfoxidation has been proposed (Leavitt and Penner, 1979) and recent evidence seems to support such a hypothesis.

The sulfoxidation of thiocarbamate herbicides in corn is believed to be an enzymatic reaction catalyzed by specific monooxygenases (EC 1.14.14.1) (Hubbell and Casida, 1977). These enzymes are hemoproteins containing cytochrome P-450 as a central feature in their function and they are also known as mixed-function oxidases (Hatzios and Penner, 1982). Treatment of corn seedlings with the monooxygenase inhibitors piperonyl butoxide and SKF-525A was reported to synergize the activity of the herbicide EPTC on corn, even in the presence of the safener dichlormid, indicating that oxidation of EPTC might play a key role in the protective action of this safener (Dutka and Komives, 1983; Hatzios, 1983b). The antioxidant propyl gallate, the herbicide tebuthiuron, and the air pollutant ozone have also been reported to synergize the activity of EPTC on corn in the presence of the safener dichlormid (Hatzios, 1981, 1983b). Similarly,it was shown that in the presence of the safener cyometrinil, the activity of the herbicide metolachlor on grain sorghum was synergized by ozone, piperonyl butoxide and propyl gallate as well as the herbicide tebuthiuron (Hatzios, 1983a, 1984d). These results combined with the observation that cyometrinil fails to protect sorghum against metolachlor under extremely wet soil conditions (Ketchersid and Merkle, 1981; Ketchersid *et al.*, 1981; Leek and Penner, 1981), indicate that the protective action of this safener may also involve the enhancement of a biological oxidation system (possibly a monooxygenase) that detoxifies the herbicide metolachlor in grain sorghum. The sulfoxidation of thiolactic acid conjugates of chloroacetanilide herbicides (Blattmann *et al.*,

1986) might be such an oxidative reaction which could play a key role in the metabolism of chloroacetanilide herbicides and the action of safeners. Additional oxidative reactions involved in the metabolism of chloroacetanilide herbicides by soil microorganisms include alkyl or aryl hydroxylations (Krause *et al.*, 1985). A safener-induced enhancement of oxidative herbicide metabolism (ring hydroxylation) has been implicated in studies demonstrating the protection of oats and wheat from diclofop-methyl by NA (Taylor and Loader, 1984) and of corn from chlorsulfuron injury by NA and cyometrinil (Sweetser, 1985). The involvement of enhanced oxidative metabolism in the safening of corn against chlorsulfuron by NA has been disputed, however, by Frear *et al.* (1987).

On the basis of the above information, it seems likely that safeners could induce the activity or the *de novo* synthesis of monooxygenase systems which may be involved in the metabolic detoxication of chloroacetanilide and thiocarbamate herbicides in protected grass crops such as corn and grain sorghum. Additional support for this hypothesis comes from a recent report by Ketchersid *et al.* (1985), who showed that several herbicide safeners can act as synergists of certain insecticides known to be metabolized by monooxygenase enzymes in insects or plants. They showed that the safeners cyometrinil, flurazole, and oxabetrinil synergized the activity of the insecticide propoxur on flies and the epoxidation of the insecticide aldrin to dieldrin in grain sorghum coleoptiles. More recently, Brooks *et al.* (1987) reported that oxime ether safeners and flurazole enhanced the activity of monooxygenases from grain sorghum shoot tissues which catalyzed the metabolism of metolachlor. However, further research is definitely needed to establish this hypothesis as a viable mechanism for the protective action of herbicide safeners. Our limited information on plant monooxygenase enzymes is an additional obstacle that needs to be overcome as part of future research on this topic.

c. A Model for the Biochemical Mechanisms of Action of Herbicide Safeners. Taking into consideration the similarities that exist in the reported effects of thiocarbamate and chloroacetanilide herbicides as well as the interactive effects of safeners or synergists with herbicides, Fedtke and Trebst (1987) proposed a model attempting to explain the action of these herbicides and the protective action of safeners. The central feature of this model is the function of monooxygenase enzymes involved in plant metabolism as well as in the detoxication of herbicides. The model proposes that thiocarbamate and chloroacetanilide herbicides act by inhibiting the activities of monooxygenase enzymes which are important for normal plant metabolism. Safeners are viewed as inhibitors of the activity of monooxygenases which are not vital for normal plant cell life. Inhibition of these enzymes by safeners is postulated to induce the activity of monooxygenases, which are target sites for herbicide inhibition or catalyze the oxidative herbicide metabolism. Therefore, monooxygenases that participate in the oxidative metabolism of herbicides are available at high levels and contribute to enhanced detoxication and crop protection. A slightly modified version of the model proposed by Fedtke and Trebst (1987) is presented in Fig. 4.

The contribution of the GSH conjugation pathway in safener action is also recognized in this model but, in my view, the glutathione-mediated metabolism of selected herbicides could also be induced separately by herbicides or safeners rather than being induced only after the induction of the monooxygenase system. A major drawback of the model of Fedtke and Trebst stems from the fact that most of its data were generated with the use of a bioassay measuring oat root growth to evaluate the individual and interactive effects of herbicides, safeners, and synergists. Additional work with corn or grain sorghum coleoptiles and root systems is undoubtedly needed to substantiate the validity of this model. As mentioned earlier (Table II), thiocarbamate herbicides do not affect root growth of grass crops. Furthermore, our present knowledge on the contribution of oxidative metabolism in the detoxication of chloroacetanilide herbicides in higher plants is obscure. The recent report by Blattmann *et al.* (1986) shows that sulfoxidation of thiolactic acid conjugates of these herbicides in corn is important but it occurs after the initial conjugation of these herbicides to GSH.

B. Interactions of Herbicides and Safeners at the Site of Action

The exact mechanisms by which chloroacetanilide and thiocarbamate herbicides exert their phytotoxic action are unknown (Fedtke, 1982). The symptomology of these herbicides is not indicative of an acute blockage of a central metabolic reaction in susceptible plants, but it is rather consistent with a chronic loss of vital metabolic components (Wilkinson, 1978). The lack of the involvement of a single target site and a specific mode of action for chloroacetanilide and thiocarbamate herbicides is further supported by their characterization as herbicides with low risk for weed resistance development (LeBaron, 1986). In spite of their continuous use in crop production during the past two to three decades, weed biotypes with resistance to thiocarbamate and chloroacetanilide herbicides have not been observed (LeBaron, 1986).

Shoot deformations and growth inhibitions caused by these herbicides could result from their reported effects on cell division and cell elongation in tissues of treated grass plants (Deal and Hess, 1980; Deal *et al.*, 1980; Fedtke, 1982). Some of the symptoms (e.g., stunting) caused by thiocarbamate and chloroacetanilide herbicides on corn and grain sorghum seedlings are similar to those observed in grasses treated with classical growth retardants (Wilkinson, 1978). In fact, the stunting effects of EPTC and alachlor on corn could be prevented by exogenous applications of gibberellin (GA), but the leaf deformations (twisting and rolling) were not prevented (Donald and Fawcett, 1975; Wilkinson, 1978). These interactions further demonstrate the complexity of the phytotoxic action of these herbicides and the lack of a single target site.

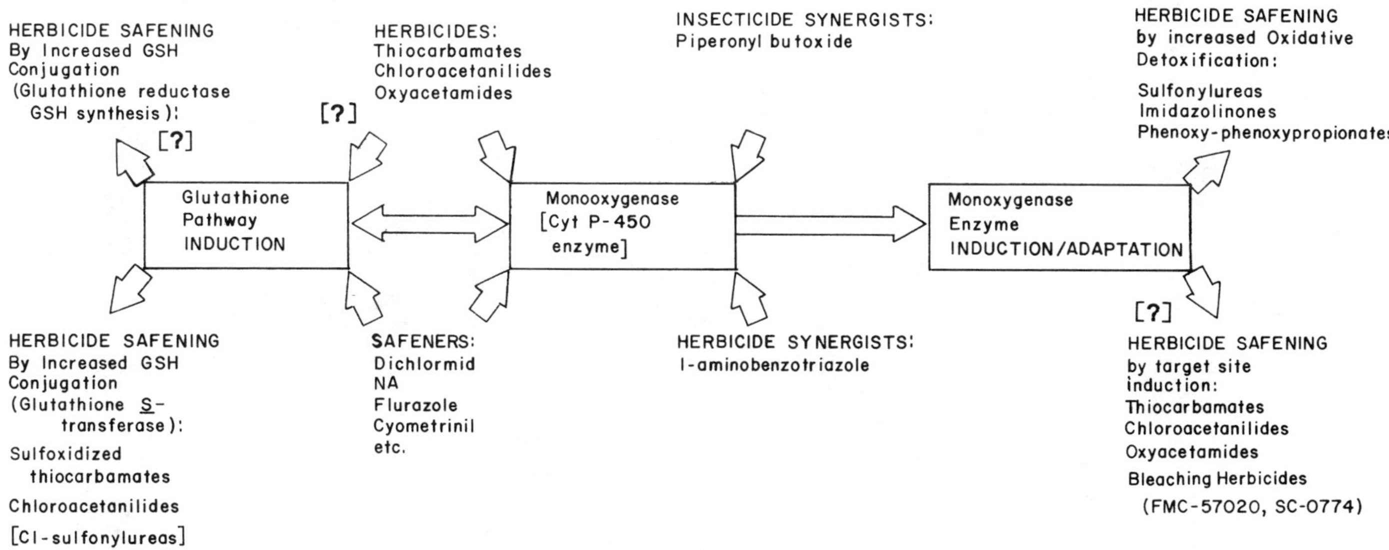

Fig. 4. The model of Fedtke and Trebst (1987) describing the interactive mechanisms of action of selected herbicides, safeners, and synergists. Question marks [?] indicate postulated actions which are not currently conclusive. (Modified from the original model, provided kindly by Dr. Fedtke).

At the level of cellular metabolism, physiological and biochemical processes that are affected by chloroacetanilide and/or thiocarbamate herbicides include lipid synthesis (Ebert, 1982; Ebert and Ramsteiner, 1984; Fedtke, 1982; Mann and Pu, 1968; Warmund *et al.*, 1985; Wilkinson and Smith, 1975), terpenoid synthesis (Wilkinson, 1981a,b, 1982, 1983, 1985), protein synthesis (Deal *et al.*, 1980; Dhillon and Anderson, 1972; Duke *et al.*, 1975; Jaworski, 1969; Mann *et al.*, 1965; Moreland *et al.*, 1969; Rao and Duke, 1976), nucleic acid (RNA) synthesis (Jaworski, 1969; McFarland and Hess, 1985), and membrane function and ion transport (Balke, 1979; Mellis *et al.*, 1982; Wilkinson and Smith, 1976).

Antagonistic interactions between herbicides and safeners at the site of action are possible and they may contribute to the mechanism of the protective action of herbicide safeners. However, because of the complexity of the phytotoxic action of thiocarbamate and chloracetanilide herbicides it seems more likely that safeners may antagonize the action of these herbicides by eliciting opposite effects at different steps of the same physiological process (physiological antagonism) rather than interacting antagonistically with these herbicides on a common target site (competitive antagonism).

1. Competitive Antagonism

This mechanism occurs when a safener acts reversibly at the same site as the herbicide (Fig. 5). By reacting with a receptor site within the cell, the safener may prevent the binding of a given herbicide to its target site and the formation of a herbicide–receptor complex that may be necessary for herbicidal activity. In contrast to the situation with animal systems, where the formation of hormone–receptor complexes is well established, our understanding of the formation of herbicide–receptor complexes in plant cells is very limited. Recent research, however, has demonstrated the existence of specific soluble and/or membrane-bound receptors for auxin-type herbicides (Stoddart and Venis, 1980). Enzymes,

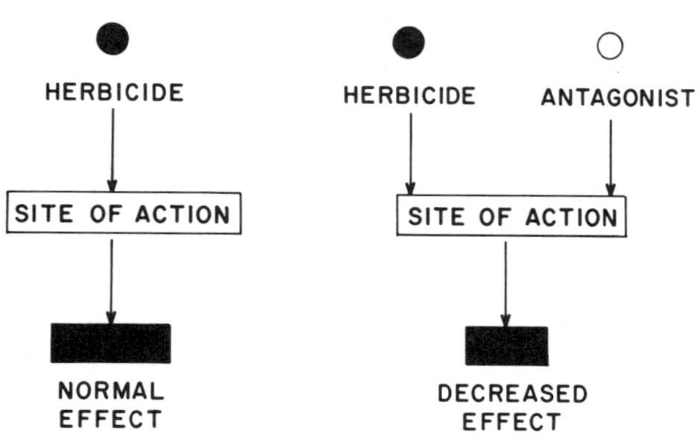

Fig. 5. Competitive antagonism as a mechanism of safener action. (After Hatzios and Penner, 1985).

proteins, nucleic acids, coenzymes and other small endogenous molecules (e.g., thiols) have all been recognized as potential receptor sites for biologically active substances in mammalian systems (Aldridge and Reiner, 1972). The potential involvement of a receptor site specific for chloroacetanilide herbicides has been postulated by Chang (1983), but much more research is needed to substantiate the existence or the nature of such a receptor.

Because of the structural similarity of chloroacetamide safeners to thiocarbamate and choloroacetanilide herbicides it has been proposed that such safeners may act as herbicidally inactive competitive inhibitors of these herbicides at some common site of action (Stephenson and Chang, 1978). The potential involvement of lipid biosynthesis as a common target site for the interactions of thiocarbamate and chloroacetanilide herbicides with safeners has been proposed by several investigators. Ezra *et al.* (1983b) and Sagaral (1978) showed that the EPTC-induced inhibition of the incorporation of [^{14}C]acetate into neutral and polar lipids of corn cells or protoplasts could be prevented by the simultaneous addition of dichlormid. Similar results were reported by Wilkinson and Smith (1975, 1976), who showed that the safeners NA and dichlormid can reverse the effects of EPTC and other thiocarbamate herbicides on fatty acid synthesis of spinach (*Spinacia olearacea* L.) chloroplasts. A significant antagonistic interaction between metolachlor and oxabetrinil on the incorporation of [^{14}C]acetate into lipids of sorghum protoplasts has also been reported by Zama and Hatzios (1987). However, counteraction of the effects of metolachlor by oxabetrinil in that study was partial and did not appear to be sufficient to account for the entire protective action of this safener. Recently, Warmund *et al.* (1985) reported that total lipid synthesis or fatty acid composition of grain sorghum seedlings were not affected by the alachlor or flurazole.

Other researchers have also disputed the involvement of lipid synthesis in the interactions of safeners with thiocarbamate and chloroacetanilide herbicides and instead they emphasized epicuticular wax formation as a potential site for such interactions. Leavitt and Penner (1978a) were first to suggest that dichlormid may act as a safener by preventing an EPTC-induced aggregation of the epicuticular wax layer of corn rather than by overcoming a blocking of lipid biosynthesis by this herbicide. Such a mechanism was confirmed later by Görög *et al.* (1982) not only with dichlormid, but also with other dichloracetamide safeners such as R-28725 and AD-67. The step of EPTC inhibition is the elongation of the C_{30} fatty acids to C_{32} derivatives (Barta *et al.*, 1983). The effects of EPTC on epicuticular wax biosynthesis of corn were counteracted by micromolar concentrations of the safeners dichlormid and MG-191 (Barta *et al.*, 1983). A similar mechanism has also been proposed for the protective action of the safener cyometrinil. According to Ebert (1982), cyometrinil protects grain sorghum from metolachlor injury by preventing a metolachlor-induced loss of cuticular integrity in sorghum plants. More specifically, cyometrinil prevents metolachlor from inhibiting the synthesis of C_{28} to C_{30} chain alcohols, which are the dominating constituents of the epicuticular wax of sorghum leaves (Ebert and Ramsteiner, 1984). As mentioned earlier, the counteraction of these herbicidal effects by safeners results in a decrease of the transpiration rate in protected plants and causes an indirect reduction of the uptake and translocation of the soil-applied thiocarbamate and chloroacetanilide herbicides.

If the competitive inhibition theory of safener action is correct, one would expect safeners to act as antagonists of chemically similar herbicides on all susceptible plants. The pronounced botanical specificity of herbicide/safener combinations, however, obscures the potential contribution of this mechanism in the protective action of safeners. Additional evidence against this mechanism comes from the fact that none of the currently available herbicide safeners is effective in protecting grass crops against injury from the herbicide dalapon (Parker, 1983). This herbicide induces morphological symptoms on grass crops nearly identical to those caused by thiocarbamates and chloroacetanilide herbicides and it is a strong inhibitor of lipid synthesis and epicuticular wax formation in treated grass crops (Fedtke, 1982). It is of interest to note, however, that the safener dichlormid offers partial protection to corn and grain sorghum against injury from the herbicide sethoxydim (Hatzios, 1984a,b). Sethoxydim is a strong inhibitor of lipid synthesis in higher plants (Burgstahler *et al.*, 1986; Hatzios, 1982). This inhibition appears to result from a strong inhibitory effect of sethoxydim on acetyl-CoA carboxylase (EC 6.4.1.2) (Burton *et al.*, 1987). In laboratory studies, the dichloroacetamide safeners dichlormid and R-28725 antagonized the effects of sethoxydim on lipid synthesis of grain sorghum protoplasts (Hatzios and Moon, 1985). Devlin and Koszanski (1987a,b) showed that the safener dichlormid is also effective in protecting corn against injury from the recently introduced bleaching herbicides clomazone (FMC-57020) and SC-0774. The metabolism of these herbicides in corn or other plants has not been studied and it remains to be seen whether dichlormid protects corn against clomazone and SC-0774 by enhancing their metabolic detoxication. Nevertheless, these results show that dichlormid is a more versatile safener than previously thought and may act by more than one mechanism.

Apart from lipid synthesis and epicuticular wax formation, other active sites also may be involved in the competitive inhibition of the effects of thiocarbamate and chloroacetanilide herbicides by safeners. Rao and Kahn (1975) reported that the alachlor-induced inhibition of polyribosome formation in barley (*Hordeum vulgare* L.) roots was reversed by dichlormid, probably by a competition with the herbicide for this active site. However, protein synthesis, measured by the incorporation of [^{14}C]leucine into corn cells or grain sorghum protoplasts, did not appear to be an active site for a competitive antagonism between the herbicides EPTC and metolachlor and the safeners dichlormid and oxabetrinil, respectively (Ezra and Gressel, 1982; Zama and Hatzios, 1987). Other metabolic enzymes that have been shown to be inhibited by the herbicides EPTC and chlorsulfuron, but stimulated by the safeners NA and dichlormid include polyphenoloxidase (EC 1.10.3.2) (Wilkinson, 1978) and acetolactate synthase (EC 4.1.3.18) (Rubin and Casida, 1985).

Recently, Molin *et al.* (1986) proposed that some aspect of CoA metabolism may be the site of action of chloroacetanilide herbicides since CoA and its intermediates are needed for terpenoid, anthocyanin, lipid, and lignin syntheses, which are all inhibited by these herbicides. Studies on the nature of the potential interactions of chloroacetanilide herbicides and safeners on this site are not yet available. The degradative enzyme arylamidase (EC 3.5.1.13) has also been reported as a potential target site for the interaction of thiocarbamate herbicides and oxime ether safeners. A 50% reversal of the inhibitory activity of EPTC on arylamidase activity extracted from lettuce was observed by Hoagland (1984) when several oxime ether safeners were used in a 1:1 ratio with the herbicide. These results, however, are difficult to assess in terms of the protective action of these safeners since arylamidase is involved only in the metabolic detoxication of the herbicide propanil. On the basis of present knowledge, amidases are not involved in the action or metabolism of thiocarbamate herbicides in corn plants.

The competitive inhibition theory also falls short in explaining how structurally unrelated compounds are effective as crop safeners against injury from chloroacetanilide, thiocarbamate, or other groups of herbicides. In such cases, a physiological antagonism seems more appropriate for consideration rather than a competitive antagonism.

2. Physiological Antagonism

This type of antagonism is observed when the herbicide and the safener counteract each other by acting at different sites and producing opposite effects on the same physiological process (Fig. 6). Each agent is unhindered in its ability to react with its receptor site and elicit a characteristic response, but the combined effect of the two chemicals is the net result of their opposing effects on the same process. Thus, physiological antagonism is quite different from

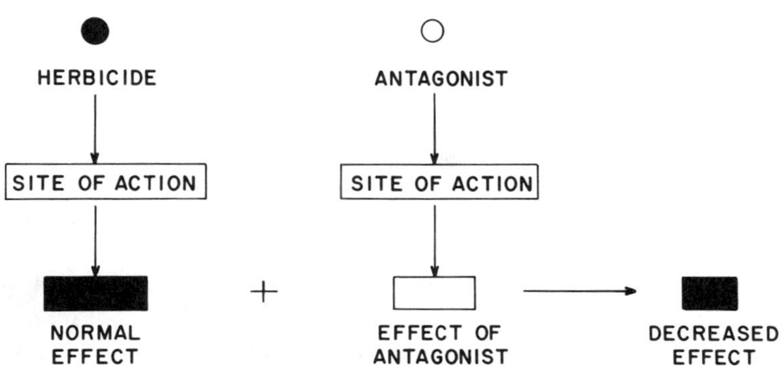

Fig. 6. Physiological antagonism as a mechanism of safener action. (After Hatzios and Penner, 1985).

competitive antagonism, where the two chemicals act on the same site and one of them actually prevents the other from exerting its effect (Fig. 5).

The involvement of endogenous growth hormones in the activity of thiocarbamate and chloroacetanilide herbicides has been suggested by their reported inhibitory effects on cell division and cell elongation of treated grass plants (Deal and Hess, 1980; Deal *et al.* 1980; Fedtke, 1982). Indeed, the responses of corn and grain sorghum to these herbicides resemble a gibberellin deficiency and, in a series of reports, Wilkinson (1981a,b, 1982, 1983) has documented the interference of thiocarbamate and chloroacetanilide herbicides with GA precursor biosynthesis in grass plants. The EPTC- or metolachlor-induced inhibition of the incorporation of [^{14}C]mevalonate into kaurene was reversed by dichlormid and cyometrinil in cell-free preparations from sorghum seedlings (Wilkinson 1981a,b, 1982, 1983). The reversal of thiocarbamate and chloroacetanilide effects on terpenoid biosynthesis of grass plants by selected herbicide safeners appears to be an example of a physiological rather than competitive antagonism. Support for this conclusion comes from the work of Wilkinson (1982) who showed that the herbicide alachlor interferes with terpenoid or GA biosynthesis of grass plants at multiple sites. Pretreatment of sorghum seeds with the safener cyometrinil altered the pattern of alachlor-induced effects on these biosynthetic processes, but failed to completely reverse the effects of this herbicide. An extensive review of the involvement of terpenoid biosynthesis in the actions of herbicides and safeners is given by Wilkinson in Chapter 10, this volume. Corn seedlings treated with EPTC exhibit increased peroxidase (EC 1.11.1.7) activity (Harvey *et al.*, 1975). Dichlormid reduced peroxidase activity and reversed the responses induced by EPTC on corn (Harvey *et al.*, 1975). GA also causes an inhibition of peroxidase activity (Birecka and Galston, 1970) and cessation of cell elongation is correlated with increased peroxidase activity (Gardiner and Cleland, 1974). Thus, an inhibition of GA synthesis by EPTC would inhibit shoot growth of treated grasses and the EPTC-induced increase in peroxidase activity would cause the coleoptilar tissues to become static and lignified (Wilkinson, 1983). These effects fit the response phenology of grass seedlings to thiocarbamate herbicides and they could explain their herbicidal activity (Wilkinson, 1983).

In contrast, chloroacetanilide herbicides have been reported to inhibit lignin synthesis (Ebert, 1980; Hickey and Krueger, 1974; Molin *et al.*, 1986; Paradies *et al.*, 1981) rather than increasing lignification as the thiocarbamate herbicides do. Molin *et al.* (1986) proposed that a reduction of lignin content would result in weaker secondary walls, which may account for the brittle appearance of the leaves and shoots of grass seedlings treated with chloroacetanilide herbicides. Hickey and Krueger (1974) proposed that safeners such as NA may have a loosening effect on the coleoptile, which prevents the growth retardation and brittleness of grass seedlings treated with chloroacetanilide herbicides.

Apart from its association with increased lignification, peroxidase is important because it metabolizes the auxin indole-3-acetic acid (IAA) (Smith *et al.*, 1982). Therefore, an EPTC-induced inhibition of GA synthesis coupled with increased peroxidase activity would result not only in a deficiency of GA required for growth, but also in a deficiency of IAA, which is needed for H^+ extrusion to loosen cell walls for cell enlargement (Wilkinson, 1983).

Therefore, it seems likely that the protective action of some safeners may involve interactions with plant hormones. Moreover, it is possible that some safeners such as NA may mimic the effects of endogenous growth hormones when they protect plants against herbicide injury. Endogenous hormones such as IAA and GA were found to offer partial, but not practical protection against EPTC injury to corn in early screening investigations for potential safeners conducted by Hoffman (1978) and Gray *et al.* (1982). These early observations were crucial for the subsequent development of NA as the first commercially developed safener by Hoffman (1969). Recently, Frear *et al.* (1987) showed that NA possesses auxinlike properties since it increased the geotropism of corn roots. Frear *et al.* (1987) also showed that several auxins including naphthalene acetic acid (NAA), indolebutyric acid (IBA), IAA, and phenylacetic acid, were effective safeners of corn seedlings against growth inhibitions caused by the herbicide chlorsulfuron. NAA, 2,6-D, 2,4,6-T, and 4-methyl-3-chloroaniline also protected rice against injury from the herbicide benthiocarb (Wakabayashi and Matsunaka, 1982). The synthetic auxin 2,4-D is an active antagonist of EPTC effects on corn (Beste and Schreiber, 1972). More recently, the antagonistic interactions of auxinlike herbicides and the herbicide diclofop-methyl on grasses have attracted considerable attention (Taylor and Loader, 1984). The antagonistic interactions of auxinlike herbicides and the thiocarbamates seem to result from a physiological antagonism on RNA synthesis. Thiocarbamate herbicides inhibit RNA synthesis whereas 2,4-D and other auxins stimulate RNA synthesis (Beste and Schreiber, 1972).

The EPTC- and metolachlor-induced changes in membrane permeability of sugar beet (*Beta vulgaris* L.) roots and onion (*Allium cepa* L.) roots could be prevented by dichlormid and NA, respectively (Bujtas, 1978; Mellis *et al.*, 1982). Antagonistic interactions between NA and chlorsulfuron on membrane permeability of corn roots were recently reported by Crowley and Prendeville (1985). However, because metolachlor, even at high concentrations, did not influence the membrane permeability of corn roots (Mellis *et al.*, 1982), the contribution of such a mechanism in the protective action of NA against metolachlor injury to corn under field conditions is questionable. GA_3 did not prevent the effects of supraoptimal EPTC concentrations on betacyanin efflux from sugar beet tonoplasts (Wilkinson, 1985). This influence of EPTC was shown to be via membrane disruption, which annulled any possible GA influence by eliminating the functionality of the membranes required for the development of a GA response. Changes in membrane permeability did not appear to be involved in the protective action of cyometrinil against metolachlor injury to grain sorghum (Ketchersid *et al.*, 1982).

The potential involvement of safener-induced transitory reductions in the metabolic activity of grass plants in the protective action of safeners is also worth mentioning. In a recent study, Zama and Hatzios (1987) showed that the metabolic activity of sorghum seedlings grown from seeds pretreated with oxabetrinil was significantly lower (two- to threefold) than that of unsafened seedlings. Although it is conceivable that the reduced metabolic activity of safened sorghum could render these seedlings less susceptible to subsequent treatments with the herbicide metolachlor, these safener effects seem to be unrelated to their protective action against herbicide injury.

Oxime safeners such as cyometrinil and oxabetrinil were also shown to interfere with respiration during the early stages of seed germination of grain sorghum (Ketchersid and Merkle, 1983). Cyometrinil was found to be a strong inhibitor of seed germination of grain sorghum (Davidson *et al.*, 1978; Warmund *et al.*, 1981). These effects of oxime ether safeners on seed respiration could possibly result from their potential interference with the cyanide-resistant alternative pathway of respiration and lipoxygenase (EC 1.12.11.12) activity in germinating seeds. Lipoxygenase oxidizes unsaturated fatty acids to conjugated hydroxyperoxy acids (Axelrod, 1974). Substituted benzohydroxamic acids such as SHAM (salicylhydroxamic acid) and the antioxidant propyl gallate are very strong inhibitors of lipoxygenase (Peterman and Siedow, 1983). Oxime ether safeners structurally resemble SHAM (Fig. 7) and it appears that these effects of safeners on respiration could mimic those of SHAM and be unrelated to their action as safeners against injury from chloroacetanilide herbicides. However, it is of interest to note that at very high dosages (6000 ppm), soil-applied SHAM protected sorghum against metolachlor (M. L. Ketchersid, Texas A & M University, personal communication). Furthermore, the carboxamide fungicide, carboxin, which is also an inhibitor of the cyanide-resistant alternative pathway of respiration in several fungi and higher plants (Kuhn, 1984), has been reported to act as a safener of selected crops against injury from carbamate herbicides (Miller and Nalewaja, 1980). Undoubtedly more research is needed to elucidate

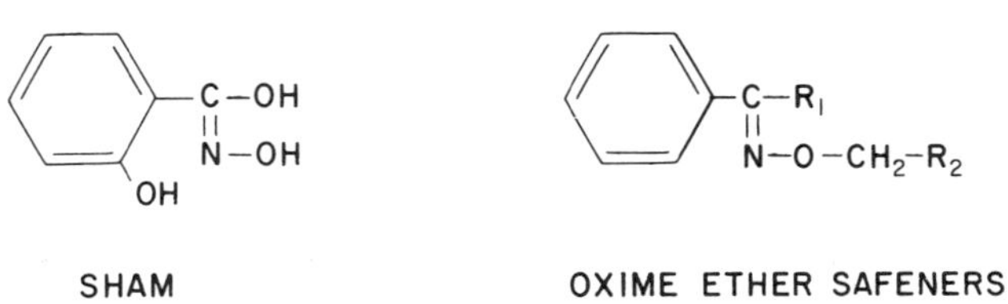

Fig. 7. Chemical structures of SHAM (salicylhydroxamic acid) and substituted oxime ether safeners.

the potential involvement of seed respiration and lipoxygenase activity in the action of chloroacetamide herbicides and the protective action of oxime ether safeners.

IV. MECHANISMS OF SAFENER ACTION: MOLECULAR ASPECTS

The ability of safeners to induce the levels or the activity of GSTs and of other metabolic enzymes in higher plants is now well documented and was discussed earlier. However, our understanding of the molecular mechanisms of safener action is incomplete. Although it is reasonable to expect that safeners could act as gene inducers regulating the expression of mRNAs coding for polypeptides involved in herbicide detoxication, direct evidence supporting such a hypothesis is limited.

A number of selected plant enzymes catalyzing the metabolic detoxication of specific herbicides in higher plants have been isolated and partially characterized (Hatzios and Penner, 1982; Lamoureaux and Frear, 1979). In most cases, however, the genes coding for these enzymes have not been studied (Hatzios, 1987). A recent report by Shah *et al.* (1986) is the first successful attempt at cloning and sequencing of a plant gene coding for the enzyme GST I in corn, which is involved in the detoxication of chloroacetanilide herbicides. Nucleotide sequence analysis of cDNA and genomic clones yielded the complete amino acid sequence and provided the exon–intron map of this GST I gene (Shah *et al.*, 1986). A polymorphism based on the restriction fragment length of the GST I gene was observed in different corn lines. An inbred corn line ('Missouri 17') contained a single gene for GST I, whereas a hybrid line ('Pioneer 3780A') contained two genes (Shah *et al.*, 1986). Furthermore, the amino acid sequence of the corn GST I gene showed no apparent homology with the published sequences of animal GSTs (Shah *et al.*, 1986).

The expression of the GST I gene can be regulated chemically with the use of exogenously applied safeners. Wiegand *et al.* (1986) showed that flurazole may act at the transcriptional level to induce a three- to fourfold increase in the steady state level of mRNA encoding for the GST I gene in corn tissues grown from flurazole-treated seeds. However, the possibility that the flurazole-induced elevation of the GST I mRNA could also result from an increase of the half-life of this mRNA was not ruled out (Wiegand *et al.*, 1986). The characteristics of the flurazole-regulated cDNA clone are summarized in Table III.

The safener-induced elevation of mRNA in treated plants is in agreement with earlier reports demonstrating that treatment with xenobiotics can induce an increase in GST isozyme levels in mammalian or insect systems (Benson *et al.*, 1982; Ding and Pickett, 1985; Hayaoka and Dauterman, 1982, 1983; Pearson *et*

Table III Characteristics of the Flurazole-Regulated cDNA Clone Isolated from Corn Tissue[a]

Tissue	Etiolated corn shoots (Pioneer 3780A)
cDNA insert size (bp)	0.74
mRNA size (kb)	1.1
Source of poly A^+-RNA used for construction of the cDNA library	Shoots from 8-day-old corn seedlings grown from seeds treated with flurazole
Level of expression of the mRNA in the intact tissue	0.15% of total poly A^+-RNA
Fold increase of mRNA as a result of treatment with flurazole	Three- to fourfold

[a] Compiled from data reported by Shah *et al.* (1986) and by Wiegand *et al.* (1986).

al., 1983). Certain xenobiotics such as phenobarbital and 3-methylcholanthrene are known to act at the transcriptional level and activate the genes coding for specific subunits of the GST dimer in rat liver (Ding and Pickett, 1985). The possible induction of GST isozymes that exhibit greater reactivity toward metolachlor has been reported recently by Gronwald *et al.* (1987) in etiolated grain sorghum seedlings treated with certain herbicide safeners. Reports on the molecular mechanisms of these safener-mediated inductions of GSTs are not yet available.

Apart from the results of Wiegand *et al.* (1986), the potential involvement of gene activation in the molecular action of herbicide safeners is supported indirectly by the observation of Frear *et al.* (1987) that the action of certain safeners (e.g., NA) may mimic the action of endogenous or synthetic auxins. Natural or synthetic auxins are known to regulate the expression of several genes in higher plants (Guilfoyle, 1986; Theologis, 1986) and the possibility that herbicide safeners may act through a similar mechanism at the molecular level is worthy of examination. Nucleophilic reagents such as hydroxamic acids and oximes are also effective reactivators of esterase enzymes inhibited by carbamate and organosulfur compounds (Aldridge and Reiner, 1972; Engelhard *et al.*, 1967).

Assuming that herbicide safeners act at the transcriptional level and in analogy to the negative and positive control models proposed for the regulation of auxin-inducible genes by Theologis (1986), the following control models can be postulated for the molecular mechanisms of action of herbicide safeners.

The negative control model (Fig. 8) assumes the existence of a shortlived protein repressor (R) which interacts with the promoter region of a gene coding for metabolic plant enzymes (e.g., GSTs) which detoxify herbicides and prevents its transcription. Both herbicides and safeners can initiate the transcription of this gene, but through different mechanisms of regulation. The herbicide may alter the concentration of the repressor and release its control over transcription by

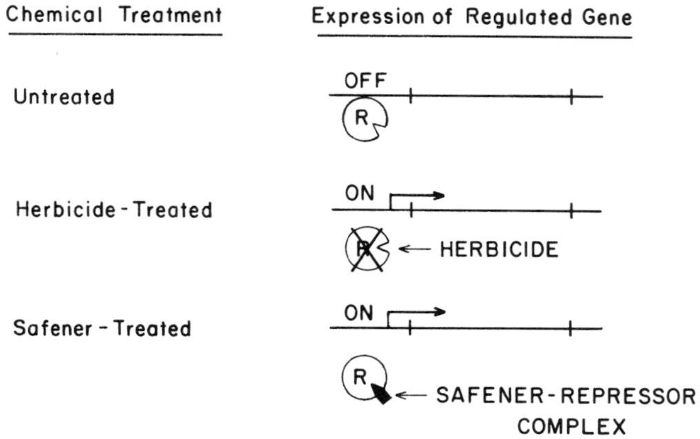

Fig. 8. Negative control model for the regulation of plant genes by herbicides and safeners.

acting as an inhibitor of protein synthesis. Chloroacetanilide and thiocarbamate herbicides could fit into this role since both have been reported to inhibit protein synthesis in higher plants (Fedtke, 1982; Jaworski, 1969; McFarland and Hess, 1985). These herbicides have also been reported to elevate the activity of GSTs by catalyzing their conjugation to GSH (Ezra *et al.*, 1985; A. V. Komives *et al.*, 1985a; Stephenson *et al.*, 1983). Safeners on the other hand could bind to the repressor protein and inactivate it.

The direct positive control model (Fig. 9) is based on the existence of an activator molecule (A) which is a protein with a short half-life and is inactive in the absence of the herbicide or the safener. When the herbicide or the safener is present, they bind to the activator and activate it by altering its functionality. The activator then turns on the regulated gene.

The indirect positive control model (Fig. 10) assumes the existence of an activator molecule (A) which in the absence of the safener interacts with a

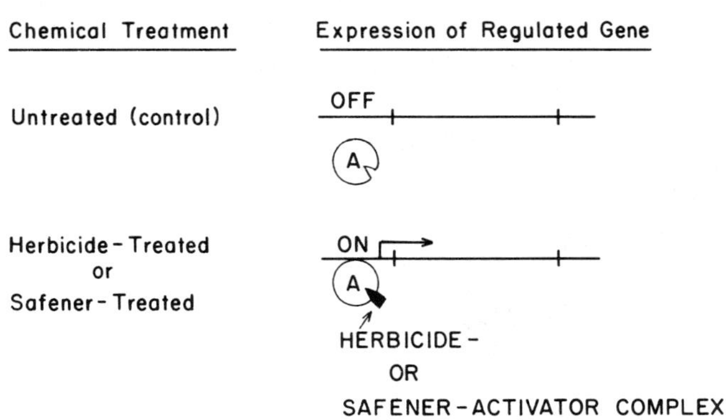

Fig. 9. Direct positive control model for the regulation of plant genes by herbicides and safeners.

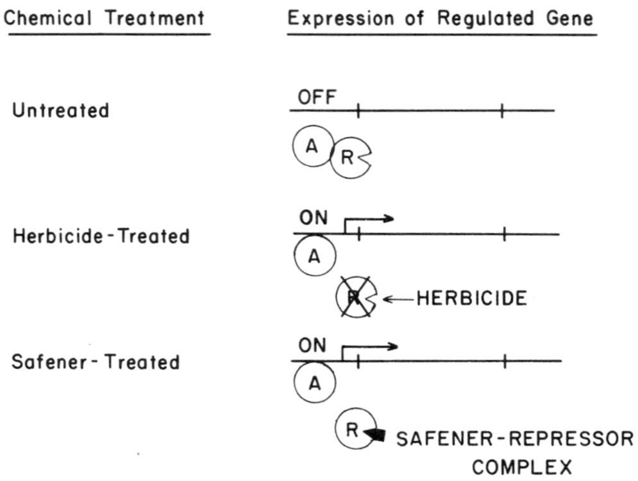

Fig. 10. Indirect positive control model for the regulation of plant genes by herbicides and safeners.

repressor protein molecule (R) and forms an inactive complex (A–R). The safener binds to the repressor and releases the activator (A), which can then interact with the promoter region of the regulated gene and turn on its transcription. The possibility, however, that a herbicide may turn on the gene for a specific metabolic enzyme whereas a safener turns on the gene for an isozyme of this protein, should also be considered.

If safeners indeed act at a transcriptional level by regulating the expression of selected plant genes then their molecular mechanism of action should meet several criteria. According to Theologis (1986), the main criteria which characterize the auxin-induced regulation of gene expression include: an induction of mRNA which is rapid, specific, and unaffected by protein synthesis inhibitors. Wiegand *et al.* (1986) showed that the induction of mRNA by flurazole is very specific, but studies on the rapidity of the mRNA induction by safeners as well as the effect of protein synthesis inhibitors on this effect of safeners are not available. In a recent study, however, Sweetser (1985) showed that the increased metabolism of sulfonylurea herbicides following treatment of corn with NA was strongly inhibited by the protein synthesis inhibitor cycloheximide. Undoubtedly, further research is needed to elucidate the molecular mechanisms of action of herbicide safeners as well as the validity of the proposed models for the potential safener-induced regulation of gene expression in higher plants.

V. CONCLUDING REMARKS

Research on the physiological and biochemical mechanisms of action of herbicide safeners has attracted considerable attention during the past 10 years. However, the exact mechanisms of action of these chemical agents have not been elucidated.

Because of the chemical and botanical specificity of the currently marketed safeners, it seems safe to conclude that safeners do not act by a single mechanism. A series of multilevel interactions between safeners and antagonized herbicides appears more likely for explaining the protective action of herbicide safeners. Some safeners are also likely to act by different mechanisms depending on the nature and specificity of particular crop–herbicide–safener combinations.

In general, safeners are believed to act either as "bioregulators" influencing the amount of a given herbicide that reaches its target site in an active form or as "antagonists" of herbicidal effects at a common site of action. A safener-induced enhancement of herbicide detoxication in protected plants seems to be the major mechanism involved in the protective action of the currently developed safeners. Most safeners enhance the conjugation of thiocarbamate and chloroacetanilide herbicides with glutathione by elevating the levels of reduced glutathione (GSH) or by inducing the activity and *de novo* synthesis of specific glutathione-*S*-transferases which catalyze this reaction in protected plants. Oxidative reactions catalyzed by specific monooxygenase enzymes are also currently viewed as possible factors contributing to the metabolism of these herbicides in grass plants and to the protective action of herbicide safeners.

Specific steps of lipid and terpenoid metabolism of grass plants appear to be the most likely target sites for a competitive or physiological antagonism between safeners and chloroacetanilide or thiocarbamate herbicides. The existing correlation of the effects of some safeners with the activity of endogenous phytohormones further demonstrates the complexity of the interactions of safeners and antagonized herbicides and the need for conducting additional research to further comprehend the protective action of herbicide safeners.

Our knowledge of the molecular aspects of the mechanisms of action of herbicide safeners is just beginning to unfold. It is now evident that several safeners induce specific mRNAs in grass plants protected from thiocarbamate and chloroacetanilide herbicide injury. The potential involvement of "gene activation" in the molecular action of herbicide safeners is quite likely and could provide the key for the elucidation of the exact mechanisms of the protective action of these chemical regulators. Undoubtedly, a fuller understanding of the exact mechanisms of action of herbicide safeners awaits further comprehensive investigations.

Acknowledgments

I express my appreciation to Dr. Carl Fedtke of Bayer AG for sharing the model that he proposed for explaining the interactive effects of herbicides, safeners, and synergists. The author's research on physiological aspects of herbicide–safener interactions has been funded by Ciba-Geigy Corporation, BASF Chemical Company, and the Jeffress Foundation, Richmond, Virginia.

References

Adams, C. A., Blee, E., and Casida, J. E. (1983). Dichloroacetamide herbicide antidotes enhance sulfate metabolism in corn roots. *Pestic. Biochem. Physiol.* **19**, 350–360.

Ahrens, W. H., and Davis, D. E. (1978). Seed protectant effects on metolachlor absorption and translocation. *Proc., South. Weed Sci. Soc.* **31**, 19.

Aldridge, W. N., and Reiner, E. (1972). "Enzyme Inhibitors and Substrates." North-Holland Publ. Co. Amsterdam.

Axelrod, B. (1974). Lipoxygenase. *ACS Symp. Ser.* **136**, 324–328.

Balke, N. E. (1979). Inhibition of ion absorption by acetanilide herbicides. *Abstr. Weed Sci. Soc. Am.* **19**, 93.

Barta, I. Cs., Komives, T., and Dutka, F. (1983). Effects of EPTC and its antidotes on epicuticular wax of corn. *Fat Sci., Proc. ISF Congr., 16th,* pp. 277–283.

Benson, A. M., Hunkeler, M. J., Morrow, J. F., Pearson, W. R., Talalay, T., and Windle, J. J. (1982). Characterization of mouse hepatic glutathione-*S*-transferases. Increased synthesis in response to antioxidants. *Fed. Proc., Fed. Am. Soc. Exp. Biol.* **41**, 638.

Beste, C. E., and Schreiber, M. M. (1972). RNA synthesis as the basis for EPTC and 2,4-D antagonism. *Weed Sci.* **20**, 8–11.

Birecka, H., and Galston, A. W. (1970). Peroxidase ontogeny in a dwarf pea stem as affected by gibberellin and decapitation. *J. Exp. Bot.* **21**, 735–745.

Blair, A. M., Parker, C., and Kasasian, L. (1976). Herbicide protectants and antidotes—A review. *Pestic. Artic. News Summ.* **22**, 65–74.

Blattmann, P., Gross, D., Kriemler, H. P., and Ramsteiner, K. (1986). Identification of thiolactic acid type conjugates as major degradation products in the glutathione dependent metabolism of the α-chloroacetamide herbicides metolachlor (DUAL), dimetachlor (TERIDOX) and pretilachlor (RIFIT). *IUPAC Int. Congr. Pesticide Chem., 6th,* Abstr. No. 7A-02.

Breaux, E. J. (1986). Identification of the initial metabolites of acetochlor in corn and soybean seedlings. *J. Agric. Food Chem.* **34**, 884-888.

Brooks, R. L., Ketchersid, M. L., and Merkle, M. G. (1987). Effects of the sorghum seed safeners on sorghum mixed-function oxidases. *Proc., South. Weed Sci. Soc.* **40**, 328.

Bujtas, G.(1978). Effect of an EPTC herbicide and its antidote on the membrane permeability of beet root cells. *Novenytermeles* **27**, 223–229.

Burgstahler, R. J., Retzlaff, G., and Lichtenthaler, H. K. (1986). Mode of action of sethoxydim: Effects on the plant's lipid metabolism. *IUPAC Int. Congr. Pestic. Chem., 6th* Abstr. No. 3B-11.

Burton, J. D., Gronwald, J. W., Somers, D. A., Connelly, J. A., Gengenbach, B. G., and Wyse, D. L. (1982). Inhibition of plant acetyl-coenzyme A carboxylase by the herbicide sethoxydim and haloxyfop. *Biochem. Biophys. Res. Commun.* **148**, 1039–1044.

Carringer, R. D., Rieck, C. E., and Bush, L. P. (1978a). Metabolism of EPTC in corn (*Zea mays*). *Weed Sci.* **26**, 157-160.

Carringer, R. D., Rieck, C. E., and Bush, L. P. (1978b). Effect of R-25788 on EPTC metabolism in corn (*Zea mays*). *Weed Sci. , 167-171.*

Chang, F. Y., Stephenson, G. R., and Bandeen, J. D. (1973). Comparative effects of three EPTC antidotes. *Weed Sci.* **21**, 292-295.

Chang, F. Y., Stephenson, G. R., and Bandeen, J. D. (1974). Effects of *N,N*-diallyl-2,2-dichloroacetamide on ethyl *N,N*-di-*n*-propylthiocarbamate uptake and metabolism by corn seedlings. *J. Agric. Food Chem.* **22**, 245–248.

Chang, T. S. (1983). Structure–activity relationship of oximes and related compounds as seed safeners. Ph.D. Dissertation, Texas A&M University, College Station.

Christ, R. A. (1981). The effects of CGA-43089 as a safener of metolachlor in sorghum (*Sorghum bicolor*) (recordings of elongation rates on single sorghum leaves). *Weed Res.* **21**, 1–9.

Christ, R. A. (1985). Effects of CGA-123407 as a safener for pretilachlor in rice (*Oryza sativa* L.). Recordings of elongation rates of single rice leaves. *Weed Res.* **25**, 193–200.

Crowley, J., and Prendeville, G. N. (1985). Effect of chlorsulfuron and 1,8-naphthalic anhydride on uptake of ^{45}Ca in maize. *Weed Res.* **25**, 341–345.

Davidson, W. E., Gagnon, S. A., Chrustensen, M. D., and Dorr, J. E. (1978). A new herbicide safener which permits effective grass control in sorghum. *Proc. West. Soc. Weed Sci.* **32**, 123–129.

Deal, L. M., and Hess, F. D. (1980). An analysis of the growth inhibitory characteristics of alachlor and metolachlor. *Weed Sci.* **28**, 168–175.

Deal, L. M., Reeves, J. T., Larkins, B. A., and Hess, F. D. (1980). Use of an *in vitro* protein synthesizing system to test the mode of action of chloroacetamides. *Weed Sci.* **28**, 334–340.

Devlin, R. M., and Koszanski, Z. K. (1987a). R-25788 as an antidote for SC-0774 in corn. *Proc. Annu. Meet. Northeast. Weed Sci. Conf.* **41**, 74–77.

Devlin, R. M., and Koszanski, Z. K. (1987b). Effect of FMC-57020 on plant growth and pigment synthesis. *Proc. Annu. Meet. Northeast. Weed Sci. Conf.* **41**, 95–99.

Dhillon, N. S., and Anderson, J. L. (1972). Morphological, anatomical and biochemical effects of propachlor on seedling growth. *Weed Res.* 12, 182–189.

Diesperger, H., and Sandermann, H., Jr. (1979). Soluble and microsomal glutathione *S*-transferase activities in pea (*Pisum sativum* L.) seedlings. *Planta* **146**, 643–648.

Ding, V. D. H., and Pickett, C. B. (1985). Transcriptional regulation of rat liver glutathione *S*-transferase genes by phenobarbital and 3-methylchlolanthrene. *Arch. Biochem. Biophys.* **240**, 553–560.

Donald, W. W., and Fawcett, R. S. (1975). Prevention of EPTC injury to hybrid corn with the antidote R-29148 and exogenous GA_3. *Proc. North* Cent. Weed Control Conf. **20**, 28-29.

Donald, W. W., and Fawcett, R. S. (1976). Prevention of EPTC injury to corn: A developmental and anatomical study. *Abstr. Weed Sci. Soc. Am.* **16**, 70.

Duessing, J. H. (1984). Genetic analysis of herbicide resistance. *In* "Biochemical and Physiological Mechanisms of Herbicide Action" (S. O. Duke, ed.), pp. 73–81. South. Sect. Am. Soc. Plant Physiol., Tallahassee, Florida.

Duke, W. B., Slife, F. W., Hanson, J. B., and Butler, H. S. (1975). An investigation on the mechanism of action of propachlor. *Weed Sci.* **23**, 142-147.

Dutka, F., and Komives, T. (1983). On the mode of action of EPTC and its antidotes. *In* "Pesticide Chemistry: Human Welfare and the Environment" (J. Miyamoto and P. C. Kearney, eds.), Vol. 3, pp. 213–218. Pergamon, Oxford.

Dutka, F., and Komives, T. (1987). MG-191: A new selective herbicide antidote. *In* "Pesticide Science and Biotechnology" (R. Greenhalgh and T. R. Roberts, eds.), pp. 201–204. Blackwell, Oxford.

Ebert, E. (1980). Herbicidal effects of metolachlor (2-chloro-*N*-(2-ethyl-6-methylphenyl)-*N*-(2-methoxy-1-methylethyl)acetamide) at the cellular level in sorghum. *Pestic. Biochem. Physiol.* **13**, 277–286.

Ebert, E. (1982). The role of waxes in the uptake of metolachlor into sorghum in relation to the protectant CGA-43089. *Weed Res.* **22**, 305–311.

Ebert, E., and Ramsteiner, K. (1984). Influence of metolachlor and the metolachlor protectant CGA-43089 on the biosynthesis and epicuticular waxes on the primary leaves of *Sorghum bicolor* Moench. *Weed Res.* **24**, 383–389.

Engelhard, N., Prchal, K., and Nenner, M. (1967). Acetylcholinesterase. *Angew. Chem., Int. Ed. Engl.* **6**, 615–618.

Ezra, G., and Gressel, J. (1982). Rapid effects of a thiocarbamate herbicide and its dichloroacetamide protectant on macromolecular synthesis and gluthathione levels in maize cell cultures. *Pestic. Biochem. Physiol.* **17**, 48–58.

Ezra, G., and Stephenson, G. R. (1986). The physiology of metazachlor and safener 145 138 interactions in corn. *IUPAC Int. Congr. Pestic. Chem., 6th* Abstr. No. 3D-12.

Ezra, G., Krochmal, E., and Gressel, J. (1982). Competition between a thiocarbamate herbicide and herbicide protectants at the level of uptake into maize cells in culture. *Pestic. Biochem. Physiol.* **18**, 107–112.

Ezra, G., Flowers, H. M., and Gressel, J. (1983a). Rapid multilevel interactions of a thiocarbamate herbicide and its protectant in maize cell culture. *In* "Pesticide Chemistry: Human Welfare and the Environment" (J. Miyamoto and P. C. Kearney, eds.), Vol. 3, pp. 225–231. Pergamon, Oxford.

Ezra, G., Gressel, J., and Flowers, H. M. (1983b). Effects of the herbicide EPTC and the protectant DDCA on incorporation and distribution of (2-^{14}C)acetate into major lipid fractions on maize cell suspension cultures. *Pestic. Biochem. Physiol.* **19**, 225–234.

Ezra, G., Rusness, D. G., Lamoureux, G. L., and Stephenson, G. R. (1985). The effect of CDAA (*N,N*-diallyl-2-chloroacetamide) pretreatments on subsequent CDAA injury to corn (*Zea mays* L.). *Pestic. Biochem. Physiol. 23*, 108-115.

Fedtke, C. (1981). Wirkung von herbiziden und herbizid-antidots auf den glutathiongehalt in mais- und sokabohnenwurzeln. Z. Pflanzenkr. (Pflanzenpathol.) Pflanzenschutz, Sonderh. **9**, 141–146.

Fedtke, C. (1982). "Biochemistry and Physiology of Herbicide Action." Springer-Verlag, Berlin and New York.

Fedtke, C. (1985). Einsatzmoglichkeiten und wirkungsweise von herbizidantidots. *Z. Pflanzenkr. Pflanzenschutz*. **92**, 654–664.

Fedtke, C., and Trebst, A. (1987). Advances in understanding herbicide modes of action. *In* "Pesticide Science and Biotechnology" (R. Greenhalgh and T. R. Roberts, eds.), pp. 161–168. Blackwell, Oxford.

Frear, D. S., and Swanson, H. R. (1970). Biosynthesis of *S*-(4-ethylamino-6-isopropylamino-2-*s*-triazino)glutathione. Partial purification and properties of a glutathione *S*-transferase from corn. *Phytochemistry* **9**, 2123–2132.

Frear, D. S., Swanson, H. R., and Mansager, E. R. (1987). 1,8-Naphthalic anhydride/auxin protection against chlorsulfuron inhibition of corn seedling growth. *In* "Pesticide Science and Biotechnology" (R. Greenhalgh and T. R. Roberts, eds.), pp. 499–503. Blackwell, Oxford.

Fuerst, E. P., and Gronwald, J. W. (1986). Induction of rapid metabolism of metolachlor in sorghum (*Sorghum bicolor*) shoots by CGA-92194 and other antidotes. *Weed Sci.* **34**, 354–361.

Gardiner, M. G., and Cleland, R. (1974). Peroxidase changes during the cessation of elongation in *Pisum sativum* stems. *Phytochemistry* **13**, 1095–1105.

Görög, K., Muschinek, G. Y., Mustardy, L. A., and Faludi-Daniel, A. (1982). Comparative studies of safeners for the prevention of EPTC injury in maize. *Weed Res.* **22**, 27–33.

Gray, R. A., and Joo, G. K. (1978). Site of uptake and action of thiocarbamate herbicides and herbicide antidotes in corn seedlings. *In* "Chemistry and Action of Herbicide Antidotes" (F. M. Pallos and J. E. Casida, eds.), pp. 67–84. Academic Press, New York.

Gray, R. A., Green, L. L., Hoch, P. E., and Pallos, F. M. (1982). The evolution of practical crop safeners. *Proc. Br. Crop Prot. Conf.—Weeds* **2**, 431-437.

Gressel, J. (1986). Modes and genetics of herbicide resistance in plants. *In* "Pesticide Resistance: Strategies and Tactics for Management" (E. H. Glass, ed.), pp. 54–73. National Academy Press, Washington, D.C.

Gressel, J., Ezra, G., and Jain, S. M. (1982). *In* "Genetic and Chemical Manipulation of Crops to Confer Tolerance to Chemicals" (J. D. McLaren, ed.), pp. 79–91. Butterworth, London.

Gronwald, J. W., Fuerst, E. P., Eberlein, C. V., and Egli, M. A. (1987). Effect of herbicide antidotes on glutathione content and glutathione *S*-transferase activity of sorghum shoots. *Pestic. Biochem. Physiol.* **29**, 66-76.

Guddewar, M. B., and Dauterman, W. C. (1979). Purification and properties of a glutathione-*S*-transferase from corn which conjugates *s*-triazine herbicides. *Phytochemistry* **18**, 735-740.

Guilfoyle, T. J. (1986). Auxin-regulated gene expression in higher plants. *CRC Crit. Rev. Plant Sci.* **4**, 247-276.

Guneyli, E. (1971). Factors affecting the action of 1,8-naphthalic anhydride in corn treated with *S*-ethyl dipropylthiocarbamate (EPTC). *Diss. Abstr. Int. B* **32**, *1957-1958.*

Gupta, O. P. (1976). Adsorbents and antidotes offer a breakthrough in weed control. World Crops **28**, 134–138.

Halliwell, B. (1984). "Chloroplast Metabolism: The Structure and Function of Chloroplasts in Green Leaves." Oxford Univ. Press, London and New York.

Hardy, R. W. (1986). Biotechnology in pesticide resistance development. *In* "Pesticide Resistance: Strategies and Tactics for Management" (E. H. Glass, ed.), pp. 130–141. National Academy Press, Washington, D.C.

Harvey, B. M. R., Chang, F. Y., and Fletcher, R. A. (1975). Relationship between *S*-ethyl dipropylthiocarbamate injury and peroxidase activity in corn seedlings. *Can. J. Bot.* **53**, 225–230.

Hatzios, K. K. (1981). Synergistic interactions of tebuthiuron with EPTC + R-25788 and butylate + R-25788 in corn (*Zea mays*). *Weed Sci.* **29**, 601–604.

Hatzios, K. K. (1982). Effects of sethoxydim on the metabolism of isolated leaf cells of soybean (*Glycine max* (L.) Merr.). *Plant Cell Rep.* **1**, 87–90.

Hatzios, K. K. (1983a). Effects of the safener CGA-43089 on response of sorghum (*Sorghum bicolor*) to metolachlor combined ozone or antioxidants. *Weed Sci.* **31**, 280–284.

Hatzios, K. K. (1983b). Interactions of the herbicides EPTC and EPTC + R-25788 with ozone or antioxidants in corn. *J. Agric. Food Chem.* **31**, 1187–1191.

Hatzios, K. K. (1983c). Herbicide antidotes: Development, chemistry and mode of action. *Adv. Agron.* **36**, 265–316.

Hatzios, K. K. (1984a). Potential safeners for protecting sorghum (*Sorghum bicolor* (L.) Moench) against chlorsulfuron, fluazifop-butyl and sethoxydim. *Weed Res.* **24**, 249-254.

Hatzios, K. K. (1984b). Interactions between selected herbicides and protectants on corn (*Zea mays*). *Weed Sci.* **32**, 51-58.

Hatzios, K. K. (1984c). Biochemical and physiological mechanisms of action of herbicide antidotes. *In* "Biochemical and Physiological Mechanisms of Herbicide Action" (S. O. Duke, ed.), pp. 7–30. South. Sect. Am. Soc. Plant Physiol., Tallahassee, Florida.

Hatzios, K. K. (1984d). Interactions of tebuthiuron with chloroacetanilide herbicides on corn (*Zea mays* L.) seedlings safened or unsafened with the antidote R-25788. *Zizaniology* **1**, 153–158.

Hatzios, K. K. (1987). Biotechnology applications in weed management: Now and in the future. *Adv. Agron.* **41**, 325–375.

Hatzios, K. K., and Moon, P. A. (1985). Combined effects of sethoxydim and chloroacetamide safeners on the metabolism of sorghum protoplasts and soybean cells. *Proc., South. Weed Sci. Soc.* **38**, 462.

Hatzios, K. K., and Penner, D. (1982). "Metabolism of Herbicides in Higher Plants." Burgess, Minneapolis, Minnesota.

Hatzios K. K., and Penner, D. (1985). Interactions of herbicides with other agrochemicals on higher plants. *Rev. Weed Sci.* **1**, 1–63.

Hayaoka, T., and Dauterman, W. C. (1982). Induction of glutathione-*S*-transferase by phenobarbital and pesticides in various house fly strains and its effect on toxicity. *Pestic. Biochem. Physiol.* **17**, 113–120.

Hayaoka, T., and Dauterman, W. C. (1983). The effect of phenobarbital induction on glutathione conjugation of diazinon in susceptible and resistant house flies. *Pestic. Biochem. Physiol.* **19**, 334–340.

Hickey, J. S., and Krueger, W. A. (1974). Alachlor and 1,8-naphthalic anhydride effects on corn coleoptiles. *Weed Sci.* **22**, 250–252.

Hilton, J. L., and Pillai, P. (1986). L-2-Oxothiazolidine-4-carboxylic acid protection against tridiphane toxicity. *Weed Sci.* **34**, 669–675.

Hilton, J. L., and Pillai, P. (1987). L-Thiazolidine-4-carboxylic acid, a new herbicide protectant. *Abstr. Weed Sci. Soc. Am.* **27**, 67.

Hoagland, R. E. (1984). Reversal of EPTC inhibition of arylamidase activity by oximes. *Abstr. Weed Sci. Soc. Am.* **24**, 86.

Hoffman, O. L. (1969). Chemical antidotes for EPTC on corn. *Abstr. Weed Sci. Soc. Am.* **9**, 12.

Hoffman, O. L. (1978). Herbicide antidotes: Concept to crop. CHEMTECH **8**, 488–492.

Holm, R. E., and Szabo, S. S. (1974). Increased metabolism of a pyrrolidine urea herbicide in corn by a herbicide antidote. *Weed Res.* **14**, 119–122.

Horváth, L., and Pulay, Á. (1980). Metabolism of EPTC in germinating corn: Sulfone as the true carbamoylating agent. *Pestic. Biochem. Physiol.* **14**, 265–270.

Hubbell, J. P., and Casida, J. E. (1977). Metabolic fate of the *N,N*-dialkylcarbamoyl moiety of thiocarbamate herbicides in rats and corn. *J. Agric. Food Chem.* **25**, 404–413.

Jackson, L. A., Kapusta, G., and Yopp, J. H. (1986). Absorption and distribution of flurazole and acetochlor in grain sorghum. *Pestic. Biochem. Physiol.* **25**, 373-380.

Jaworski, E. G. (1969). Analysis of the mode of action of herbicidal α-chloroacetamides. *J. Agric. Food Chem.* **17**, 165-170.

Jordan, L. S., and Jolliffe, V. A. (1971). Protection of plants from herbicides with 1,8-naphthalic anhydride as illustrated with sorghum. *Bull. Environ. Contam. Toxicol.* **6**, 417–421.

Ketchersid, M. L., and Merkle, M. G. (1981). The effect of CGA-43089 on absorption and metabolism of metolachlor in grain sorghum. *Abstr. Weed Sci. Soc. Am.* **21**, 100.

Ketchersid, M. L., and Merkle, M. G. (1983). Effects of protectants on sorghum seedling respiration. Proc. South. Weed Sci. Soc. **36**, 384.

Ketchersid, M. L., Norton, K., and Merkle, M. G. (1981). Influence of soil moisture on the safening effect of CGA-43089 in grain sorghum. (*Sorghum bicolor*). *Weed Sci.* **29**, 281-287.

Ketchersid, M. L., Vietor, D. M., and Merkle, M. G. (1982). CGA-43089 effects on metolachlor uptake and membrane permeability in grain sorghum (*Sorghum bicolor*). *J. Plant Growth Regul.* **1**, 285-294.

Ketchersid, M. L., Plapp, F. W., and Merkle, M. G. (1985). Sorghum (*Sorghum bicolor*) seed safeners as insecticide synergists. *Weed Sci.* **33**, 774-778.

Komives, A. V., Komives, T., and Dutka, F. (1985). Effects of thiocarbamate herbicides on the activity of glutathione *S*-transferase in maize. *Cereal Res. Commun.* **13**, 253–257.

Komives, T., Komives, V. A., Balazs, M., and Dutka, F. (1985a). Role of glutathione-related enzymes in the mode of action of herbicide antidotes. *Proc. Br. Crop Prot. Conf.—Weeds*, pp. 1155–1162.

Komives, T., Balazs, M., Komives, V., and Dutka, F. (1985b). Effects of herbicide antidotes on reduced glutathione content and levels of l. cytochrome P-450 and glutathione *S*-transferase in *Zea mays* L. *Proc. Int. Conf. Cytochrome P-450, 5th,* pp. 451–454.

Kondo, T., Taniguchi, N., and Kawakami, Y. (1984). Significance of glutathione S-conjugate for glutathione metabolism in human erythrocytes. *Eur. J. Biochem.* **145**, 131–136.

Krause, A., Hancock, W. S., Minard, R. D., Freyer, A. S., Honeycutt, R. C. LeBaron, H. M., Paulson, D. L., Liu, S. Y., and Bollag, J. M. (1985). Microbial transformation of the herbicide metolachlor by a soil actinomycete. *J. Agric. Food Chem.* **33**, 585–589.

Kuhn, P. J. (1984). Mode of action of carboxamides. *In* "Mode of Action of Antifungal Agents" (A. P. J. Trinci and J. F. Riley, eds.), pp. 155–183. Br. Mycol. Soc., London.

Lamoureux, G. L., and Frear, D. S. (1979). Pesticide metabolism in higher plants: *In vitro* enzyme studies. *ACS Symp. Ser.* **97**, 78–128. American Chemical Society, Washington, D.C.

Lamoureux, G. L., and Frear, D. S. (1987). Current problems, trends, and developments in pesticide metabolism in plants. *In* "Pesticide Science and Biotechnology" (R. Greenhalgh and T. R. Roberts, eds.), pp. 455–462. Blackwell, Oxford.

Lamoureux, G. L., and Rusness, D. G. (1986). Xenobiotic conjugation in higher plants. *ACS Symp. Ser.* **299**, 62–105.

Lay, M.-M., and Casida, J. E. (1976). Dichloroacetamide antidotes enhance thiocarbamate sulfoxide detoxification by elevating corn root glutathione content and glutathione *S*-transferase activity. *Pestic. Biochem. Physiol.* **6**, 442-456.

Lay, M.-M., and Casida, J. E. (1978). Influence of glutathione and glutathione *S*-transferase in the action of dichloroacetamide antidotes for thiocarbamate herbicides. *In* "Chemistry and Action of Herbicide Antidotes" (F. M. Pallos and J. E. Casida, eds.), pp. 151–160. Academic Press, New York.

Lay, M.-M., and Niland, A. M. (1985). Biochemical response of inbred and hybrid corn (*Zea mays* L.) to R-25788 and its distribution with EPTC in corn seedlings. *Pestic. Biochem. Physiol.* **23**, 131–140.

Lay, M.-M., Hubbell, J. P., and Casida, J. E. (1975). Dichloroacetamide antidotes for thiocarbamate herbicides: Mode of action. *Science* **189**, 287-289.

Leavitt, J. R. C., and Penner, D. (1978a). Prevention of EPTC-induced epituticular wax aggregation on corn (*Zea mays*) with R-25788. *Weed Sci.* **26**, *47-50.*

Leavitt, J. R. C., and Penner, D. (1978b). Protection of corn (*Zea mays*) from acetanilide herbicidal injury with the antidote R-25788. *Weed Sci.* **26**, 653–659.

Leavitt, J. R. C., and Penner, D. (1979). *In vitro* conjugation of glutathione and other thiols with acetanilide herbicides and EPTC sulfoxide and the action of the herbicide antidote R-25788. *J. Agric. Food Chem.* **27**, 533–536.

LeBaron, H. M. (1986). Resistance of weeds to herbicides. Presentation to the Illinois Custom Spray Training School.

LeBaron, H. M., and Gressel, J., eds. (1982). "Herbicide Resistance in Plants." Wiley, New York.

Leek, G. L., and Penner, D. (1981). Environmental conditions affecting efficacy of CGA-43089 as an antidote for sorghum. *Abstr. Weed Sci. Soc. Am.* **21**, 101.

McFarland, J. E., and Hess, F. D. (1985). Herbicidal activity of acetanilides parallels alkylation potential. *Abstr. Weed Sci. Soc. Am.* **25**, 72.

Mann, J. D., and Pu, M. (1968). Inhibition of lipid synthesis by certain herbicides. *Weed Sci.* **16**, 197-198.

Mann, J. D., Jordan, L. S., and Day, B. E. (1965). A survey of herbicides for their effects upon protein synthesis. *Plant Physiol.* **40**, 840-843.

Martell, A. E. (1971). Principles of complex formation. *In* "Organic Compounds in Aquatic Environments" (S. D. Faust and J. V. Hunter, eds.), pp. 259–263. Dekker, New York.

Marton, A. F., Appokovacs, A. V., Komives, T., Fodor-Csobra, K., and Dutka, F. (1978). Comparison of the uptake and translocation of EPTC in a susceptible and a resistant plant. *Proc. Hung. Annu. Meet. Biochem.* **18**, 111-112.

Mazelis, M., and Creveling, R. K. (1978). 5-Oxoprolinase (L-pyroglutamate hydrolase) in higher plants. *Plant Physiol.* **62**, 798-801.

Meister, A. (1984). New aspects of glutathione biochemistry and transport. Selective alteration of glutathione metabolism. *Fed. Proc., Fed. Am. Soc. Exp. Biol.* **43**, 3031–3042.

Mellis, J. M., Pillai, P., Davis, D. E., and Truelove, B. (1982). Metolachlor and alachlor effects on membrane permeability and lipid synthesis. *Weed Sci.* **30**, 399–404.

Miaullis, J. B., Thomas, V. M., Gray, R. A., Murphy, J. J., and Hollingworth, R. M. (1978). Metabolism of R-25788 (*N,N*-diallyl-2,2-dichloroacetamide) in corn plants, rats, and soil. *In* "Chemistry and Action of Herbicide Antidotes" (F. M. Pallos and J. E. Casida, eds.), pp. 109–131. Academic Press, New York.

Mikami, N., Baba, Y., Miyashita, T., Matsuda, T., and Miyamoto, J. (1986). Purification and characterization of glutathione S-transferase in higher plants. *IUPAC Int. Congr. Pestic. Chem., 6th* Abstr. No. 7A-15.

Miller, S. D., and Nalewaja, J. D. (1980). Herbicide antidotes with triallate. *Agron. J.* **72**, 662–664.

Molin, W. T., Anderson, E. J., and Porter, C. A. (1986). Effects of alachlor on anthocyanin and lignin synthesis in etiolated sorghum (*Sorghum bicolor* (L.) Moench) mesocotyls. *Pestic. Biochem. Physiol.* **25**, 105–111.

Moreland, D. E., Malhorta, S. S., Gruenhagen, R. D., and Shokrah, E. H. (1969). Effects of herbicides on RNA and protein synthesis. *Weed Sci.* **17**, 556–563.

Mozer, T. J., Tiemeier D. C., and Jaworski, E. G. (1983). Purification and characterization of corn glutathione-*S*-transferase. *Biochemistry* **22**, *1068–1072.*

Murphy, J. J. (1972). Effect of 1,8-naphthalic anhydride on the uptake of *S*-ethyl *N,N*-dipropylthiocarbamate (EPTC) by *Zea mays*. *Chem. Biol. Interact.* **5**, 284–286.

Nyffeler, A., Gerber, H. R., and Hensley, J. R.(1980). Laboratory studies on the behavior of the herbicide safener CGA-43089. *Weed Sci.* **28**, 6–10.

Pallos, F. M., and Casida, J. E., eds. (1978). "Chemistry and Action of Herbicide Antidotes." Academic Press, New York.

Paradies, I., Ebert, E., and Elstner, E. F.(1981). Metolachlor (2-chloro-*N*-(2-ethyl-6-methyphenyl)-*N*-(2-methoxy-1-methylethyl)acetamide) and the metolachlor safener CGA-43089 (α-(cyanomethoximino)-benzoacetonitrile) in sorghum seedlings: Correlations between morphological effects and ethylene formation. *Pestic. Biochem. Physiol.* **15**, 209–212.

Parker, C. (1983). Herbicide antidotes—A review. *Pestic. Sci.* **14**, 40–48.

Pearson, W. R., Windle, J. J., Morrow, J. F., Benson, A. M.and Talalay, P. (1983). Increased synthesis of glutathione-*S*-transferases in response to anticarcinogenic antioxidants. *J. Biol. Chem.* **258**, 2052–2062.

Peterman, T. K., and Siedow, J. N. (1983). Structural features required for inhibition of soybean lipoxygenase-2 by propyl gallate. *Plant Physiol.* **71**, 55-58.

Rao, V. S., and Duke, W. B. (1976). Effect of alachlor, propachlor, and prynachlor on GA_3-induced production of protease and α-amylase. *Weed Sci.* **24**, 616-618.

Rao, V. S., and Khan, A. A. (1975). Protection of barley from alachlor injury. *Abstr. Weed Sci. Soc. Am.* **15**, 66-67.

Rennenberg, H. (1982). Glutathione metabolism and possible biological roles in higher plants. *Phytochemistry* **21**, 2771-2781.

Rennenberg, H., Birk, C., and Shcaer, B. (1982). Effect of *N,N*-diallyl-2,2-dichloroacetamide (R-25788) on efflux and synthesis of glutathione in tobacco suspension cultures. *Phytochemistry* **21**, 5-8.

Rubin, B., and Casida, J. E. (1985). R-25788 effects on chlorsulfuron injury and acetohydroxyacid synthase activity. *Weed Sci.* **33**, 462–468.

Rubin, B., Kirino, O., and Casida, J. E. (1985). Chemistry and action of *N*-phenylmaleamic acids and their progenitors as selective herbicide antidotes. *J. Agric. Food Chem.* **33**, 489–494.

Sagaral, E. (1978). Toxicity, selectivity, uptake, distribution, and site of action of EPTC in corn (*Zea mays* L.) as affected by a herbicide antidote. *Diss. Abstr. Int. B* **39**, 5174.

Shah, D. M., Hironaka, C. H., Wiegand, R. C., Harding, E. I., Krivi, G. G., and Tiemeier, D. C.(1986). Structural analysis of a maize gene coding for glutathione-*S*-transferase involved in herbicide detoxification. *Plant Molec. Biol.* **6**, 203–211.

Smith, A. M., Morrison, W. L., and Milham, P. J. (1982). Oxidation of indole-3-acetic acid by peroxidase. Involvement of reduced peroxidase and compound III with superoxide as a product. *Biochem. J. 21*, 4414–4424.

Stephenson, G. R., and Chang, F. Y. (1978). Comparative activity and selectivity of herbicide antidotes. *In* "Chemistry and Action of Herbicide Antidotes" (F. M. Pallos and J. E. Casida, eds.), pp. 35–61. Academic Press, New York.

Stephenson, G. R., and Ezra, G. (1982). The mode of action of herbicide safeners. *Proc. Br. Crop Prot. Conf.—Weeds,* **2**, 451–459.

Stephenson, G. R., and Ezra, G. (1983). Herbicide antidotes—A new era in selective chemical weed control. *In* "Plant Growth Regulating Chemicals" (L. G. Nickell, ed.), Vol. 2, pp. 193–212. CRC Press, Boca Raton, Florida.

Stephenson, G. R., and Ezra, G. (1985). Use of subtoxic pretreatments to improve crop tolerance to herbicides. *ACS Symp. Ser.* **276** 69–84.

Stephenson, G. R., Bunce, J. J., Makowski, R. I., and Curry, J. C. (1978). Structure–activity relationships for *S*-ethyl *N,N*-dipropylthiocarbamate (EPTC) antidotes in corn. *J. Agric. Food Chem.* **26**, 137–140.

Stephenson, G. R., Ali, A., and Ashton, F. M. (1983). Influence of herbicides and antidotes on the gluthathione levels of maize seedlings. *In* "Pesticide Chemistry: Human Welfare and the Environment" (J. Miyamoto and P. C. Kearney, eds.), Vol 3, pp. 219–224. Pergamon Press, Oxford.

Stoddart, J. L., and Venis, M. A. (1980). Molecular and subcellular aspects of hormone action. *Encycl. Plant Physiol. New Ser.* **9**, 445–510.

Sweetser, P. B. (1985). Safening of sulfonylurea herbicides to cereal crops: Mode of herbicide antidote action. *Proc. Br. Crop Prot. Conf.—Weeds*, **3**, 1147-1154.

Taylor, H. F., and Loader, M. P. C. (1984). Research on the control of wild oats and weeds by herbicide mixtures. *Outlook Agric.* **13**, 58–68.

Theologis, A. (1986). Rapid gene regulation by auxin. *Annu. Rev. Plant Physiol.* **37**, 407–438.

Thiessen, E. P. (1978). Barban plus naphthalic anhydride for the selective control of wild oats in oats. M. S. Thesis, University of Guelph, Guelph, Ontario.

Turner, D. J., and Loader, M. C. P. (1978). Complexing agents as herbicide additives. *Weed Res.* **18**, 199–207.

Wakabayashi, K., and Matsunaka, S. (1982). Crop safening in Japan. *Proc. Br. Crop Prot. Conf.—Weeds* **2**, 439-450.

Warmund, M. R., England, D. C., and Kerr, H. D. (1981). Anatomical site of action of MON-4606 in milo, *Sorghum bicolor*. *Proc.—North Cent. Weed Control Conf.* **36**, 123–124.

Warmund, M. R., Kerr, H. D., and Peters, E. J. (1985). Lipid metabolism in grain sorghum (*Sorghum bicolor*) treated with alachlor plus flurazole. *Weed Sci.* **33**, 25–28.

Wiegand, R. C., Shah, D. M., Mozer, T. J., Harding, E. I., Diaz-Collier, J., Saunders, C., Jaworski, E. G., and Tiemeier, D. C. (1986). Messenger RNA encoding a gluthathione-*S*-transferases responsible for herbicide tolerance in maize is induced in response to safener treatment. *Plant Molec. Biol.* **7**, 235-243.

Wilkinson, R. E. (1978). Physiological response of lipid components to thiocarbamates and antidotes. *In* "Chemistry and Action of Herbicide Antidotes" (F. M. Pallos and J. E. Casida, eds.), pp. 85–108. Academic Press, New York.

Wilkinson, R. E. (1981a). Metolachlor influence on growth and terpenoid synthesis. *Pestic. Biochem. Physiol.* **16**, 63–71.

Wilkinson, R. E. (1981b). Metolachlor (2-chloro-*N*-(2-ethyl-6-methylhenyl)-*N*-(2-methoxy-1-methylethyl)acetamide) inhibition of gibberellin precursor biosynthesis. *Pestic. Biochem. Physiol.* **16**, 199–205.

Wilkinson, R. E. (1982). Alachlor influence on sorghum growth and gibberellin precursor biosynthesis. *Pestic. Biochem. Physiol.* **17**, 177–184.

Wilkinson, R. E. (1983). Gibberellin precursor biosynthesis inhibition by EPTC and reversal by R-25788. *Pestic. Biochem. Physiol.* **19**, 321–329.

Wilkinson, R. E. (1985). Gibberellin influence on membrane disruption by thiocarbamate herbicides. *Pestic. Biochem. Physiol.* **24**, 103-111.

Wilkinson, R. E., and Smith, A. E. (1975). Reversal of EPTC induced fatty acid synthesis inhibition. *Weed Sci.* **23**, 90-92.

Wilkinson, R. E., and Smith, A. E. (1976). EPTC altered beet disc betacyanin efflux and fatty acid synthesis. *Weed Sci.* **24**, 235-237.

Winkle, M. E., Leavitt, J. R. C., and Burnside, O. C. (1980). Acetanilide–antidote combinations for weed control in corn (*Zea mays*) and sorghum (*Sorghum bicolor*). *Weed Sci.* **28**, 699-704.

Wright, T. H., Rieck, C. E., and Thompson, L. (1973). Comparative butylate metabolism in corn hybrids. *Proc., South. Weed Sci. Soc.* **26**, 383.

Zaitlin, M., Day, P., and Hollaender, A. (1986). "Biotechnology in Plant Science: Relevance to Agriculture in the Eighties." Academic Press, Orlando, Florida.

Zama, P., and Hatzios, K. K. (1986). Effects of CGA-92194 on the chemical reactivity of metolachlor with glutathione and metabolism of metolachlor in grain sorghum (*Sorghum bicolor*). *Weed Sci.* **34**, 834–841.

Zama, P., and Hatzios, K. K. (1987). Interactions between the herbicide metolachlor and the safener CGA-92194 at the levels of uptake and macromolecular synthesis in sorghum leaf protoplasts. *Pestic. Biochem. Physiol.* **27**, 86–96.

Chapter 4

Influence of Herbicide Safeners on Herbicide Metabolism

John W. Gronwald
USDA–ARS, Plant Science
Research Unit and Department of
Agronomy and Plant Genetics
University of Minnesota
St. Paul, Minnesota 55108

I. INTRODUCTION

The thiocarbamate herbicides, EPTC[1] and butylate, applied preplant incorporated, provide adequate control of grass and some broadleaf weeds in corn (*Zea mays* L.) (Anderson, 1983). It is often desirable to use higher rates of these herbicides to control shattercane [*Sorghum bicolor* (L.) Moench], wooly cupgrass [*Eriochloa villosa* (thunb.) Kunth], wild proso millet (*Panicum miliaceum* L.), and certain perennial grasses. However, injury to corn may occur (Burnside *et al.*, 1971; Chang *et al.*, 1972). The chloroacetanilide herbicides, metolachlor and alachlor, applied preemergent, provide acceptable control of grass weeds in grain sorghum [*Sorghum bicolor* (L.) Moench], but the tolerance of grain sorghum to these herbicides is marginal and substantial injury may result (Spotanski and Burnside, 1973; Ellis *et al.*, 1980; Winkle *et al.*, 1980). For both

[1] Mention of a trademark, vendor, or proprietary product does not constitute a guarantee or warranty of the product by the U.S. Department of Agriculture or the University of Minnesota and does not imply its approval to the exclusion of other products or vendors that may also be suitable.

of the above herbicide class–crop combinations, safeners (also referred to as antidotes or protectants) have been developed which reduce injury to crop species without impairing herbicide efficacy on weed species.

The chemical structures of safeners used to protect corn against thiocarbamate and/or sorghum from chloroacetanilide herbicides are given in Fig. 1. Naphthalic anhydride, which was first introduced in 1969, is no longer used commercially. This safener, applied as a seed treatment, will protect against thiocarbamate injury in corn (Burnside *et al.*, 1971) and chloroacetanilide injury in sorghum (Jordan and Jolliffe, 1971; Spotanski and Burnside, 1973; Fuerst and Gronwald, 1986). However, the safening action of naphthalic anhydride is not limited to these herbicide class–crop combinations. Naphthalic anhydride is considered to be a broad-spectrum safener which provides varying degrees of protection to several grass species against a number of herbicides (Hatzios, 1983c; Hatzios, 1984; Parker, 1983). Other examples of herbicide–crop combinations where naphthalic anhydride will confer protection are: barban–wheat (*Triticum aestivum* L.), perfluridone–corn, chlorsulfuron–corn, molinate–rice (*Oryza sativa* L.), cisanilide–sorghum, and preemergent applications of sethoxydim or fluazifop-butyl in corn. Dichlormid, formerly referred to as R-25788, is used commercially to protect corn from injury by EPTC or butylate (Pallos *et al.*, 1975). It is effective when applied either as a seed treatment or as a tank mix with EPTC or butylate (Chang *et al.*, 1973). Dichlormid is similar to naphthalic anhydride in that it will provide some protection against several herbicide classes but unlike naphthalic anhydride the safening action of dichlormid is rather specific for corn (Stephenson and Chang, 1978). Dichlormid provides good protection to corn against thiocarbamate (Burnside *et al.*, 1971; Chang *et al.*, 1972; Pallos *et al.*, 1975) and chloroacetanilide (Leavitt and Penner, 1978) herbicides. It will also confer some protection to corn against injury by chlorsulfuron and preemergent applications of sethoxydim or fluazifop-butyl (Hatzios, 1983c; Hatzios, 1984; Parker, 1983). CGA-92194 {α-[(1,3-dioxolan-2-yl-methoxy) imino] benzeneacetonitrile}, an oxime ether, and flurazole, a substituted thiazolecarboxylate, are used commercially to protect sorghum from chloroacetanilide injury (Hatzios, 1983c). Cyometrinil was the first oxime ether used to protect sorghum from chloroacetanilides but has since been replaced by CGA-92194. These safeners are applied as seed-dressings.

The mechanism by which the safeners described above increase herbicide tolerance in crop species is not understood. It has been postulated that they confer protection by one or more of the following mechanisms: (1) reducing uptake and/or translocation of the herbicide; (2) competing with the herbicide for binding at the site of action; and (3) enhancing herbicide metabolism (Hatzios, 1983c). There is increasing evidence that safeners which protect corn from thiocarbamate herbicides and sorghum from chloroacetanilide herbicides act by enhancing herbicide metabolism. This review will be limited to discussing the effects of safeners on EPTC metabolism in corn and chloroacetanilide (alachlor,

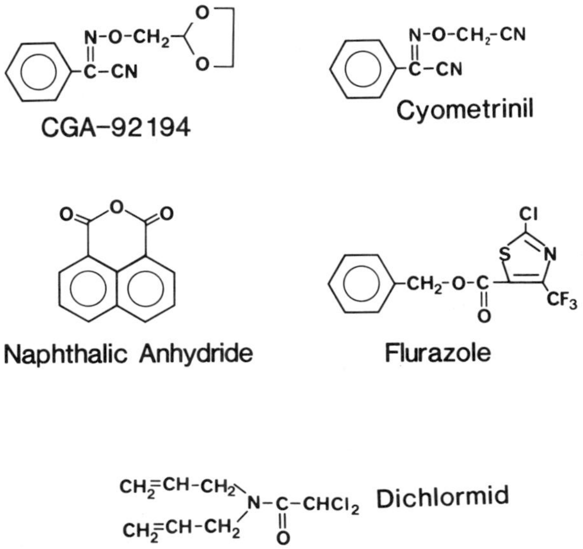

Fig. 1. Chemical structures of safeners.

metolachlor, acetochlor) metabolism in grain sorghum. The metabolism of a particular herbicide class (thiocarbamate or chloroacetanilide) will be discussed, followed by a review of the literature concerning the influence of safeners on metabolism. The chapter ends with a general discussion which includes both an overview of research in the area as well as interpretative speculation.

II. METABOLISM OF THIOCARBAMATE HERBICIDES IN PLANTS

Most studies concerning metabolism of thiocarbamate herbicides have centered on EPTC metabolism in corn. The proposed pathway for EPTC metabolism (Fig. 2) involves what are referred to in toxicology as phase I, II, and III reactions (Hatzios and Penner, 1982; Shimabukuro, 1985). Phase I reactions involve a bioactivation of the xenobiotic (foreign compound) which predisposes the molecule for a phase II reaction. The phase I bioactivation usually involves the insertion of a reactive polar group into the molecule. Phase II reactions result in detoxication via conjugation with an endogenous substrate. Phase III reactions, which are unique to plants, lead to the formation of a terminal metabolite(s).

In the metabolism of EPTC, the phase I reaction involves a sulfoxidation and is thought to be mediated by a microsomal monooxygenase (EC 1.14.14.1), also referred to as a mixed-function oxidase (Lay *et al.*, 1975; Hubbell and Casida, 1977). The effect of sulfoxidation is twofold; it generates the active form of the

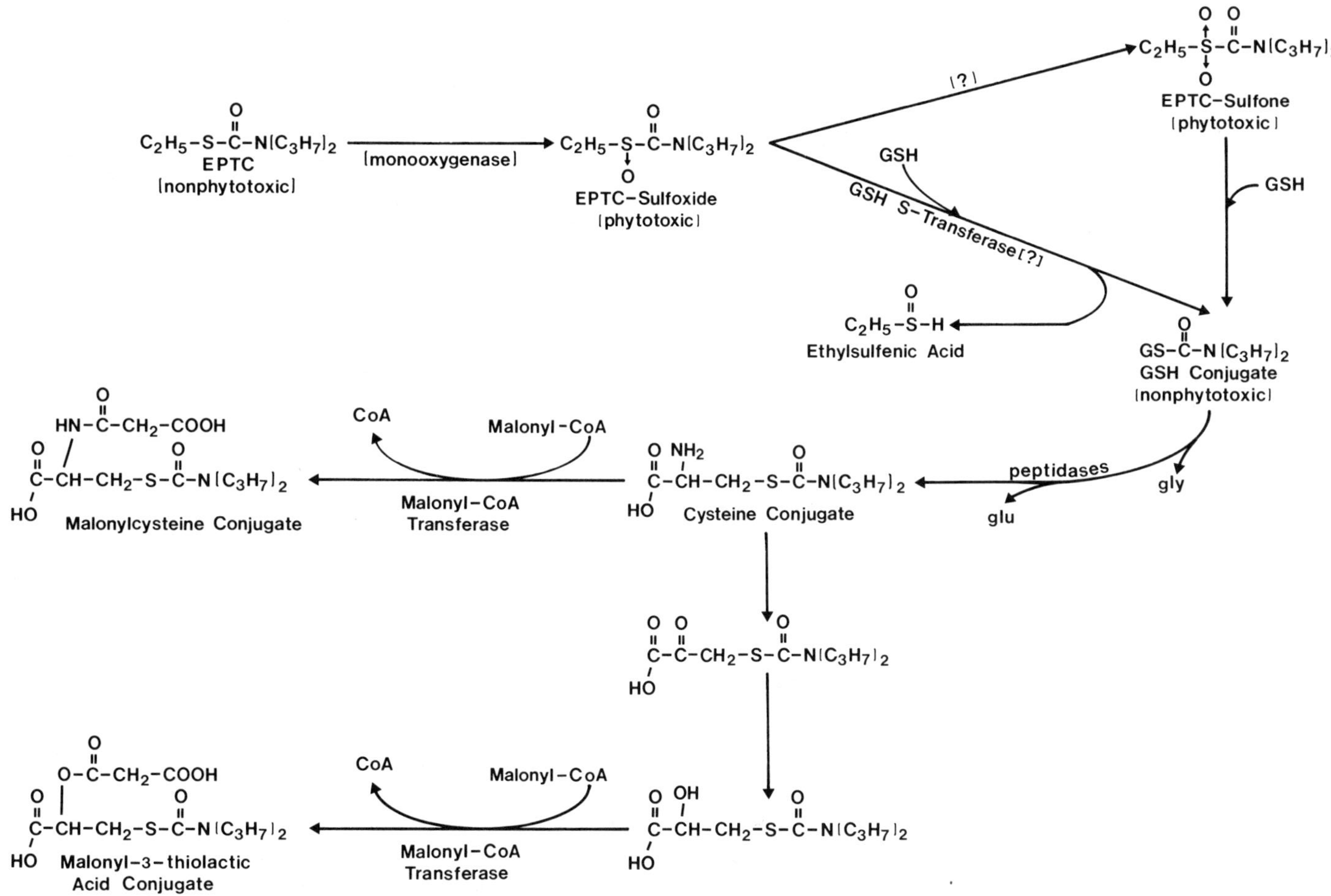

Fig. 2. Proposed pathway for the metabolism of EPTC in plants. (Reproduced in part from Lamoureux and Rusness, 1987.)

herbicide, EPTC-sulfoxide (Casida *et al.* 1974; Shimabukuro, 1985), and it converts EPTC to a substrate suitable for a phase II reaction. There is speculation that a second oxygen atom may be added to the sulfoxide to form a sulfone (Horváth and Pulay, 1980). However, this is difficult to evaluate since sulfones, if formed, are hydrolytically unstable (Casida *et al.*, 1974) and have not been detected in plant extracts (Hubbell and Casida, 1977; Horváth and Pulay, 1980). In the phase II reaction, EPTC-sulfoxide is conjugated with reduced glutathione (GSH) with the simultaneous release of ethylsulfenic acid, a volatile and unstable cleavage product of the reaction (Lay and Casida, 1976). The phase II reaction results in herbicide detoxication (Lay *et al.*, 1975; Shimabukuro, 1985). There is a lack of consensus as to whether the reaction is nonenzymatic or is catalyzed by a glutathione *S*-transferase (GST; EC 2.5.1.18) (Lay and Casida, 1976; Carringer *et al.*, 1978a; Leavitt and Penner, 1979). The GSH conjugate is metabolized by peptidases, sequentially releasing glycine and glutamic acid from the GSH moiety, yielding an *S*-cysteine conjugate of the herbicide. The metabolism of the *S*-cysteine conjugate can proceed via two pathways. In corn, soybean, and cotton (*Gossypium hirsutum* L.), malonyl-CoA transferase catalyzes the addition of malonate, forming an *N*-malonylcysteine conjugate (Lamoureux and Rusness, 1983). A recent report has also indicated that in corn and cotton, the *S*-cysteine conjugate can be metabolized to a malonyl-3-thiolactic acid conjugate (Lamoureux and Rusness, 1987). The series of reactions involved in the metabolism of the GSH conjugate of EPTC to form the *N*-malonylcysteine or the malonyl-3-thiolactic acid conjugate are considered to be phase III reactions since they result in the formation of a terminal metabolite.

It has been postulated that the sulfoxidation reaction is mediated by a microsomal NADPH-dependent cytochrome P-450 monooxygenase (Casida *et al.*, 1975; Lay *et al.*, 1975; Hubbell and Casida, 1977). This enzyme system is known to occur in plants and, based on rather limited studies, has properties similar to the extensively investigated mammalian (rat liver) NADPH-dependent cytochrome P-450 monooxygenase. The mammalian cytochrome P-450 monooxygenase is a multicomponent enzyme complex located in the endoplasmic reticulum (Coon and Persson, 1980). The complex consists of a cytochrome P-450 and a flavoprotein, NADPH-cytochrome *c* (P-450) reductase (EC 1.6.2.4), which serves to transfer electrons from NADPH to cytochrome P-450. Enzyme activity requires molecular oxygen and NADPH. The basic reaction catalyzed by a monooxygenase involves the insertion of an atom of oxygen into the substrate molecule. The other atom of molecular oxygen is reduced to water. In plants, cytochrome P-450 monooxygenases have been implicated to play a role in gibberellin biosynthesis (Murphy and West, 1969; Hedden *et al.*, 1978), ω-hydroxylation of fatty acids (Soliday and Kolattukudy, 1977; Benveniste *et al.*, 1982), *trans*-cinnamic acid hydroxylation (Potts *et al.*, 1974; Young and Beevers, 1976; Rich and Lamb, 1977), and the N-demethylation of substituted phenylurea herbicides (Frear *et al.*, 1969).

Although a cytochrome P-450 monooxygenase is postulated to catalyze the sulfoxidation of thiocarbamate herbicides in plants, this has not yet been demonstrated *in vitro*. There are, however, two pieces of indirect evidence which suggest that plant monooxygenases catalyze this reaction. The first is based on analogy with the metabolism of EPTC in mouse liver. When EPTC is applied to either corn plants (Hubbell and Casida, 1977) or mouse liver (Casida *et al.*, 1975), EPTC-sulfoxide is generated. In the case of mouse liver, it has been demonstrated that the reaction is catalyzed by a NADPH-dependent microsomal enzyme, presumably a cytochrome P-450 monooxygenase. The second piece of evidence is the synergism observed when certain insecticide synergists are applied with EPTC to corn. Synergism occurs when the toxicity of two compounds applied together is greater than would be expected from the sum of their individual effects. Insecticide synergists are compounds demonstrated to inhibit cytochrome P-450 monooxygenases in mammals and insects (Casida, 1970; Wilkinson, 1976). When added to insecticides these compounds increase insecticide potency by impairing the ability of insects to detoxify insecticides via cytochrome P-450-dependent monooxygenase reactions. Komives and Dutka (1980) reported that applying the insecticide synergist SKF 525-A(2-(diethylamino)ethyl 2,2-diphenyl pentanoate) or piperonyl butoxide to EPTC-treated corn results in a synergistic response. These authors concluded that the formation of the sulfoxide is important in the detoxication of EPTC and that insecticide synergists, by inhibiting monooxygenase activity, prevent sulfoxide formation and thus cause greater injury.

The sulfoxide of EPTC is found only as a transitory metabolite in corn treated with EPTC because it is rapidly conjugated with GSH (Lay and Casida, 1976; Carringer *et al.*, 1978a). While there is general agreement that this conjugation occurs, there is controversy as to whether it is primarily a nonenzymatic or enzymatic reaction. There are reports that nonenzymatic conjugation of EPTC-sulfoxide and GSH will occur spontaneously in neutral aqueous media (Carringer *et al.*, 1978a; Leavitt and Penner, 1979). Carringer *et al.* (1978a) were unable to detect GST activity in corn root homogenates capable of catalyzing the GSH conjugation of EPTC-sulfoxide. On the basis of this finding and their observation that GSH conjugation of EPTC-sulfoxide can occur spontaneously *in vitro*, these authors concluded that the *in vivo* reaction is primarily nonenzymatic. In contrast, it has been reported that little or no nonenzymatic conjugation of EPTC-sulfoxide and GSH occurs *in vitro* (Horváth and Pulay, 1980) and that corn roots and shoots contain a GST which catalyzes the conjugation of EPTC-sulfoxide with GSH (Lay and Casida, 1976; Lay and Niland, 1985).

III. EFFECT OF SAFENERS ON THIOCARBAMATE METABOLISM

The first study of the effects of safeners on herbicide metabolism was conducted by Güneyli (1971), who investigated the effect of naphthalic anhydride on EPTC metabolism in corn seedlings. Naphthalic anhydride was found to enhance the rate of EPTC metabolism and it was postulated that the safener conferred protection by activating enzymes involved in EPTC detoxication.

The vast majority of studies concerning the effect of safeners on EPTC metabolism have involved examining the effects of dichlormid on EPTC metabolism in corn. Chang *et al.* (1974) reported that dichlormid enhanced EPTC metabolism in corn. However, the authors suggested that the magnitude of enhancement did not seem adequate to account for the degree of protection conferred by the safener. Other reports have clearly shown that treatment with dichlormid enhances metabolism of EPTC in corn (Lay and Casida, 1976; Hubbell and Casida, 1977; Carringer *et al.*, 1978b).

There has been considerable research into the mechanism whereby dichlormid enhances EPTC metabolism. Most investigations have focused on the effect of dichlormid on the GSH conjugation of EPTC-sulfoxide. Lay and Casida (1976) reported that dichlormid increased the levels of both cofactor (GSH) and enzyme (GST) involved in the GSH conjugation of EPTC-sulfoxide. A 24-hr treatment with 3 to 30 ppm dichlormid caused a two- to threefold increase in GST activity and GSH concentration in corn roots (Fig. 3). It was postulated that the dichlormid-induced increase in GSH and GST resulted in enhanced detoxication of EPTC-sulfoxide formed by the phase I monooxygenase reaction, and thereby conferred protection. The importance of GSH and GST in the metabolism of EPTC and the mechanism of action of dichlormid was also suggested by the results obtained with oats (*Avena sativa* L.) (Lay *et al.*, 1975). In contrast to corn, oats are very susceptible to EPTC and not protected by dichlormid. It was found that oat roots contain low levels of GSH and GST and that dichlormid caused little or no increase in these parameters. Additional support for the hypothesis that dichlormid-induced increases in GSH and GST are involved in

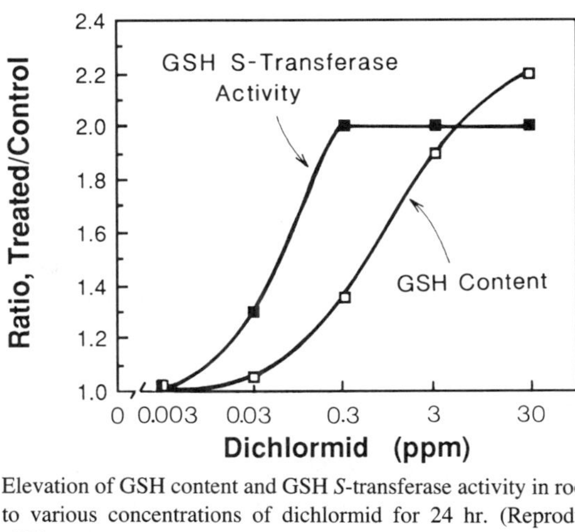

Fig. 3. Elevation of GSH content and GSH *S*-transferase activity in roots of corn seedlings exposed to various concentrations of dichlormid for 24 hr. (Reproduced from Lay and Casida, 1978.)

safening action was provided by Lay and Casida (1976). In a comparative study with 30 compounds (mostly dichloroacetamide analogs of dichlormid), a reasonably good correlation was found between the ability of a compound to confer protection against EPTC injury in corn and its ability to enhance GSH and GST in corn roots. In a similar study, Stephenson *et al.* (1983) evaluated a series of structurally related chloroacetamide compounds and found that safening activity was correlated with ability to increase GSH content in corn roots.

The ability of dichlormid to increase GSH content in corn is well established (Lay and Casida, 1976; Carringer *et al.*, 1978b; Ezra and Gressel, 1982; Stephenson *et al.*, 1983). This effect, however, is not limited to corn. Dichlormid will also increase GSH content in heterotrophically grown suspension cultures of tobacco (*Nicotiana tabacum* L.) (Rennenberg *et al.*, 1982), citrus (*Citrus sinensis* L.) leaves (Guy *et al.*, 1984), pigweed (*Amaranthus retroflexus* L.) callus culture (Ezra and Gressel, 1982), and sorghum shoots (Gronwald *et al.*, 1987).

Three mechanisms have been proposed to explain the increase in GSH content in response to dichlormid treatment. Carringer *et al.* (1978b) reported that dichlormid caused an *in vitro* activation of glutathione synthetase (EC 6.3.2.3), an enzyme catalyzing the conjugation of γ-glutamylcysteine with glycine, and proposed that dichlormid acted similarly *in vivo*. Rennenberg *et al.* (1982) found that dichlormid enhanced the specific activity of enzymes [γ-glutamylcysteine synthetase (EC 6.3.2.2) and/or glutathione synthetase] involved in GSH synthesis in tobacco cells grown in suspension culture. Enhanced enzyme activity was attributed to an induction effect, not the result of direct enzyme activation as proposed by Carringer *et al.* (1978b). A report by Adams *et al.* (1983) suggested that dichlormid enhances GSH content in corn roots by promoting sulfate assimilation. Dichlormid was reported to increase the activity of ATP sulfurylase (EC 2.7.7.4), an enzyme involved in the early stages of sulfate assimilation.

Although it has been clearly established that treatment with dichlormid increases GSH content in corn, certain studies have raised questions regarding the importance of the increase to safener activity. First, Ezra and Gressel (1982) reported that in corn suspension cells an increase in GSH was not detected until 12 hr after treatment with dichlormid. Yet, most of the EPTC applied to the suspension cells had been metabolized after only 8 hr. Second, naphthalic anhydride protects against EPTC injury in corn (Burnside *et al.*, 1971; Hatzios, 1983c), but this safener has little or no effect on GSH content of corn roots (Lay and Casida, 1976). Finally, dichlormid will protect against EPTC injury in the corn inbred 'MO 17 Ht', but has no effect on root GSH content. Conversely, dichlormid is not effective in conferring protection against EPTC injury in certain corn hybrids ('DeKalb XL-55A','XL-67','XL-379') or the inbred 'Oh-551', but will enhance GSH levels (Lay and Niland, 1985).

The role of GST in the GSH conjugation of EPTC-sulfoxide has not been clearly established for either untreated or dichlormid-treated corn seedlings. Carringer *et al.* (1978a) reported that GST activity capable of catalyzing the GSH conjugation of EPTC-sulfoxide was not detectable in corn roots. In contrast, Lay and Casida (1976) reported that such activity exists in both roots and shoots of corn seedlings and that treatment with dichlormid enhanced the level of activity by approximately twofold. The increase in GST in response to dichlormid treatment was thought to be due to an induction response, since the safener did not cause a direct activation of the enzyme. Treatment with dichlormid has also been reported to increase GST activity, as evaluated with CDNB (1-chloro-2,4-dinitrobenzene) as substrate, in etiolated shoots of corn (Mozer *et al.*, 1983) and sorghum seedlings (Gronwald *et al.*, 1987). CDNB is not a herbicide but is a substrate frequently used to measure GST activity in biological systems (Habig and Jakoby, 1981). In etiolated corn shoots, the dichlormid-induced increase in GST activity, as measured with CDNB, was the result of an increase in constitutive GST activity and the induction of a novel GST isozyme (Mozer *et al.*, 1983; Wiegand *et al.*, 1986).

The effects of dichlormid on the sulfoxidation of EPTC have not been as extensively investigated as its effects on the GSH conjugation of EPTC-sulfoxide. Leavitt and Penner (1979) hypothesized that dichlormid protects corn from EPTC injury by enhancing the rate of EPTC sulfoxidation and there is indirect evidence to support this hypothesis. Hatzios (1983a) examined the effect of insecticide synergists (piperonyl butoxide, *n*-propyl gallate) and ozone on the safening action of dichlormid in corn. As discussed earlier, insecticide synergists are considered to act as inhibitors of cytochrome P-450 monooxygenases. In the case of ozone, it has been suggested that this molecule may inhibit cytochrome P-450 monooxygenases by interfering with the interaction of molecular oxygen with the enzyme complex (Rubin *et al.*, 1980). It was found that insecticide synergists or ozone interact synergistically in corn grown in the presence of EPTC and dichlormid, i.e., in the presence of insecticide synergists or ozone, EPTC injury in dichlormid-treated corn was greater than the additive injurious effects of EPTC and insecticide synergist or ozone treatment alone. It was suggested that dichlormid may act by enhancing monooxygenase activity responsible for catalyzing the formation of EPTC-sulfoxide and that the synergistic effect observed when insecticide synergists or ozone are applied with dichlormid is the result of the inhibition of this enzyme system.

IV. METABOLISM OF CHLOROACETANILIDE HERBICIDES IN PLANTS

Based on available evidence, the pathway for the metabolism of chloroacetanilide herbicides in plants is shown in Fig. 4. Metabolism of chloroacetanilides involves phase II and III reactions. Although there has been speculation that a phase I bioactivation may be involved in the initial metabolism of the chloroacetanilide herbicides (Hatzios, 1983b; Ketchersid *et al.*, 1985), this was not

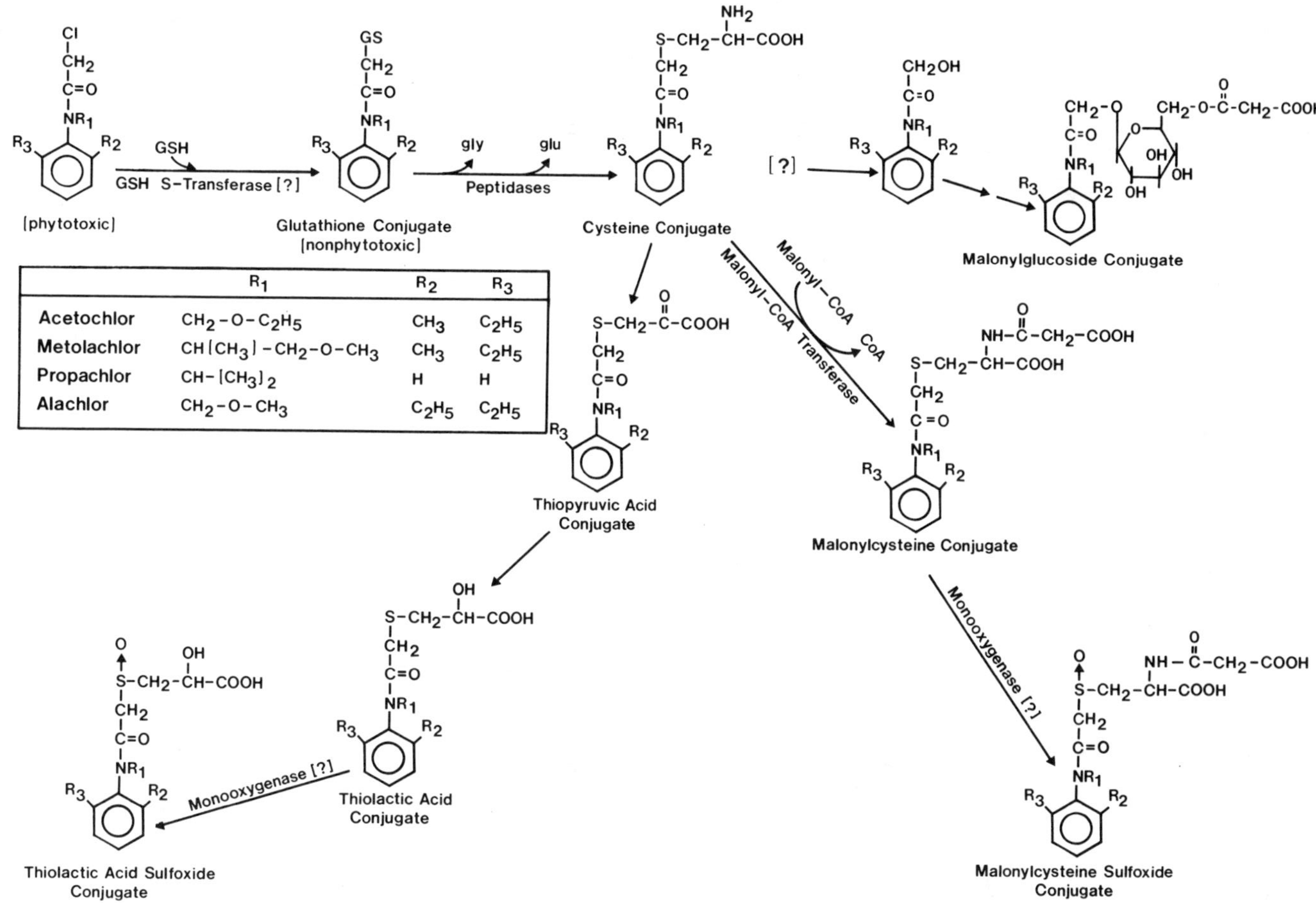

	R_1	R_2	R_3
Acetochlor	$CH_2-O-C_2H_5$	CH_3	C_2H_5
Metolachlor	$CH(CH_3)-CH_2-O-CH_3$	CH_3	C_2H_5
Propachlor	$CH-(CH_3)_2$	H	H
Alachlor	CH_2-O-CH_3	C_2H_5	C_2H_5

Fig. 4. Proposed pathway for the metabolism of chloroacetanilides in plants. [Reproduced in part from Lamoureux and Frear (1987) and G. L. Lamoureux, USDA-ARS, Metabolism and Radiation Research Laboratory, Fargo, North Dakota, personal communication.]

confirmed in recent reports by Breaux (1986, 1987). The initial metabolite formed in both tolerant and susceptible species is the GSH conjugate (Lamoureux *et al.*, 1971; Breaux, 1986, 1987; Fuerst and Gronwald, 1986) with certain exceptions. In soybeans [*Glycine max* (L.) Merr.] and certain other legumes [kidney bean (*Phaseolus vulgaris* L.), lima bean (*Phaseolus limensis* Macfad.), white clover (*Trifolium repens* L.)], homoglutathione (γ-glutamylcysteinyl-β-alanine) rather than GSH (γ-glutamylcysteinylglycine) is the predominant nonprotein thiol (Rennenberg, 1982) and the homoglutathione conjugate is formed in these species (Breaux, 1986, 1987). The formation of the GSH or homoglutathione conjugate detoxifies the herbicide (Shimabukuro, 1985). Further metabolism of the GSH conjugate involves the sequential release of glycine and glutamic acid from GSH as a result of peptidase activity. The resulting *S*-cysteine conjugate is then N-malonated via malonyl-CoA transferase to form the *N*-malonylcysteine conjugate of the chloroacetanilide which can be oxidized to form the *N*-malonylcysteine sulfoxide conjugate (Lamoureux and Rusness, 1983; Lamoureux and Frear, 1987). The *S*-cysteine conjugate can also be metabolized to a *S*-thiolactic acid sulfoxide conjugate. Presumably, this pathway involves the formation of a thiopyruvic acid intermediate. The glutathione conjugate of chloroacetanilides can also be metabolized to an *O*-malonylglucoside conjugate, but little is known about the metabolic pathway (Lamoureux and Frear, 1987).

The role of GST in the formation of the GSH conjugate of the chloroacetanilides is disputed. Nonenzymatic conjugation of chloroacetanilide herbicides with GSH is known to occur *in vitro* (Lamoureux *et al.*, 1971; Leavitt and Penner, 1979; Gronwald *et al.*, 1987) and is strongly dependent on pH (Leavitt and Penner, 1979). Leavitt and Penner (1979) reported that the *in vitro* rate of nonenzymatic conjugation was limited at physiological pH. Nevertheless, these authors concluded that nonenzymatic conjugation may make a significant contribution to conjugate formation *in vivo*. The conclusion of Leavitt and Penner (1979) regarding the importance of nonenzymatic conjugation rests in part on their inability to detect GST activity in plant extracts capable of catalyzing the formation of a GSH–chloroacetanilide conjugate. More recent studies indicate that this enzyme activity exists in plants. GSTs capable of catalyzing the formation of a GSH–chloroacetanilide conjugate have been reported in etiolated corn (Mozer *et al.*, 1983) and sorghum (Gronwald *et al.*, 1987) shoots, as well as corn cell culture (Edwards and Owen, 1986). In corn cell culture, two isozymes of GST were detected, one of which exhibited greater specificity for metolachlor.

V. EFFECT OF SAFENERS ON CHLOROACETANILIDE METABOLISM

The majority of studies concerning the effects of safeners on the metabolism of chloroacetanilide herbicides have involved examining the effect of oxime ethers (cyometrinil, CGA-92194) on the metabolism of metolachlor or alachlor in grain sorghum. A number of earlier studies suggested that cyometrinil had little or no effect on the metabolism of chloroacetanilide herbicides in grain sorghum. Winkle *et al.* (1980) reported that the metabolism of [^{14}C]metolachlor in sorghum seedlings was unaffected by seed treatment with cyometrinil. Ketchersid *et al.* (1982) found that seed treatment with cyometrinil reduced metolachlor content in sorghum shoots. However, this effect was primarily attributed to the ability of cyometrinil to reduce metolachlor uptake, not to its ability to enhance metabolism. Leek (1981) reported that [^{14}C]metolachlor was metabolized at the same rate or less rapidly in cyometrinil-treated sorghum seedlings as compared to untreated seedlings. A recent study by Jackson *et al.* (1986) indicated that flurazole, a thiazolecarboxylic acid safener, had little or no effect on the metabolism of acetochlor in sorghum seedlings. After 3 days, essentially all of the [^{14}C]acetochlor applied via the roots had been converted to polar metabolites in both treated and untreated seedlings.

The above reports conflict with two recent reports which indicate that cyometrinil and flurazole do indeed enhance metabolism of chloroacetanilide herbicides in sorghum. Fuerst and Gronwald (1986) showed that seed treatment with naphthalic anhydride, dichlormid, CGA-92194, cyometrinil, and flurazole resulted in accelerated metolachlor metabolism in etiolated sorghum shoots. After a 100 min incubation in [^{14}C]metolachlor, the percentage of absorbed radiolabel that was metabolized in shoots of untreated and CGA-92194-treated seedlings was 41 and 90%, respectively (Fig. 5). Concomitant with the decrease

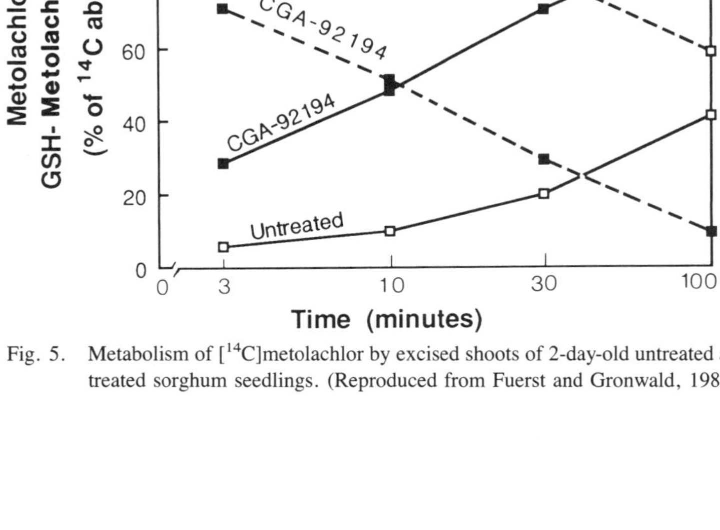

Fig. 5. Metabolism of [^{14}C]metolachlor by excised shoots of 2-day-old untreated and CGA-92194-treated sorghum seedlings. (Reproduced from Fuerst and Gronwald, 1986.)

in metolachlor was an increase in the GSH conjugate of metolachlor in both untreated and CGA-92194-treated tissue. Zama and Hatzios (1986) also reported that CGA-92194 enhanced metolachlor metabolism in sorghum seedlings. Additional evidence that safeners protect sorghum against chloroacetanilide injury by promoting herbicide metabolism was provided by Fuerst and Gronwald (1986). In a comparative study involving five safeners (naphthalic anhydride, dichlormid, CGA-92194, cyometrinil, flurazole), a strong correlation was found between the ability of a safener to protect sorghum against injury by metolachlor and its ability to enhance the formation of a GSH–metolachlor conjugate *in vivo*.

The work by Fuerst and Gronwald (1986) and Zama and Hatzios (1986) suggests that safeners act by promoting an existing pathway of detoxication, i.e., GSH conjugation. It has also been proposed that cyometrinil may confer protection by promoting oxidative metabolism of chloroacetanilide herbicides (Hatzios, 1983b). Ozone and some rates of insecticide synergists were reported to interact synergistically in the presence of metolachlor to impair the safening action of cyometrinil. The results were interpreted to indicate that cyometrinil promotes oxidative metabolism of chloroacetanilide herbicides. However, as discussed earlier, most evidence to date suggests that oxidative metabolism is not involved in the initial stages of chloroacetanilide metabolism in either untreated (Lamoureux *et al.*, 1971; Breaux, 1986, 1987) or safener-treated (Fuerst and Gronwald, 1986) plants.

Although there is increasing evidence that safeners protect sorghum from chloroacetanilide herbicide injury by promoting the formation of a GSH conjugate, there is no consensus as to whether this is the result of a safener-induced increase in cofactor (GSH) or enzyme (GST) or both. Leavitt and Penner (1979) suggested that dichlormid conferred protection against chloroacetanilide injury in corn by enhancing GSH levels leading to an accelerated rate of nonenzymatic conjugation. Gronwald *et al.* (1987) recently evaluated the effect of four safeners (dichlormid, naphthalic anhydride, CGA-92194, flurazole) on GSH content in etiolated sorghum shoots. In general, there was not a strong correlation between the ability of a safener to increase GSH content and ability to confer protection against injury by chloroacetanilide herbicides. Dichlormid increased GSH content whereas CGA-92194 and naphthalic anhydride had little or no effect on GSH content. Yet dichlormid was much less effective than CGA-92194 or naphthalic anhydride in conferring protection.

It has also been proposed that safeners may promote nonenzymatic GSH conjugation of chloroacetanilides by a direct activation of the reaction. Zama and Hatzios (1986) reported that CGA-92194 promoted the nonenzymatic conjugation of metolachlor with GSH *in vitro* and speculated that CGA-92194 may act in a similar fashion *in vivo*. However, Gronwald *et al.* (1987) were unable to confirm this *in vitro* activation of nonenzymatic conjugation by CGA-92194.

There are reports which indicate that enhancement of GST activity may be important in safener protection against chloroacetanilide injury in corn (Mozer *et al.*, 1983) and sorghum (Fuerst and Gronwald, 1986; Zama and Hatzios, 1986). Mozer *et al.* (1983) reported that dichlormid, naphthalic anhydride, cyometrinil, and flurazole enhanced GST activity in etiolated corn seedlings. In the case of dichlormid and flurazole, the enhanced activity was attributed to an increase in the constitutive GST, referred to as GST I, and the appearance of an induced isozyme of GST, referred to as GST II. Both GST I and II were capable of catalyzing the conjugation of alachlor with GSH *in vitro*. However, the induced GST II isozyme exhibited a greater selectivity toward alachlor. It was postulated that the safener-induced increase in GST activity enhanced the formation of a GSH–alachlor conjugate *in vivo* and thereby conferred protection. However, it has yet to be demonstrated that dichlormid or flurazole protects corn from chloroacetanilide injury by enhancing herbicide detoxication via GSH conjugation. Nevertheless, this is a reasonable assumption based on studies conducted with etiolated sorghum shoots (Fuerst and Gronwald, 1986).

Evidence that safeners confer protection against chloroacetanilide injury in sorghum by increasing GST activity was provided by Zama and Hatzios (1986) and Gronwald *et al.* (1987). Zama and Hatzios (1986) found that the atrazine synergist, tridiphane, which is thought to act by inhibiting GST activity (Lamoureux and Rusness, 1986), reduced the enhanced formation of the GSH–metolachlor conjugate in CGA-92194-treated sorghum seedlings. Gronwald *et al.* (1987) examined the effect of four safeners (CGA-92194, flurazole, dichlormid, naphthalic anhydride) on GST activity in etiolated sorghum shoots. A strong correlation was found between the ability of a safener to enhance GST activity, as evaluated with metolachlor as substrate, and ability to confer protection. Although Gronwald *et al.* (1987) did not determine whether safener action involved induction of a novel isozyme of GST, as is the case in etiolated corn seedlings (Mozer *et al.*, 1983), the results were interpreted to suggest that this may be the case. As compared to the GST activity in untreated sorghum shoots, the GST activity in CGA-92194-treated shoots exhibited a greater reactivity toward metolachlor (as compared with CDNB), which was particularly evident at low concentrations of the herbicide.

Enhanced GST activity in safener-treated corn or sorghum shoots appears to be the result of an induction response involving *de novo* protein synthesis. The appearance of a novel isozyme of GST in safener-treated, etiolated corn seedlings was interpreted to indicate that safeners elicit an induction response (Mozer *et al.*, 1983). Gronwald *et al.* (1987) reported that the increase in GST activity in sorghum shoots treated with CGA-92194 was not due to a direct activation of enzyme activity by the safener. Recently, some light has been shed on the induction of GST activity in etiolated corn seedlings. Wiegand *et al.* (1986) reported that treatment with flurazole caused a three- to fourfold enhancement of mRNA levels coding for a subunit of GST I. It was postulated that flurazole increased mRNA levels by either increasing the half-life of the GST I mRNA or by activating the transcription of a gene coding for the GST I subunit.

VI. DISCUSSION

Herbicide tolerance for most plant species is a function of the ability of the plant to metabolize the herbicide at a rate sufficient to prevent irreversible damage (Hatzios and Penner, 1982; Shimabukuro, 1985). Although there is yet no consensus on the mechanism of action of herbicide safeners, there is increasing evidence that safeners which protect corn from injury by thiocarbamate herbicides and sorghum from chloroacetanilide herbicides act by enhancing the rate of herbicide metabolism.

Most studies concerning the effects of dichlormid on EPTC metabolism in corn have concluded that this safener enhances EPTC metabolism (Chang *et al.*, 1974; Lay and Casida, 1976; Carringer *et al.*, 1978b). However, the report by Chang *et al.* (1974) concluded that the degree of enhancement of metabolism did not seem adequate to explain the degree of protection conferred by the safener. In the case of the chloroacetanilides, a number of studies have concluded that safeners which protect sorghum against this herbicide class have little or no effect on herbicide metabolism (Winkle *et al.*, 1980; Leek, 1981; Ketchersid *et al.*, 1982; Jackson *et al.*, 1986). These reports conflict with two recent studies which indicate that safeners promote metabolism of chloroacetanilides in sorghum (Fuerst and Gronwald, 1986; Zama and Hatzios, 1986).

A. Factors to Consider in Herbicide Metabolism Studies

Inconsistencies in the literature with regard to the effects of safeners on herbicide metabolism may be due to the manner in which the studies were conducted and the plant tissue used. Certain factors should be kept in mind when planning experiments to investigate the effect of safeners on metabolism of thiocarbamate and chloroacetanilide herbicides in corn and sorghum. First, corn (Lay and Casida, 1976; Carringer *et al.*, 1978b) and sorghum (Fuerst and Gronwald, 1986; Jackson *et al.*, 1986; Zama and Hatzios, 1986) have a significant endogenous capacity to metabolize thiocarbamate and chloroacetanilide herbicides, respectively. Second, the shoot, as opposed to the root, is more sensitive to these herbicides and is also more responsive to the protective action of the safener (Dawson, 1963; Appleby *et al.*, 1965; Jordan and Jolliffe, 1971; Gray and Joo, 1978; Ellis *et al.*, 1980; Nyffeler *et al.*, 1980; Ketchersid *et al.*, 1981; Fuerst and Gronwald, 1986). Third, the critical period of plant exposure is during the preemergent phase of growth when the coleoptile is emerging through the soil layer containing the herbicide (Güneyli, 1971; Donald and Fawcett, 1976; Nyffeler *et al.*, 1980). This is also the period when safening action is most critical.

If it is assumed that safeners act by enhancing the rate of herbicide metabolism, it is particularly important to evaluate the kinetics of metabolism in untreated and safener-treated tissue over an appropriate time course. Depending on the amount of radiolabeled herbicide applied this may require short-term studies. This is important because both untreated and safener-treated tissues have the ability to metabolize the herbicide and eventually the untreated tissue will metabolize the herbicide to the same extent as the safened tissue. For example, in an early study concerning the effects of cyometrinil on metolachlor metabolism in sorghum (Winkle *et al.*, 1980), metabolite levels were not examined until 3 weeks after application of radiolabeled metolachlor. It is therefore not surprising that differences in metabolite levels were not observed. Essentially all of the radiolabeled metolachlor had been converted to metabolite in both untreated and safened plants.

Given that the coleoptile, during the preemergent phase of growth, is most susceptible to thiocarbamate and chloroacetanilide herbicides, and also more responsive to safening action, it would appear logical to conduct metabolism studies with this tissue, yet this is frequently not the case. It is quite surprising that no studies have been conducted to examine the effects of dichlormid on EPTC metabolism in etiolated corn shoots. It would seem that using the appropriate tissue would be of critical importance when trying to establish a correlation between the protection conferred by a safener and its effect on the rate of herbicide metabolism. In most studies in which the effects of dichlormid on EPTC metabolism have been examined, roots and/or shoots of photosynthetically active plants have been utilized. This was the case in the study conducted by Chang *et al.* (1974), in which it was concluded that enhancement of EPTC metabolism in roots and leaves of 1- to 2-week-old corn plants did not seem adequate to explain protection conferred by dichlormid. Perhaps if EPTC metabolism had been evaluated in etiolated shoots that had been exposed to dichlormid during the preemergent phase of growth, a greater enhancement of metabolism by the safener would have been apparent.

It would also seem appropriate that etiolated shoots be used when examining the effects of safeners on the levels of enzymes and cofactors that may influence the rate of herbicide metabolism. Unfortunately, the majority of studies concerning the effects of dichlormid on GSH content and GST activity have been conducted with corn roots (Lay and Casida, 1976; Carringer *et al.*, 1978b; Stephenson *et al.*, 1983; Lay and Niland, 1985). Roots of grasses are not sensitive to EPTC and in fact low levels of thiocarbamate herbicides often promote root growth (Gray and Weierich, 1969).

B. Effects of Safeners on GSH and Herbicide-Metabolizing Enzymes

In those studies where safener-induced enhancement of thiocarbamate and chloroacetanilide metabolism has been reported, enhanced metabolism appears to be due to the ability of safeners to increase the levels of enzymes (GST, monooxygenase) and/or the cofactor GSH involved in herbicide metabolism.

The proposed role of safeners in enhancing monooxygenase activity is largely speculative. Although it has been suggested that safeners may promote oxidative metabolism of chloroacetanilides by stimulating monooxygenase activity (Hatzios, 1983b), there is no evidence that oxidative metabolism is involved in the initial metabolism of this herbicide class in either untreated or safener-treated plants. All studies to date suggest that a GSH conjugate is the initial metabolite formed (Lamoureux *et al.*, 1971; Breaux, 1986, 1987; Fuerst and Gronwald, 1986) and that a phase I activation by a monooxygenase is not required.

With regard to the metabolism of EPTC, it has been proposed that the initial metabolite, EPTC-sulfoxide, is generated by monooxygenase activity (Lay *et al.*, 1975; Hatzios and Penner, 1982). It has also been postulated that dichlormid may act by promoting sulfoxidation of EPTC and thereby enhance the overall rate of EPTC metabolism (Hatzios, 1983a). The evidence for this is indirect and based on the synergism observed when ozone or insecticide synergists are applied to corn seedlings treated with EPTC and dichlormid. While dichlormid may indeed enhance cytochrome P-450 monooxygenase activity resulting in accelerated EPTC sulfoxidation, the evidence obtained with insecticide synergists and ozone is equivocal. The interpretation of the synergistic response obtained with ozone and insecticide synergists is based on two assumptions. One of these is that a cytochrome P-450 monooxygenase is involved in the sulfoxidation of EPTC. Another is that insecticide synergists inhibit this monooxygenase. It is, however, by no means certain that a cytochrome P-450 monooxygenase is responsible for the sulfoxidation of EPTC. Recent reports have indicated that peroxygenase hemoproteins in plants are capable of catalyzing sulfoxidation of certain xenobiotics (Doerge and Corbett, 1984; Blee and Durst, 1986, 1987). Peroxygenase hemoproteins differ from cytochrome P-450 hemoproteins in that the oxygen atom incorporated into the substrate is derived from H_2O_2 or other hydroperoxides. Whether this type of hemoprotein is responsible for the sulfoxidation of EPTC has not been investigated. If it is assumed that a cytochrome P-450 monooxygenase is responsible for the sulfoxidation of EPTC, it is uncertain whether it would be inhibited by insecticide synergists or ozone. It should be noted that Frear *et al.* (1969) reported that the insecticide synergist SKF 525-A was not an effective inhibitor of the microsomal cytochrome P-450 monooxygenase involved in the N-demethylation of phenylurea herbicides in cotton. Regarding the effect of ozone on monooxygenase activity, there is no evidence that this molecule inhibits cytochrome P-450 monooxygenase activity in either mammalian or plant tissue. An alternative explanation for the synergism observed in the presence of ozone may relate to the effects of this molecule on plant GSH levels. Ozone exposure has been shown to reduce GSH content in sensitive bean (*Phaseolus vulgaris* L.) cultivars (Guri, 1983).

There is strong evidence that safeners which protect against injury by thiocarbamate and chloroacetanilide herbicides increase GSH levels and/or GST activity (Lay and Casida, 1976; Carringer *et al.*, 1978b; Ezra and Gressel, 1982;

Mozer *et al.*, 1983; Gronwald *et al.*, 1987). The relative importance of the increase in cofactor or enzyme is not resolved. If the *in vivo* conjugation of EPTC-sulfoxide and chloroacetanilides is primarily nonenzymatic then the enhancement of GSH content would be most important. If, on the other hand, conjugation *in vivo* is largely enzyme-mediated, then increases in both GSH and GST could contribute to enhanced detoxication.

As regards the importance of antidote-induced enhancement of GSH content, there are reports indicating a reasonably good correlation between the ability of dichlormid and other dichloroacetamides to confer protection against EPTC injury in corn and ability to increase root GSH content (Lay and Casida, 1976; Stephenson *et al.*, 1983). On the other hand, there are reports which question the importance of enhanced GSH content. For example, the ability of dichlormid to increase GSH content in corn roots does not always correlate with safening activity in certain inbreds or hybrids (Lay and Niland, 1985). Also, naphthalic anhydride protects corn from injury by EPTC, yet it has no effect on corn root GSH content (Lay and Casida, 1976). In the case of safeners which protect sorghum against chloroacetanilide injury, a correlation was not observed between safening activity and ability to enhance GSH content of sorghum shoots (Gronwald *et al.*, 1987).

There is considerable controversy as to whether GSH conjugation of EPTC-sulfoxide and chloroacetanilides *in vivo* is catalyzed by GST. Some investigators have concluded that conjugation *in vivo* is primarily nonenzymatic (Carringer *et al.*, 1978a; Leavitt and Penner, 1979). This conclusion is based on (1) the lack of detectable enzyme activity capable of catalyzing GSH conjugation of EPTC-sulfoxide (Carringer *et al.*, 1978a) or chloroacetanilides (Leavitt and Penner, 1979) in crude plant extracts and (2) the demonstration of a spontaneous nonenzymatic reaction between GSH and the herbicide substrates (EPTC-sulfoxide, chloroacetanilides) *in vitro*. However, the basis for this conclusion is equivocal. It is known that all GST substrates will react with GSH nonenzymatically at a finite rate *in vitro* (Jakoby and Habig, 1980). Furthermore, other investigators have detected GST activity capable of catalyzing GSH conjugation of EPTC-sulfoxide (Lay and Casida, 1976; Lay and Niland, 1985) and chloroacetanilides (Gronwald *et al.*, 1987) in plant extracts and found that antidotes increased activity.

C. Induction Response

There is indirect evidence which suggests that the increase in GSH levels and GST activity in response to safener treatment involves an induction process (i.e., *de novo* protein synthesis) as opposed to a direct activation of enzyme activity (Lay and Casida, 1976; Rennenberg *et al.*, 1982; Mozer *et al.*, 1983; Wiegand *et al.*, 1986; Gronwald *et al.*, 1987). The mechanism whereby safeners trigger an induction response leading to enhanced GSH levels and/or GST activity is not understood. In mammals it is known that pretreatment with certain chemicals that act as substrates for xenobiotic-detoxifying enzymes (cytochrome P-450 monooxygenase, GST) will serve as inducers of these enzymes (Nebert and Gelboin, 1968; Guthenberg *et al.*, 1980; Eisen, 1986). In corn, pretreatment with certain herbicides results in enhanced metabolism of the same herbicide when it is applied at a later date (Jachetta and Radosevich, 1981; Ezra *et al.*, 1985). Jachetta and Radosevich (1981) reported that pretreating corn with atrazine resulted in enhanced metabolism via GSH conjugation when the herbicide was applied later. This enhanced metabolism was attributed to the ability of atrazine pretreatment to increase the levels of GST catalyzing the formation of a GSH–atrazine conjugate. Pretreating corn plants with low levels of the chloroacetamide CDAA protected corn from the herbicide when it was applied later at higher rates (Ezra *et al.*, 1985). The protection conferred was attributed to the ability of the treatment to enhance GSH conjugation of CDAA by increasing levels of GSH and GST. In another study, Stephenson *et al.* (1983) reported that pretreating corn roots with herbicides metabolized via GSH conjugation (EPTC, alachlor, propachlor, atrazine, barban) increased root GSH content.

As an explanation for the effects of herbicide pretreatment on GSH content and GST activity, it has been proposed that herbicides which serve as substrates for GSH conjugation elevate the levels of cofactor (GSH) and enzyme (GST) required for their own metabolism (Stephenson *et al.*, 1983; Ezra *et al.*, 1985). Stephenson *et al.* (1983) postulated that safeners confer protection to certain herbicides that are conjugated with GSH because they exhibit structural similarities to these herbicides and as a result can act as inducers of GSH and GST involved in herbicide metabolism. Furthermore, these authors hypothesized that the advantage of the safener over the herbicide is that the safener can act as an effective inducer of GSH and GST without having herbicidal activity. The structural similarities between certain safeners and the herbicides they protect against are not always readily apparent; for example, the oxime ethers used to protect sorghum against chloroacetanilide injury. It may not be necessary that safeners mimic the herbicide structurally in order to confer protection. Instead, the ability of the safener to serve as a substrate for the enyzme system(s) involved in detoxifying the herbicide may be of primary importance.

Assuming safeners are acting to elicit a substrate induction of GSH and GST, and perhaps monooxygenases, they must elicit a rapid induction. The examples of induction of GSH and GST by herbicides described above involved a pretreatment. Under field conditions safeners are not applied as pretreatments, although safener seed treatments could be considered a form of pretreatment. To date, the kinetics of enhancement of herbicide metabolism in response to safener treatment have been examined in only one study. Sweetser (1985) reported that treatment with naphthalic anhydride increased the rate of chlorsulfuron metabolism in corn leaves by 3.6-fold. Maximum enhancement of chlorsulfuron metabolism was attained approximately 4 hr after treatment with the safener. These results suggest that naphthalic anhydride is capable of eliciting a rapid induction of an enzyme(s) involved in chlorsulfuron metabolism in corn.

D. Cross-protection

Many of the safeners in Fig. 1 confer protection to more than one herbicide class, i.e., they exhibit cross-protection. This is true of dichlormid, which protects corn against injury by thiocarbamates, chloroacetanilides, and, to a lesser extent, the sulfonylureas (Hatzios, 1983c; Parker, 1983). It is difficult to explain cross-protection in terms of an effect of dichlormid on herbicide uptake or site of action since these herbicide classes exhibit quite distinct chemistries and mechanisms of action. A more probable explanation is that dichlormid enhances herbicide metabolism and in order for this to be accepted, one would have to hypothesize that dichlormid enhances more than one enzyme system involved in herbicide detoxication. In mammalian tissues, treatment with certain xenobiotics (phenobarbital, 3-methylcholanthrene) will induce increases in both cytochrome P-450 monooxygenase (Eisen, 1986) and GST activity (Kaplowitz *et al.*, 1975; Ding and Pickett, 1985). Cross-protection in corn could be explained if dichlormid acts by inducing an increase in both monooxygenase and GST activity. Metabolism of the thiocarbamates in corn involves both sulfoxidation and GSH conjugation (Fig. 2). Hence, the protection conferred by dichlormid could be due to the ability of the safener to increase both the GST/GSH and monooxygenase enzyme systems. There is in fact considerable evidence that dichlormid increases GSH content and GST activity in corn roots and shoots. It has also been postulated that dichlormid enhances monooxygenase activity (Leavitt and Penner, 1979; Sweetser, 1985). The initial metabolism of the chloroacetanilides does not involve a monooxygenase reaction (Breaux, 1986, 1987) and in this case the ability of dichlormid to increase the GST/GSH enzyme system may account for enhanced metabolism. Metabolism of the sulfonylurea, chlorsulfuron, in grasses involves a monooxygenase reaction followed by glycosylation (Sweetser *et al.*, 1982). Sweetser (1985) recently proposed that the protection conferred by dichlormid against sulfonylurea injury in corn is due to the ability of the safener to induce an increase in cytochrome P-450 monooxygenase activity.

E. Analogy with Effects of Xenobiotics in Mammals

In many respects, herbicide safeners may be considered as relatively benign xenobiotics which are capable of eliciting an induction of enzymes involved in herbicide metabolism. The effects of safeners on herbicide-detoxifying enzymes in plants appear to be similar to the effects of xenobiotics on xenobiotic-detoxifying enzymes in mammals. Approximately 30 years ago, it was found that pretreating mammals with certain xenobiotics resulted in enhanced metabolism of the xenobiotics when they were applied later (Conney *et al.*, 1957). Moreover, it was found that pretreatment with xenobiotics with no apparent structural similarities to the xenobiotic substrate whose metabolism rate was being measured also elicited a similar response. Later investigations revealed that xenobiotic pretreatment increased the rate of xenobiotic metabolism by enhancing the levels of the enzymes involved (Conney *et al.*, 1959; Conney, 1967; Nebert and Gelboin, 1968). The enhancement was the result of an induction response involving *de novo* protein synthesis. Two major xenobiotic-detoxifying enzymes induced by xenobiotic treatment in mammals were cytochrome P-450 monooxygenase (Eisen, 1986) and GST (Guthenberg *et al.*, 1980; Pickett *et al.*, 1984). It was also shown that certain xenobiotics (phenobarbital, 3-methylcholanthrene) were capable of enhancing the activity of both enzyme systems (Alvares *et al.*, 1967; Kaplowitz *et al.*, 1975; Sundheimer *et al.*, 1983; Ding and Pickett, 1985; Pike *et al.*, 1985). Recent studies have indicated that the enhancement of GST and cytochrome P-450 monooxygenase activity by xenobiotics involves an increase in mRNA levels coding for these proteins (Hardwick *et al.*, 1983; Ding and Pickett, 1985; Pike *et al.*, 1985). It is thought that the xenobiotics act by augmenting gene transcription; however, it has not been ruled out that they may also increase the half-life of mRNA coding for xenobiotic-metabolizing enzymes.

References

Adams, C. A., Blee, E., and Casida, J. E. (1983). Dichloroacetamide herbicide antidotes enhance sulfate metabolism in corn roots. *Pestic. Biochem. Physiol.* **19**, 350–360.

Alvares, A. P., Schilling, G., Levin, W., and Kuntzman, R. (1967). Studies on the induction of CO-binding pigments in liver microsomes by phenobarbital and 3-methylcholanthrene. *Biochem. Biophys. Res. Commun.* **29**, 521–526.

Anderson, W. P. (1983). "Weed Science: Principles", 2nd ed. West Publ. Co., St. Paul, Minnesota.

Appleby, A. P., Furtick, W. R., and Fang, S. C. (1965). Soil placement studies with EPTC on *Avena sativa*. *Weed Res.* **5**, 115–122.

Benveniste, I., Salaün, J-P., Simon, A., Reichhart, D., and Durst, F. (1982). Cytochrome P-450-dependent ω-hydroxylation of lauric acid by microsomes from pea seedlings. *Plant Physiol.* **70**, 122–126.

Blee, E., and Durst, F. (1986). Oxidation of an organosulfur xenobiotic by microsomes from soybean cotyledons. *Biochem. Biophys. Res. Commun.* **135**, 922–927.

Blee, E., and Durst, F. (1987). Hydroperoxide-dependent sulfoxidation catalyzed by soybean microsomes. *Arch. Biochem. Biophys.* **254**, 43–52.

Breaux, E. J. (1986). Identification of the initial metabolites of acetochlor in corn and soybean seedlings. *J. Agric. Food Chem.* **34**, 884–888.

Breaux, E. J. (1987). Initial metabolism of acetochlor in tolerant and susceptible seedlings. *Weed Sci.* **35**, 463–468.

Burnside, O. C., Weeks, G. A., and Fenster, C. R. (1971). Protecting corn from herbicide injury by seed treatment. *Weed Sci.* **19**, 565–568.

Carringer, R. D., Rieck, C. E., and Bush, L. P. (1978a). Metabolism of EPTC in corn (*Zea mays*). *Weed Sci.* **26**, 157–160.

Carringer, R. D., Rieck, C. E., and Bush, L. P. (1978b). Effect of R-25788 on EPTC metabolism in corn (*Zea mays*). *Weed Sci.*, **26**, 167–171.

Casida, J.E. (1970). Mixed-function oxidase involvement in the biochemistry of insecticide synergists. *J. Agric. Food Chem.*, **18**, 753–772.

Casida, J.E., Gray, R. A., and Tilles, H. (1974). Thiocarbamate sulfoxides: Potent, selective, and biodegradable herbicides. *Science* **184**, 573–574.

Casida, J. E., Kimmel, E. C., Ohkawa, H., and Ohkawa, R. (1975). Sulfoxidation of thiocarbamate herbicides and metabolism of thiocarbamate sulfoxides in living mice and liver enzyme systems. *Pestic. Biochem. Physiol.* **5**, 1–11.

Chang, F. Y., Bandeen, J. D., and Stephenson, G. R. (1972). A selective antidote for prevention of EPTC injury in corn. *Can. J. Plant Sci.* **52**, 707–714.

Chang, F. Y., Bandeen, J. D., and Stephenson, G. R. (1973). *N,N*-diallyl-α,α-dichloroacetamide as an antidote for EPTC and other herbicides in corn. *Weed Res.* **13**, 399–406.

Chang, F.Y., Stephenson, G. R., and Bandeen, J. D. (1974). Effects of *N,N*-diallyl-2,2,-dichloroacetamide on ethyl *N,N*-di-*n*-propylthiocarbamate uptake and metabolism by corn seedlings. *J. Agric. Food Chem.* **22**, 245–248.

Conney, A. H. (1967). Pharmacological implications of microsomal enzyme induction. *Pharmacol. Rev.* **19**, 317–366.

Conney, A. H., Miller, E. C., and Miller, J. A. (1957). Substrate-induced synthesis and other properties of benzpyrene hydroxylase in rat liver. *J. Biol. Chem.* **228**, 753–766.

Conney, A. H., Gillette, J. R., Inscoe, J. K., Trams, E. R., and Posner, H. S. (1959). Induced synthesis of liver microsomal enzymes which metabolize foreign compounds. *Science* **130**, 1478–1479.

Coon, M. J., and Persson, A. V. (1980). Microsomal cytochrome P-450: A central catalyst in detoxication reactions. *In* "Biochemical Pharmacology and Toxicology" (W. B. Jakoby, ed.), Vol. 1, pp. 117–134. Academic Press, New York.

Dawson, J. H. (1963). Development of barnyard grass seedlings and their response to EPTC. *Weeds* **11**, 60–66.

Ding, V. D. H., and Pickett, C. B. (1985). Transcriptional regulation of rat liver glutathione *S*-transferase genes by phenobarbital and 3-methylcholanthrene. *Arch. Biochem. Biophys.* **240**, 553–559.

Doerge, D. R., and Corbett, M. D. (1984). Oxidation of organosulfur compounds by the microsomal fraction of germinating pea seeds (*Pisum sativum*). *Biochem. Biophys. Res. Commun.* **121**, 1001–1005.

Donald, W. W., and Fawcett, R. S. (1976). Prevention of EPTC injury to corn: A developmental and anatomical study. *Abstr. Weed Sci. Soc. Am.*, p. 70.

Edwards, R., and Owen, W. J. (1986). Comparison of glutathione *S*-transferases of *Zea mays* responsible for herbicide detoxification in plants and suspension-cultured cells. *Planta* **169**, 208–215.

Eisen, H. J. (1986). Induction of hepatic P-450 isozymes. *In* "Cytochrome P-450: Structure, Mechanism, and Biochemistry" (P. R. Ortiz de Montellano, ed.), pp. 315–344. Plenum, New York.

Ellis, J. F., Peek, J. W., Boehle, J., Jr., and Müller, G. (1980). Effectiveness of a new safener for protecting sorghum (*Sorghum bicolor*) from metolachlor injury. *Weed Sci.* **28**, 1–5.

Ezra, G., and Gressel, J. (1982). Rapid effects of a thiocarbamate herbicide and its dichloroacetamide protectant on macromolecular syntheses and glutathione levels in maize cell cultures. *Pestic. Biochem. Physiol.* **17**, 48–58.

Ezra, G., Rusness, D. G., Lamoureux, G. L., and Stephenson, G. R. (1985). The effect of CDAA (*N,N*-diallyl-2-chloroacetamide) pretreatments on subsequent CDAA injury to corn (*Zea mays* L.) *Pestic. Biochem. Physiol.* **23**:108–115.

Frear, D. S., Swanson, H. R., and Tanaka, F. S. (1969). *N*-Demethylation of substituted 3-(phenyl)-1-methylureas: Isolation and characterization of a microsomal mixed function oxidase from cotton. *Phytochemistry* **8**, 2157–2169.

Fuerst, E. P., and Gronwald, J. W. (1986). Induction of rapid metabolism of metolachlor in sorghum (*Sorghum bicolor*) shoots by CGA-92194 and other antidotes. *Weed Sci.* **34**, 354–361.

Gray, R. A., and Joo, G. K. (1978). Site of uptake and action of thiocarbamate herbicides and herbicide antidotes in corn seedlings. *In* "Chemistry and Action of Herbicide Antidotes" (F. M. Pallos and J. E. Casida, eds.), pp. 67–84. Academic Press, New York.

Gray, R. A., and Weierich, A. J. (1969). Importance of root, shoot, and seed exposure on the herbicidal activity of EPTC. *Weed Sci.* **17**, 223–229.

Gronwald, J. W., Fuerst, E. P., Eberlein, C. V., and Egli, M. A. (1987). Effect of herbicide antidotes on glutathione content and glutathione *S*-transferase activity of sorghum shoots. *Pestic. Biochem. Physiol.* **29**, 66–76.

Güneyli, E. (1971). Factors affecting the action of 1,8-naphthalic anhydride in corn treated with *S*-ethyldipropylthiocarbamate (EPTC). Ph.D. Dissertation, Oregon State University, Corvallis.

Guri, A. (1983). Variation in glutathione and ascorbic acid content among selected cultivars of *Phaseolus vulgaris* prior to and after exposure to ozone. *Can. J. Plant Sci.* **63**, 733–737.

Guthenberg, C., Morgenstern, R., DePierre, J. W., and Mannervik, B. (1980). Induction of glutathione *S*-transferases A, B and C in rat liver cytosol by *trans*-stilbene oxide. *Biochim. Biophys. Acta* **631**, 1–10.

Guy, C. L., Carter, J. V., Yelenosky, G., and Guy, C. T. (1984). Changes in glutathione content during cold acclimation in *Cornus sericea* and *Citrus sinensis*. *Cryobiology* **21**, 443–453.

Habig, W. H., and Jakoby, W. B. (1981). Assays for differentiation of glutathione *S*-transferase. *In* "Methods in Enzymology" (W. B. Jakoby, ed.), Vol. 77, pp. 398–405. Academic Press, New York.

Hardwick, J. P., Gonzalez, F. J., and Kasper, C. B. (1983). Transcriptional regulation of rat liver epoxide hydratase, NADPH-cytochrome P-450 oxidoreductase, and cytochrome P-450 genes by phenobarbital. *J. Biol. Chem.* **258**, 8081–8085.

Hatzios, K. K. (1983a). Interactions of the herbicides EPTC and EPTC plus R-25788 with ozone and antioxidants in corn. *J. Agric. Food Chem.* **31**, 1187–1191.

Hatzios, K. K. (1983b). Effects of CGA-43089 on responses of sorghum (*Sorghum bicolor*) to metolachlor combined with ozone or antioxidants. *Weed Sci.* **31**, 280–284.

Hatzios, K. K. (1983c). Herbicide antidotes: Development, chemistry, and mode of action. *Adv. Agron.* **36**, 265–316.

Hatzios, K. K. (1984). Interactions between selected herbicides and protectants on corn (*Zea mays*). *Weed Sci.* **32**, 51–58.

Hatzios, K. K., and Penner, D. (1982). "Metabolism of Herbicides in Higher Plants." Burgess, Minneapolis, Minnesota.

Hedden, P., MacMillan, J., and Phinney, B. O. (1978). The metabolism of the gibberellins. *Annu. Rev. Plant Physiol.* **29**, 149–192.

Horváth, L., and Pulay, Á. (1980). Metabolism of EPTC in germinating corn: Sulfone as the true carbamoylating agent. *Pestic. Biochem. Physiol.* **14**, 265–270.

Hubbell, J.P., and Casida, J. E. (1977). Metabolic fate of the *N,N*-dialkylcarbamoyl moiety of thiocarbamate herbicides in rats and corn. *J. Agric. Food Chem.* **25**, 404–413.

Jachetta, J. J., and Radosevich, S. R. (1981). Enhanced degradation of atrazine by corn (*Zea mays*). *Weed Sci.* **29**, 37–44.

Jackson, L. A., Yopp, J. H., and Kapusta, G. (1986). Absorption and distribution of flurazole and acetochlor in grain sorghum. *Pestic. Biochem. Physiol.* **25**, 373–380.

Jakoby, W. B., and Habig, W. H. (1980). Glutathione transferases. *In* "Biochemical Pharmacology and Toxicology" (W. B. Jakoby, ed)., Vol. 2, pp. 63–94. Academic Press, New York.

Jordan, L. S., and Jolliffe, V. A. (1971). Protection of plants from herbicides with 1,8-naphthalic anhydride as illustrated with sorghum. *Bull. Environ. Contam. Toxicol.* **6**, 417–421.

Kaplowitz, N., Kuhlenkamp, J., and Clifton, G. (1975). Drug induction of hepatic glutathione *S*-transferases in male and female rats. *Biochem. J.* **146**, 351–356.

Ketchersid, M. L., Norton, K., and Merkle, M. G. (1981). Influence of soil moisture on the safening effect of CGA-43089 in grain sorghum (*Sorghum bicolor*). *Weed Sci.* **29**, 281–287.

Ketchersid, M. L., Vietor, D. M., and Merkle, M. G. (1982). CGA-43089 effects on metolachlor uptake and membrane permeability in grain sorghum (*Sorghum bicolor*). *J. Plant Growth Regul.* **1**, 285–294.

Ketchersid, M. L., Plapp, F. W., and Merkle, M. G. (1985). Sorghum (*Sorghum bicolor*) seed safeners as insecticide synergists. *Weed Sci.* **33**, 774–778.

Komives, T., and Dutka, F. (1980). On the mode of action of EPTC and its antidotes on corn. *Cereal Res. Commun.* **8**, 627–633.

Lamoureux, G. L., and Rusness, D. G. (1983). Malonylcysteine conjugates as end-products of glutathione conjugate metabolism in plants. *In* "Pesticide Chemistry: Human Welfare and the Environment" (J. Miyamoto and P. C. Kearney, eds.), Vol. 3, pp. 295–300. Pergamon, Oxford.

Lamoureux, G. L., and Rusness, D. G. (1986). Tridiphane [2-(3,5-dichlorophenyl)-2-(2,2,2-trichloroethyl) oxirane] an atrazine synergist: Enzymatic conversion to a potent glutathione *S*-transferase inhibitor. *Pestic. Biochem. Physiol* **26**, 323–342.

Lamoureux, G. L., and Frear, D. S. (1987). Current problems, trends, and developments in pesticide metabolism in plants. *In* "Pesticide Science and Biochemistry" (R. Greenhalgh and T. R. Roberts, eds.), pp. 455–462. Blackwell Scientific, Boston.

Lamoureux, G. L., and Rusness, D. G. (1987). EPTC metabolism in corn, cotton, and soybean: Identification of a novel metabolite derived from the metabolism of a glutathione conjugate. *J. Agric. Food Chem.* **35**, 1–7.

Lamoureux, G. L., Stafford, L. E., and Tanaka, F. S. (1971). Metabolism of 2-chloro-*N*-isopropylacetanilide (propachlor) in the leaves of corn, sorghum, sugarcane, and barley, *J. Agric. Food Chem.* **19**, 346–350.

Lay, M-M., and Casida, J. E. (1976). Dichloroacetamide antidotes enhance thiocarbamate sulfoxide detoxification by elevating corn root glutathione content and glutathione *S*-transferase activity. *Pestic. Biochem. Physiol.* **6**, 442–456.

Lay, M-M., and Casida, J. E. (1978). Involvement of glutathione and glutathione *S*-transferases in the action of dichloroacetamide antidotes for thiocarbamate herbicides. *In* "Chemistry and Action of Herbicide Antidotes" (F. M. Pallos and J. E. Casida, eds.), pp. 151–160. Academic Press, New York.

Lay, M-M., and Niland, A. M. (1985). Biochemical response of inbred and hybrid corn (*Zea mays* L.) to R-25788 and its distribution with EPTC in corn seedlings. *Pestic. Biochem. Physiol.* **23**, 131–140.

Lay, M-M., Hubbell, J. P., and Casida, J. E. (1975). Dichloroacetamide antidotes for thiocarbamate herbicides: Mode of action. *Science* **189**, 287–289.

Leavitt, J. R. C., and Penner, D. (1978). Protection of corn (*Zea mays*) from acetanilide herbicidal injury with the antidote R-25788. *Weed Sci.* **26**, 653–659.

Leavitt, J. R. C., and Penner, D. (1979). *In vitro* conjugation of glutathione and other thiols with acetanilide herbicides and EPTC sulfoxide and the action of the herbicide antidote R-25788. *J. Agric. Food Chem.* **27**, 533–536.

Leek, G. L. (1981). Efficacy of CGA-43089 [α-(cyanomethoximino)-benzeneacetonitrile] as a herbicide antidote for sorghum (*Sorghum vulgare* Pers.) Ph.D. Dissertation, Michigan State University, East Lansing

Mozer, T. J., Tiemeier, D. C., and Jaworski, E. G. (1983). Purification and characterization of corn glutathione *S*-transferase. *Biochemistry* **22**, 1068–1072.

Murphy, P. J., and West, C. A. (1969). The role of mixed function oxidases in kaurene metabolism in *Echinocystis macrocarpa* Greene endosperm. *Arch. Biochem. Biophys* **133**, 395–407.

Nebert, D. W., and Gelboin, H. V. (1968). Substrate-inducible microsomal aryl hydroxylase in mammalian cell culture. I. Assay and properties of induced enzyme. *J. Biol. Chem.* **243**, 6242–6249.

Nyffeler, A., Gerber, H. R., and Hensley, J. R. (1980). Laboratory studies on the behavior of the herbicide safener CGA-43089. *Weed Sci.* **28**, 6–10.

Pallos, F. M., Gray, R. A., Arneklev, D. R., and Brokke, M. E. (1975). Antidotes protect corn from thiocarbamate herbicide injury. *J. Agric. Food Chem.* **23**, 821–822.

Parker, C. (1983). Herbicide antidotes—A review. *Pestic. Sci.* **14**, 40–48.

Pickett, C. B., Telakowski-Hopkins, C. A., Ding, G. J. F., Argenbright, L., and Lu, A. Y. H. (1984). Rat liver glutathione *S*-transferases. Complete nucleotide sequence of a glutathione *S*-transferase mRNA and the regulation of the Ya, Yb, and Yc mRNAs by 3-methylcholanthrene and phenobarbital. *J. Biol. Chem.* **259**, 5182–5188.

Pike, S. F., Shephard, E. A., Rabin, B. R., and Phillips, I. R. (1985). Induction of cytochrome P-450 by phenobarbital is mediated at the level of transcription. *Biochem. Pharmacol.* **34**, 2489–2494.

Potts, J. R. M., Weklych, R., and Conn, E. E. (1974). The 4-hydroxylation of cinnamic acid by sorghum microsomes and the requirement for cytochrome P-450. *J. Biol. Chem.* **249**, 5019–5026.

Rennenberg, H. (1982). Glutathione metabolism and possible biological roles in higher plants. *Phytochemistry* **21**, 2771–2781.

Rennenberg, H., Birk, C., and Schaer, B. (1982). Effect of *N,N*-diallyl-2,2-dichloroacetamide (R-25788) on efflux and synthesis of glutathione in tobacco suspension cultures. *Phytochemistry* **21**, 5–8.

Rich, P. R., and Lamb, C. J. (1977). Biophysical and enzymological studies upon the interaction of *trans*-cinnamic acid with higher plant microsomal cytochromes P-450. *Eur. J. Biochem.* **72**, 353–360.

Rubin, B., Leavitt, J. R. C., Penner, D., and Saettler, A. W. (1980). Interaction of antioxidants with ozone and herbicide stress. *Bull. Environ. Contam. Toxicol.* **25**, 623–629.

Shimabukuro, R. H. (1985). Detoxication of herbicides. *In* "Weed Physiology" (S. O. Duke, ed.), Vol. 2, pp. 215–240. CRC Press, Boca Raton, Florida.

Soliday, C. L., and Kolattukudy, P.E. (1977). Biosynthesis of cutin. ω-hydroxylation of fatty acids by a microsomal preparation from germinating *Vicia faba*. *Plant Physiol.* **59**, 1116–1121.

Spotanski, R. F., and Burnside, O. C. (1973). Reducing herbicide injury to sorghum with crop protectants. *Weed Sci.* **21**, 531–536.

Stephenson, G. R., and Chang, F. Y. (1978). Comparative activity and selectivity of herbicide antidotes. *In* "Chemistry and Action of Herbicide Antidotes" (F. M. Pallos and J. E. Casida, eds.), pp. 35–61. Academic Press, New York.

Stephenson, G. R., Ali, A., and Ashton, F. M. (1983). Influence of herbicides and antidotes on the glutathione levels of maize seedlings. *In* "Pesticide Chemistry: Human Welfare and the Environment" (J. Miyamoto and P. C. Kearney, eds.), Vol 3, pp. 219–224. Pergamon, Oxford.

Sundheimer, D. W., Caveness, M. B., and Goldstein, J. A. (1983). Differential metabolism of acetanilide versus ethoxycoumarin and benzo[α]pyrene by two 3-methylcholanthrene-inducible forms of rat liver cytochrome P-450. *Arch. Biochem. Biophys.* **226**, 548–557.

Sweetser, P. B. (1985). Safening of sulfonylurea herbicides to cereal crops: Mode of herbicide antidote action. *Proc. Br. Crop Prot. Conf.—Weeds* **3**, 1147–1154.

Sweetser, P. B., Schow, G. S., and Hutchison, J. M. (1982). Metabolism of chlorsulfuron by plants: Biological basis for selectivity of a new herbicide for cereals. *Pestic. Biochem. Physiol.* **17**, 18–23.

Wiegand, R. C., Shah, D. M., Mozer, T. J., Harding, E. I., Diaz-Collier, J., Saunders, C., Jaworski, E. G., and Tiemeier, D. C. (1986). Messenger RNA encoding a glutathione-*S*-transferase responsible for herbicide tolerance in maize is induced in response to safener treatment. *Plant Mol. Biol.* **7**, 235–243.

Wilkinson, C. F. (1976). Insecticide synergism. *In* "The Future for Insecticides: Needs and Prospects" (R. L. Metcalf and J. J. McKelvey, Jr., eds.), pp. 195–218. Wiley, New York.

Winkle, M. E., Leavitt, J. R. C., and Burnside, O. C. (1980). Acetanilide–antidote combinations for weed control in corn (*Zea mays*) and sorghum (*Sorghum bicolor*). *Weed Sci.* **28**, 699–704.

Young, O., and Beevers, H. (1976). Mixed function oxidases from germinating castor bean endosperm. *Phytochemistry* **15**, 379–385.

Zama, P., and Hatzios, K. K. (1986). Effects of CGA-92194 on the chemical reactivity of metolachlor with glutathione and metabolism of metolachlor in grain sorghum (*Sorghum bicolor*). *Weed Sci.* **34**, 834–841.

Chapter 5

Effects of Herbicide Safeners on Levels and Activity of Cytochrome P-450 and Other Enzymes of Corn

Tamas Komives and Ferenc Dutka
Department of Pesticide Research
Central Research Institute for Chemistry
of the Hungarian Academy of Sciences
Budapest, Hungary

I. INTRODUCTION

A knowledge of herbicide safener effects on plant enzymes is desirable from several viewpoints. Of primary importance is the understanding of the mode of protective action of safeners, but results at the enzymatic level may also explain herbicide selectivity and provide insights into mechanisms of herbicide activation/detoxication in plants. It is now well established that a number of protective systems exist in plant tissues to trap and inactivate toxic xenobiotics or to counteract their harmful effects (Shimabukuro, 1985; Dodge, 1982).

Glutathione (γ-L-glutamyl-L-cysteinyl-glycine, GSH), or in some plants homoglutathione (γ-L-glutamyl-L-cysteinyl-β-alanine), represents important components of these systems. In addition to its essential functions in sulfur transport and metabolism, in amino acid transport, and in stabilization of enzyme systems, GSH participates in the cellular antioxidant system and reacts with electrophilic chemicals to form "GSH conjugates" (Rennenberg, 1982). GSH conjugates of herbicides are thought to be biologically inactive derivatives that may undergo a variety of metabolic transformations leading to the formation of stable, so-called terminal metabolites (Shimabukuro, 1985).

Cytochrome P-450-dependent oxygenases are membrane-bound mixed function oxygenase (MFO, EC 1.14.14.1) enzymes, which utilize molecular oxygen and NADPH in the oxidative transformations of a range of endogenous and exogenous chemicals in plants (West, 1980). Oxidation of xenobiotics may lead to bioactivation, resulting in the formation of compounds that are more reactive or more toxic than the parent molecules. Conjugation with GSH may contribute to the detoxication of these bioactivated derivatives (Lay and Casida, 1976).

The interrelationships of herbicide safeners with the key enzymes of the above systems as well as with other enzymes that may be involved in herbicide phytotoxicity and/or detoxication are the subject of this chapter.

II. CYTOCHROME P-450 LEVELS AND ACTIVITY

The importance of cytochrome P-450-containing MFO enzymes in the bioactivation and/or detoxication of xenobiotics in mammalian systems is extensively documented (West, 1980). This enzyme system consists of a number of isozymes with broad substrate requirements. Some of the isozymes are selectively inducible by treatments with appropriate chemicals. These inducers form ligand complexes with cytochrome P-450 resulting initially in inhibition of enzyme activity, followed later by enzyme induction (West, 1980). By contrast, much less is known of similar enzyme systems in other living organisms. Though such enzymes have been detected in plants (West, 1980), including corn (*Zea mays* L.) (Larson and Bussard, 1986), their specificity, inducibility, and other properties are not yet defined. From the available data it seems that multiple and inducible forms of cytochrome P-45O exist in plants as well. Unlike the mammalian enzymes, plant cytochrome P-450 systems may be characterized by a high degree of substrate selectivity. They participate in important biosynthetic pathways producing gibberellins, unsaturated fatty acids, phenylpropanoids, etc. (West, 1980), as well as in the detoxication of several herbicides (Reichhart *et al.*, 1979; Cole, 1983; Cabanne *et al.*, 1985).

Involvement of oxidative reactions in safener action was first proposed by Leavitt and Penner (1979). They found that the sulfoxide derivative of EPTC reacted with GSH *in vitro*, and suggested that its detoxication in corn is rapid even without prior dichlormid treatment to increase GSH levels. The safener could therefore protect corn from EPTC injury by increasing the sulfoxidation of the herbicide. We have reached a similar conclusion (Komives and Dutka, 1980; Dutka and Komives, 1983) after comparing the phytotoxicity of EPTC and its sulfoxide and sulfone derivatives to corn seedlings.

Structural similarities between some safeners and chemicals known to interact with mammalian or insect cytochrome P-450 also support a possible involvement of this enzyme system in safener action (Ketchersid *et al.*, 1985). For example, the antibiotic chloramphenicol, a structural analog of dichlormid, and several related dichloroacetamides have been shown to be selective suicidal substrates of mammalian cytochrome P-450 isozymes. The mechanism of the suicidal action involves oxidative dechlorination of the dichloromethyl group to yield a highly reactive oxamyl chloride (Halpert *et al.*, 1985, 1986; Miller and Halpert, 1986) according to the reaction:

$$\text{R–CO–CHCl}_2 \xrightarrow[\text{MFO}]{[\text{O}]} \text{R–CO–CCl}_2\text{OH} \xrightarrow{-\text{HCl}} \text{R–CO–CO–Cl}$$

Acylation by this acyl chloride of the heme or protein moiety of cytochrome P-450 leads to enzyme inactivation. Structure–activity investigations showed that the presence of the dichloromethyl group in the molecule is essential for biological activity: compounds structurally altered at the amine portion retained enzyme inactivation capability, whereas monochloroacetyl or difluoroacetyl derivatives were inactive (Halpert *et al.*, 1985).

Results on metabolic studies of dichlormid in corn seem to support a biotransformation following the above reaction: more than 10% of the parent chemical was converted to *N,N*-diallyloxamic acid in corn seedlings (Miaullis *et al.*, 1978). It should be noted, however, that chloramphenicol has little or no activity in safening EPTC on corn, or increasing GSH levels in this species (Lay and Casida, 1976). In addition, *in vitro* experiments with Jerusalem artichoke (*Helianthus tuberosus* L.) tuber slices, analogous to those with chloramphenicol in mammalian systems (Miller and Halpert, 1986), failed to show any significant change in the cytochrome P-45O content in the presence of 200 μM dichlormid during a 72-hr study (F. Durst, personal communication, 1985).

Unfortunately, the elucidation of the nature of the possible involvement of MFOs in safener action is complicated by lack of target-oriented inhibitors or inducers of cytochrome P-450 isozymes in plants. Thus, treatments of corn and of other plants with combinations of different classes of herbicides and known inhibitors of cytochrome P-450 led to increased (Komives and Dutka, 1980; Hatzios, 1983b; Cabanne *et al.*, 1985;), unaffected (Taft, 1975), and reduced herbicide phytotoxicity (Fedtke and Trebst, 1987; I. C. Barta, unpublished results; T. Komives, unpublished results).

Studies on the participation of MFOs in the biochemical mode of action of safeners at the enzymatic level have not reached firm conclusions either. For example, the oxidation steps during the biosynthesis of the gibberellin precursor kaurenoic acid from kaurene have been shown to be catalyzed by microsomal MFO enzymes (West, 1980). Dichlormid was found to inhibit the first oxidation step (the conversion of kaurene to kaurenol) in a cell-free system from etiolated coleoptiles of grain sorghum [*Sorghum bicolor* (L.) Moench], but had no effect on the further oxidations (Wilkinson, 1983). Interestingly, CDAA, a structural analog of dichlormid with moderate safening effectiveness (Lay and Casida, 1976), behaved quite differently in the same system. It did not affect the oxidation of kaurene to kaurenol, but did inhibit the subsequent oxidations (Wilkinson, 1985).

Dichlormid and naphthalic anhydride (NA) did not change the levels of cytochrome P-450 (Komives *et al.*, 1985a) and the activity of the well-characterized cytochrome P-450-dependent plant enzyme cinnamic acid 4-hydroxylase (EC 1.14.13.11) was not significantly different in dichlormid-treated (10 μM) and untreated etiolated corn seedlings after 48 hr exposure (T. Komives, unpublished results). Similarly, pretreatment of sorghum coleoptiles with the safeners cyometrinil and flurazole did not cause a measurable effect on the activity of the MFO enzyme aldrin epoxidase (EC 1.14.14.1). In this latter system, however, the insect cytochrome P-450 inhibitor piperonyl butoxide was inhibitory only during the first 8 hr of treatment (Ketchersid *et al.*, 1985).

These above observations do not rule out the possibility of increased herbicide detoxication potency by oxidation in safener-treated plants. Rates of oxidation by MFO enzyme systems are not necessarily determined by cytochrome P-450 content (Higashi *et al.*, 1982; Hagmann *et al.*, 1983), or, alternatively, oxidative detoxication of herbicides in plants may be catalyzed by enzymes other than cytochrome P-450-dependent MFOs (Blee *et al.*, 1985; Blee and Durst, 1986).

III. CELLULAR THIOLS AND GLUTATHIONE-RELATED ENZYMES

A. Glutathione Conjugation System

GSH conjugation has been identified as the first step in the biosynthesis of mercapturic acids in mammals (Habig, 1983), and proceeds according to the reaction

$$\underset{}{RX} + \underset{\text{GSH}}{\text{Glu–Cys(SH)–Gly}} \xrightarrow{-HX} \underset{\text{GSH conjugate}}{\text{Glu–Cys(SR)–Gly}}$$

X = leaving group

The conjugation may take place nonenzymatically as a nucleophilic substitution by the mercaptide function of GSH, but may also be catalyzed by glutathione *S*-transferase (GST, EC 2.5.1.18) enzymes. GSTs represent a family of enzymes with overlapping and broad substrate specificities, which facilitate the above reactions of hydrophobic, electrophilic substrates. These conjugation reactions, with few exceptions, are considered as detoxication processes. Several GST enzymes have been described in a number of organisms, including higher plants

(Habig, 1983). Corn plants contain at least three GST isozymes differing in their separation characteristics and substrate preference (Mozer *et al.*, 1983; Edwards and Owen, 1986; Wiegand *et al.*, 1986).

The first examples of the existence of a GSH conjugation system in plants were published in 1970 when GST-catalyzed conjugation of atrazine with GSH was shown to occur in corn and grain sorghum (Shimabukuro *et al.*, 1970; Lamoureux *et al.*, 1970). Since then, GST-mediated metabolism of a number of pesticides in corn and in other plants has been described, including other herbicides to which corn exhibits high tolerance (Shimabukuro, 1985).

Pioneering work by Lay and Casida (1976) connected the GSH conjugation system with safener action. In a study on thiocarbamate herbicide and safener mode of action in corn, they showed that thiocarbamates are metabolized in this plant by sulfoxidation followed by GSH conjugation according to the reaction

$$R^1\text{–S–CO–NR}^2R^3 \xrightarrow[\text{MFO}]{\text{oxidation}} R^1\text{–SO–CO–NR}^2R^3 \xrightarrow[\text{GST}]{\text{GSH}} \text{GS–CO–NR}^2R^3$$

Sulfoxidation has been viewed as a bioactivation process, because the sulfoxide derivatives of a number of thiocarbamate herbicides exhibited higher phytotoxicity to several grass weeds than their parent molecules, and were capable of reacting with GSH to produce the *S*-carbamoylated derivatives (GSH conjugates). Consistently increased levels of GSH and GST in corn plants treated with dichlormid and related dichloroacetamide safeners led to the conclusion that the combined increases in the levels of the participants of the glutathione conjugation system are responsible for the safening action (Lay and Casida, 1976). Subsequent research in other laboratories has confirmed safener-induced increases in GSH levels and it is summarized in Table I.

The role of GST in safening has been questioned by some researchers (Carringer *et al.*, 1978; Leavitt and Penner, 1979), who suggested that the spontaneous reaction between thiocarbamate sulfoxides and GSH is rapid enough for their fast detoxication. Further research, however, supported the observations of Lay and Casida (1976). Increased levels of GST were measured with several substrates in corn and other plants, such as rice (*Oryza sativa* L.) and grain sorghum treated with a number of safeners including recently developed compounds, such as fenclorim, flurazole, MG-191, and oxabetrinil (Table I). A new isozyme of GST was found during induction of safener-treated corn seedlings (Mozer *et al.*, 1983; Wiegand *et al.*, 1986).

The microsomal fraction of corn shoot extracts was also shown to contain measurable levels of GST. In contrast to mammalian microsomal GST, this enzyme in corn is inducible: its levels are increased by dichlormid and NA (Komives *et al.*, 1985a). It has been suggested that microsomal GSTs may be effective in the detoxication of lipophilic herbicides and involved in safening against their phytotoxicity.

Table I Effects of Herbicide Safeners on the Levels of GSH and GST *in vivo*

Safener	Plant	GSH	GST	Reference
Cyometrinil	Corn	No effect	n.d[a]	Fedtke (1981)
	Sorghum	n.d.		Mozer *et al*. (1983)
	Soybean	No effect		Fedtke (1981)
Dichlormid and dichloroacetamide analogs	Corn	Increased	Increased	Lay and Casida (1976); Ezra *et al*. (1985); Komives *et al*. (1985a,b); Lay and Niland (1985); Stephenson and Ezra (1985); Ezra and Stephenson (1986)
		Increased	n.d.	Carringer *et al*. (1978); Fedtke (1981); Ezra and Gressel (1982); Adams *et al*. (1983); Stephenson *et al*. (1983)
		n.d.	Increased	Mozer *et al*. (1983); Wiegand *et al*. (1986)
	Tobacco	Increased	n.d.	Rennenberg (1982)
	Sorghum	n.d.	Increased	Mozer *et al*. (1983)
		Increased	Increased	Gronwald *et al*. (1987)
	Soybean	Increased	n.d.	Fedtke (1981)
Fenclorim	Rice	n.d.	Increased	T. Komives, unpublished results
	Corn	Increased	Increased	T. Komives, unpublished results
Flurazole	Corn	n.d.	Increased	Mozer *et al*. (1983)
		Increased	n.d.	Breaux *et al*. (1987)
	Sorghum	Increased	Increaed	Gronwald *et al*. (1987)
Maleic acid derivatives	Soybean	Increased	n.d.	Kirino *et al*. (1985)
MG-191	Corn	Increased	Increased	Dutka and Komives (1987)
	Sorghum	Increased	Increased	T. Komives, unpublished results
NA	Corn	No effect	Increased	Komives *et al*. (1985a); Frear *et al*. (1987)
		n.d.	Increased	Mozer *et al*.(1983)
		No effect	n.d.	Fedtke (1981)
		No effect	No effect	Lay and Casida (1976)
	Sorghum	n.d.	Increased	Mozer *et al*. (1983)
		No effect	Increased	Gronwald *et al*. (1987)
OTC	Corn	Increased	No effect	Hilton and Pillai (1986)
Oxabetrinil	Sorghum	No effect	Increased	Gronwald *et al*. (1987)

[a] n.d. = not determined.

A direct activation of GST by safeners was ruled out by the results of *in vitro* experiments (Lay and Casida, 1976). Recent work provided evidence that in safener-treated corn plants GST induction is controlled at the level of mRNA accumulation (Wiegand *et al.*, 1986).

Levels of GSH available for conjugation with herbicides are determined by several biochemical processes. The biosynthesis of GSH is known to be a two-step process involving the production of γ-L-glutamyl-L-cysteine in the first step, followed by addition of glycine to the C-terminal of this dipeptide. This second step is catalyzed by the enzyme glutathione synthetase (GS, EC 6.3.2.3) (Law and Halliwell, 1986) and is described by the reaction

$$\gamma\text{-L-glutamyl-L-cysteine} + \text{glycine} + \text{ATP} \xrightarrow{\text{GS}} \text{GSH} + \text{ADP} + P_i$$

Carringer *et al.* (1978) found no significant difference between GS levels of untreated and dichlormid-treated corn seedlings. They showed, however, that dichlormid increased the activity of corn GS enzyme *in vitro*, an effect not found with GS of tobacco (*Nicotiana tabacum* L.), though tobacco cell cultures also respond to dichlormid with elevated GSH content (Rennenberg, 1982).

The importance of cysteine supply in safening action was shown by feeding experiments with OTC (L-2-oxothiazolidine-4-carboxylic acid), a chemical that is enzymatically transformed to *S*-carboxy-L-cysteine, which is spontaneously decarboxylated to yield L-cysteine (Hilton and Pillai, 1986) according to the reaction,

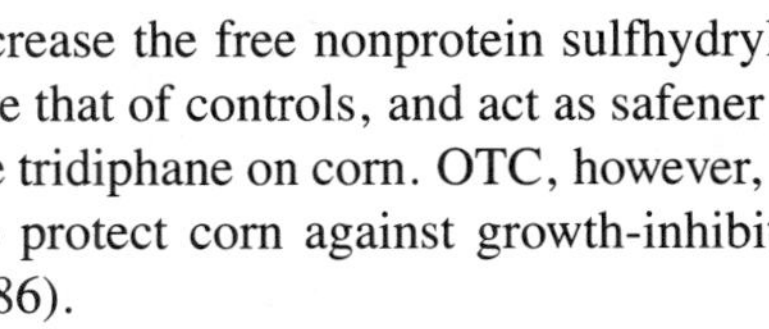

OTC was found to increase the free nonprotein sulfhydryl (nonPR–SH) content of corn seedlings above that of controls, and act as safener against the phytotoxic effects of the herbicide tridiphane on corn. OTC, however, did not alter the levels of GST, and failed to protect corn against growth-inhibitory doses of atrazine (Hilton and Pillai, 1986).

Though the cysteine required for synthesis of GSH in response to safener treatment in corn seedlings might be released from seed storage protein, data suggest that GSH biosynthesis in plants is, at least in part, controlled by the availability of inorganic sulfate for cysteine synthesis according to the pathway (Adams *et al.*, 1983)

sulfate
↓
adenosine-5′-phosphosulfate
↓
3′-phosphoadenosine-5′-phosphosulfate
↓
$X{-}S{-}SO_3^-$ (“bound sulfite”)
↓
$X{-}S{-}S^-$ (“bound sulfide”)
↓
L-cysteine

Indeed, dichlormid and several structurally related amide safeners at low concentrations slightly increased the uptake of inorganic sulfate in corn roots, and greatly enhanced its metabolism to “bound sulfide,” cysteine, and GSH (Adams *et al.*, 1983). The biosynthetic step of the above scheme influenced by dichlormid in corn occurs prior to the reduction step to form “bound sulfide”; in fact, dichlormid was shown to enhance the activity of ATP-sulfurylase (EC 2.7.7.4), the first enzyme in sulfate assimilation, which catalyzes the reaction between ATP and sulfate to yield adenosine-5′-phosphosulfate (Adams *et al.*, 1983).

The observations that several herbicides known to be metabolized by GSH conjugation in corn react with GSH nonenzymatically at a measurable rate under physiological conditions drew attention to the fact that the rate of GSH conjugation *in vivo* is determined by the combined rates of enzymatic and nonenzymatic reactions.

The nonenzymatic GSH conjugation reaction follows the rate equation

$$R = k'[\mathrm{RX}][\mathrm{GS}^-] + k''[\mathrm{RX}][\mathrm{GSH}] \qquad (1)$$

In the presence of a catalyst, Eq. (1) is modified by two additional terms

$$R_{\mathrm{cat}} = R + k'_{\mathrm{cat}}[\mathrm{catalyst}][\mathrm{RX}][\mathrm{GS}^-] + k''_{\mathrm{cat}}[\mathrm{catalyst}][\mathrm{RX}][\mathrm{GSH}] \qquad (2)$$

The elevated levels of GSH in safener-treated corn plants therefore lead to increased rates of nonenzymatic conjugations. Other assistance by the safener, however, in increasing the rate of the nonenzymatic conjugation may also occur. For example, Zama and Hatzios (1986) presented data showing that in the presence of the safener oxabetrinil (CGA-92194) the rate of nonenzymatic conjugation of GSH and metolachlor is increased in a concentration-dependent manner. The mechanism of the assistance by the safener is unknown. It may be general base catalysis by the cyano group of oxabetrinil (Komives *et al.*, 1977), but other catalytic mechanisms may also be operative.

Although the rate of the enzyme-catalyzed GSH conjugation may also be limited by the concentration of GSH, usually this rate is determined by the activity of the GST enzyme. Recent findings indicate that high levels of GST do not necessarily correspond to high rates of GSH conjugation. GST enzymes may be strongly inhibited by their reaction products, the GSH conjugates. Thus, the GSH conjugate of tridiphane was found to be a powerful inhibitor of corn GST, the GSH conjugate of propachlor was less effective, whereas the conjugate of atrazine was not inhibitory (Lamoureux and Rusness, 1986). In our laboratory, the GSH conjugate of acetochlor was found to be a strong inhibitor of corn GST, whereas the cysteine conjugate was much less effective (T. Komives and I. Jablonkai, unpublished results). Similarly, Lamoureux and Rusness (1986) found lower inhibitory activity by the degradation products of GSH conjugates. Therefore, the rate of enzymatic GSH conjugation may also be influenced by the activity of the enzymes that catalyze the decomposition of GSH conjugates to products less inhibitory to GST. It is not known whether safeners affect these decomposition processes or not.

In addition to processes that influence the synthesis and consumption of GSH, its redox state is also a major factor in determining GSH levels in tissues. GSH is known to act as an intracellular reductant, participating, via the ascorbate–dehydroascorbate redox system, in the detoxication of hydrogen peroxide (Dodge, 1982). As a result, GSH is converted to its disulfide (GSSG). High GSH/GSSG ratios in cells are maintained by reduction of GSSG to GSH by the NADPH-dependent enzyme, glutathione reductase (GR, EC 1.6.4.2) according to the reaction

$$\text{GSSG} + 2\text{NADPH} \xrightarrow{\text{GR}} 2\text{GSH} + 2\text{NADP}$$

Since oxidative stress of plant cells by herbicides, especially in photosynthesizing tissues, may result in phytotoxic symptoms because of membrane lipid peroxidation (Dodge, 1982; Lambert *et al.*, 1984), safeners may counteract such stress not only by increasing GSH levels, but also by enhancing GR enzyme activity. Increased activity of GR in dichlormid-treated corn suggests that this is a real possibility (Komives *et al.*, 1985b). GR may be a site of secondary action of herbicides that form GSH conjugates: inhibiton of mammalian GR enzyme by certain GSH conjugates has been published recently (Bilzer *et al.*, 1984). The interrelationships of the above GSH-mediated processes that may be influenced by safener treatments in corn are summarized in Fig. 1.

Cytosolic and microsomal GST enzymes in mammals may act as glutathione peroxidases (EC 6.4.11.6) by catalyzing the reaction between GSH and lipophilic hydroperoxides (Fig. 1; Reed, 1985), thereby protecting cell membrane polyunsaturated fatty acid moieties against peroxidation. Though not reported in corn, such an activity of its GST enzyme(s) may contribute to safener action.

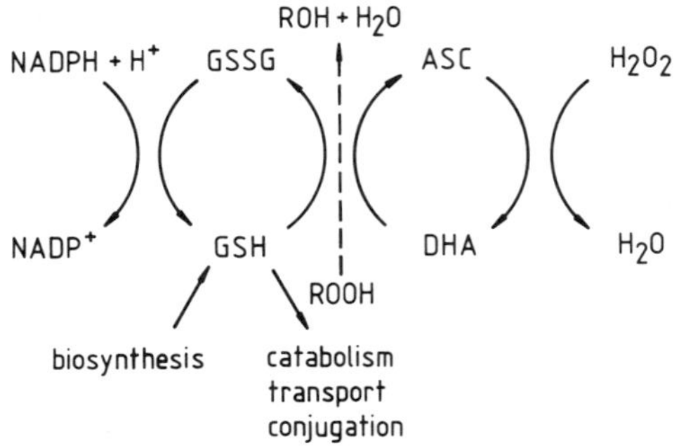

Fig. 1. An illustration of biochemical processes of corn facilitated by GSH.

B. Protein Sulfhydryl Content

Sulfhydryl groups of proteins (PR-SH) have important roles in regulating catalytic and other biological functions. At the active sites of enzymes, sulfhydryl groups may participate in substrate binding and chemical transformation, whereas other sulfhydryls of the protein backbone may influence protein structure (Reed, 1985).

Though several safeners are known to influence nonPR-SH levels in corn (see Section III,A), there are no published data on PR-SH responses to safener treatments. Here we present results of a study on the effects of amide safeners on PR-SH content of roots of 10-day old etiolated corn ('Pioneer 3747' hybrid) seedlings. Treatments were made as described earlier (Komives *et al.*, 1985b), and nonPR-SH and PR-SH levels were measured spectrophotometrically by the method of Sedlak and Lindsay (1968) using Ellman's reagent.

Exposure of corn seedlings to 10 μM dichlormid detectably increased the root PR-SH content. The observed changes occurred within 12 hr and reached 320% relative to the control levels (Fig. 2). PR-SH levels were changed in a concentration-dependent manner, increasing at low concentrations and reaching a maximum at 10 μM dichlormid (Fig. 3). The most effective compound studied at 10 μM in elevating root PR-SH content was dichlormid (Table II). These time course, dose–response, and limited structure–activity data show a close relationship between PR-SH and nonPR-SH levels of corn seedlings following safener treatment, and a good correlation of ability of a safener to elevate PR-SH and counteract thiocarbamate herbicide phytotoxicity.

Increased PR-SH contents caused by safener treatments may contribute to safening activity, because of (1) possible increases in levels of enzymes that contain sulfhydryl groups at their active site and are targets of herbicide inhibition (Rubin and Casida, 1985) and (2) higher levels of enzymes that catalyze herbicide detoxication or of ligandins (binding proteins) that trap herbicides and phytotoxic herbicide metabolites.

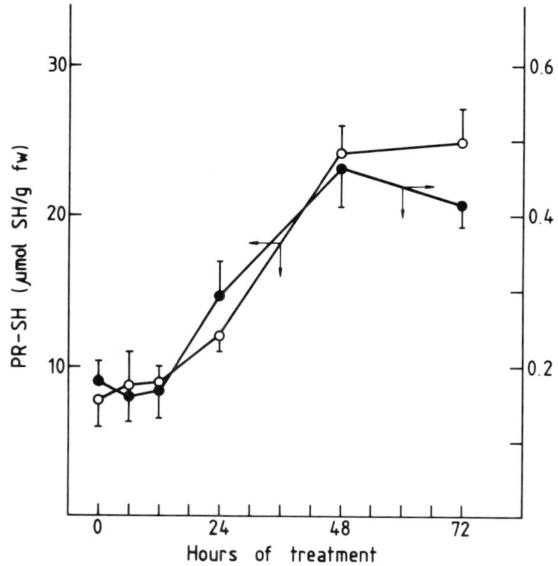

Fig. 2. Effect of exposure time on PR-SH and nonPR-SH content of roots of etiolated corn seedlings in 10 μ*M* dichlormid solutions. Results are expressed in GSH equivalents and error bars represent one SD from means of three separate experiments. Arrows indicate the coordinate axes used to plot each curve.

IV. OTHER SELECTED ENZYMES

A. Acetohydroxyacid Synthase

Sulfonylurea and imidazolinone herbicides are potent inhibitors of the enzyme acetohydroxyacid synthase (AHAS, EC 4.1.3.18) and of the biosynthesis of branched chain amino acids (LaRossa and Falco, 1984). AHAS catalyzes the condensation of pyruvate and 2-ketobutyrate as well as the self-condensation of pyruvate to yield intermediates for the ultimate biosynthesis of isoleucine, leucine, and valine as follows

$$CH_3\text{–CO–COOH} + RCH_2\text{–CO–COOH} \xrightarrow{\text{AHAS}} CH_3\text{–CO–C}(CH_2R)(OH)\text{–COOH}$$

The phytotoxicity of the above classes of herbicides to corn may be reduced by safeners (Hatzios, 1983a, 1984; Rubin and Casida, 1985; Sweetser, 1985; Barrett and Olson, 1986; Frear *et al.*, 1987). Dichlormid, for example, provided corn with protection against injury from chlorsulfuron, but it did not modify the inhibition of the AHAS enzyme by the herbicide *in vitro*. Pretreatment of corn

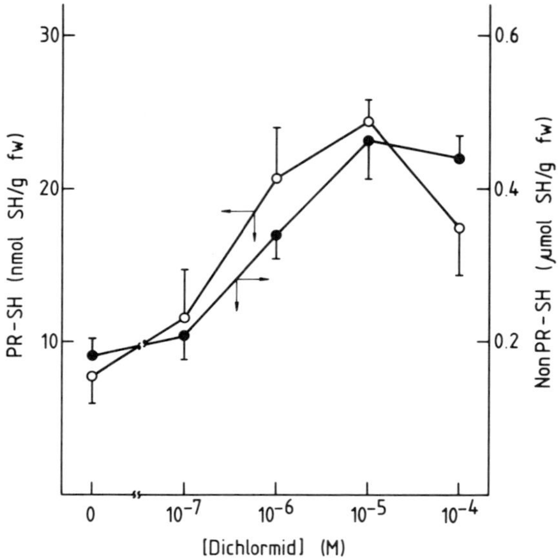

Fig. 3. Effect of dichlormid concentration on PR-SH and nonPR-SH content of roots of etiolated corn seedlings after 48 hr of exposure. Results are expressed in GSH equivalents and error bars represent one SD from means of three separate experiments. Arrows indicate the coordinate axes used to plot each curve.

with dichlormid significantly increased AHAS activity both in the roots and in the shoots. The *in vitro* response of the safener-induced enzyme to chlorsulfuron was not altered. It was suggested that dichlormid increased the biosynthesis of AHAS, and that the elevation in AHAS activity is sufficient to partially compensate for inhibition by chlorsulfuron (Rubin and Casida, 1985). Related studies on the metabolism of sulfonylureas (Sweetser, 1985) and on effects of imidazolinones on AHAS (Barrett and Olson, 1986), however, suggest that safeners protect corn against injury by these herbicides by increasing their rate of metabolic detoxication.

Table II PR-SH and nonPR-SH Content[a] of Roots of 10-Day-Old Etiolated Corn Seedlings ('Pioneer 3747' Hybrid) Treated with 10 μM of Safeners for 48 hr

Safener	Safening activity[b]	PR-SH (nmol/g fresh wt)	nonPR-SH (nmol/g fresh wt)
$H_3C{-}CO{-}N(CH_2{-}CH{=}CH_2)_2$	Inactive	8.0 ± 1.2	160 ± 32
$H_2ClC{-}CO{-}N(CH_2{-}CH{=}CH_2)_2$	Moderate	11.1 ± 1.6	299 ± 43
$Cl_2HC{-}CO{-}N(CH_2{-}CH{=}CH_2)_2$	Superior	24.3 ± 1.5	463 ± 52
$Cl_3C{-}CO{-}N(CH_2{-}CH{=}CH_2)_2$	Moderate	9.4 ± 1.4	230 ± 35

[a]Expressed in GSH equivalents.

[b]Data from Lay and Casida (1976).

B. Peroxidase

Peroxidase (PO, EC 1.11.1.7) activity in plants is shown by a number of isozyme hemoproteins. PO enzymes have a potential role in plant development, disease resistance, and wound healing, and act as catalysts in the indoleacetic acid oxidation, the hydroxylation of proline, and lignification by oxidizing their substrates utilizing hydrogen peroxide (Butt, 1980), as shown in the reaction

$$RH_2 + H_2O_2 \xrightarrow{PO} 2H_2O + R$$

A study of EPTC effects on corn seedlings revealed that EPTC stimulates PO activity in this plant and enhances lignin deposition. These effects of the herbicide were annulled by dichlormid (Harvey *et al.*, 1975). Interestingly, treatments with the safener alone decreased PO levels by 20 to 30%, but had no effect when applied *in vitro*. Since this phenomenon has not been investigated further, the relationship between PO levels and dichlormid action is unknown.

V. CONCLUDING REMARKS

Our knowledge of the effects of herbicide safeners on the levels and activity of enzymes involved in herbicide detoxication or that are target sites of herbicide action has expanded greatly in recent years. The key role of the GSH conjugation system in safening against chloroacetanilide herbicides has been clearly established by several approaches, including molecular biology (Wiegand *et al.*, 1986). The participation of this system in safener action against thiocarbamate herbicides is also well known, but its relative importance to safener effects on the sulfoxidation of these herbicides is not clear. Recent studies demonstrated synergistic interactions between thiocarbamate herbicides and tridiphane, an inhibitor of GST (Lamoureux and Rusness, 1986), on corn (Ezra and Stephenson, 1985; T. Komives, unpublished results), indicating that GST plays an active role in thiocarbamate detoxication in this species. Thus, a major question needing further clarification is the possible role of oxidative enzymes in safener action against thiocarbamates and those herbicides against which GSH and the GSH conjugation system cannot provide protection. *In vitro* studies with plant cell-free systems to detect covalent binding of safeners to cytochrome P-450, as well as to follow safener-induced changes in the oxidative transformation of herbicides, provide us with intriguing challenges for further research.

References

Adams, C. A., Blee, E., and Casida, J. E. (1983). Dichloroacetamide herbicide antidotes enhance sulfate metabolism in corn roots. *Pestic. Biochem. Physiol.* **19**, 350–360.

Barrett, M., and Olson, G. L. (1986). Protection from imazaquin injury: Studies on the mode of action of herbicide protectants. *IUPAC Int. Congr. Pestic. Chem. 6th* Abstr., No. 3D-14.

Bilzer, M., Krauth-Siegel, L., Schirmer, H., Akerboom, T. P. M., Sies, H., and Schultz, G. E. (1984). Interactions of glutathione S-conjugate with glutathione reductase. *Eur. J. Biochem.* **138**, 373–378.

Blee, E., and Durst, F. (1986). Oxidation of an organosulfur xenobiotic from soybean cotyledons. *Biochem. Biophys. Res. Commun.* **135**, 922–927.

Blee, E., Casida, J. E., and Durst, F. (1985). Sulfoxidation of methiocarb in higher plants. *Biochem. Pharmacol.* **34**, 389–390.

Breaux, E. J., Patanella, J. E., and Sanders, E. F. (1987). Chloroactanilide herbicide selectivity: Analysis of glutathione and homoglutathioine in tolerant, susceptible, and safened seedlings. *J. Agric. Food Chem.* **35**, 474–478.

Butt, V. S. (1980). Direct oxidases and related enzymes. *In* "The Biochemistry of Plants" (D. D. Davies, ed.), Vol. 2, pp. 81–123. Academic Press, New York.

Cabanne, P., Gaillardon, P., Scalla, R., and Durst, F. (1985). Aminobenzotriazole as a synergist of urea herbicides. *Proc. Br. Crop Prot. Conf.—Weeds* **3**, 1163–1170.

Carringer, R. D., Rieck, R. C. E., and Bush, B. P. (1978). Effect of R-25788 on EPTC metabolism in corn (*Zea mays*). *Weed Sci.* **26**, 167–171.

Cole, D. (1983). Oxidation of xenobiotics in plants. *Prog. Pestic. Biochem. Toxic.* **3**, 199–254.

Dodge, A. D. (1982). The role of light and oxygen in the action of photosynthetic inhibitor herbicides. *In* "Biochemical Responses Induced by Herbicides" (D. E. Moreland, J. B. St. John, and F. D. Hess, eds.), pp. 57–77. Am. Chem. Soc., Washington, DC.

Dutka, F., and Komives, T. (1983). On the mode of action of EPTC and its antidotes. *In* "Pesticide Chemistry: Human Welfare and the Environment" (J. Miyamoto and P. C. Kearney, eds.), pp. 213–218. Pergamon, Oxford.

Dutka, F., and Komives, T. (1987). MG-191: A new, selective herbicide antidote. *In* "Pesticide Science and Biotechnology" (R. Greenhalgh and T. R. Roberts, eds.), pp. 201–204. Blackwell, Oxford.

Edwards, R., and Owen, W. J. (1986). Comparison of glutathione *S*-transferases of *Zea mays* responsible for herbicide detoxication in plants and suspension cultured cells. *Planta* **169**, 208–215.

Ezra, G., and Gressel, J. (1982). Rapid effects of a thiocarbamate herbicide and its dichloroacetamide protectant on macromolecular syntheses and glutathione levels in maize cell cultures. *Pestic. Biochem. Physiol.* **17**, 48–52.

Ezra, G., and Stephenson, G. R. (1985). Comparative metabolism of atrazine and EPTC in proso millet (*Panicum miliaceum* L.) and corn. *Pestic. Biochem. Physiol.* **24**, 207–212.

Ezra, G., and Stephenson, G. R. (1986). The physiology of metazachlor and safener 145 138 interactions in corn. *IUPAC Int. Congr. Pestic. Chem., 6th* Abstr. No. 3D-12.

Ezra, G., Rusness, D. G., and Lamoureux, G. R. (1985). The effects of CDAA (*N,N*-diallyl-2-chloroacetamide) pretreatments on subsequent CDAA injury to corn (*Zea mays* L.). *Pestic. Biochem. Physiol.* **23**, 108–115.

Fedtke, C. (1981). Wirkung von herbiziden und herbizid-antidots auf den glutathiongehalt in mais und sojabohnenwurzeln. *Z. Pflanzenr. (Pflanzenpathol.) Pflanzenchutz, Sonderh.* **9**, 141–146.

Fedtke, C., and Trebst, A. (1987). Advances in understanding herbicide modes of action. *In* "Pesticide Science and Biotechnology" (R. Greenhalgh and T. R. Roberts, eds.), pp. 161–168. Blackwell, Oxford.

Frear, D. S., Swanson, H. R., and Mansager, E. R. (1987). 1,8-Naphthalic anhydride/auxin protection against chlorsulfuron inhibition of corn seedling growth. *In* "Pesticide Science and Biotechnology" (R. Greenhalgh and T. R. Roberts, eds.), pp. 499–503. Blackwell, Oxford.

Gronwald, J. W., Fuerst, E. P., Eberlein, C. V., and Egli, M. A. (1987). Effect of herbicide antidotes on glutathione content and glutathione *S*-transferase activity of sorghum shoots. *Pestic Biochem. Physiol.* **29**, 66–76.

Habig, W. H. (1983). Glutathione *S*-transferases: Versatile enzymes of detoxication. *In* "Radioprotectors and Anticarcinogens" (O. F. Nygaard and M. G. Simic, eds.), pp. 169–190. Academic Press, New York.

Hagmann, M. L., Heller, W., and Grisebach, H. (1983). Induction and characterization of a microsomal flavonoid 3′hydroxylase from parsley cell cultures. *Eur. J. Biochem.* **134**, 547–555.

Halpert, J. E., Miller, N. E., Gorsky, L. D., and Kaminsky, L. S. (1985). Suicide inactivation of rat liver cytochrome P-45O by chloramphenicol: Mechanism, isozyme-selectivity, and structural requirements. *In* "Cytochrome P-450: Biochemistry, Biophysics, and Induction" (L. Vereczkey and K. Magyar, eds.), pp. 163–166. Akadémiai Kiadó, Budapest.

Halpert, J. E., Balfour, C., Miller, N. E. , and Kaminsky, L. S. (1986). Dichloromethyl compounds as mechanism-based inactivators of rat liver cytochrome P-450 *in vitro*. *Mol. Pharmacol.* **30**, 19–24.

Harvey, B. M. R., Chang, F. Y., and Fletcher, R. A. (1975). Relationship between *S*-ethyl dipropylthiocarbamate injury and peroxidase activity in corn seedlings. *Can. J. Bot.* **53**, 225–230.

Hatzios, K. K. (1983a). Herbicide antidotes: Development, chemistry, and mode of action. *Adv. Agron.* **36**, 265–316.

Hatzios, K. K. (1983b). Effects of CGA-43089 on response of sorghum (*Sorghum bicolor*) to metolachlor combined with ozone and antioxidants. *Weed Sci.* **31**, 280–284.

Hatzios, K. K. (1984). Interactions between selected herbicides and protectants in corn (*Zea mays*). *Weed Sci.* **32**, 51–58.

Higashi, K., Karasaki, Y., and Nakashima, K. (1982). Possible multiplicity of microsomal cytochrome P-450 in higher plants. *In* "Microsomes, Drug Oxidations and Drug Toxicity" (R. Sato and R. Kato, eds.), pp. 103–104. Jpn. Sci. Soc. Press, Tokyo.

Hilton, J. L., and Pillai, P. (1986). L-2-Oxothiazolidine-4-carboxylic acid protection against tridiphane toxicity. *Weed Sci.* **34**, 669–675.

Ketchersid, M. L., Plapp, F. W., and Merkle, M. G. (1985). Sorghum (*Sorghum bicolor*) seed safeners as insecticide synergists. *Weed Sci.* **33**, 774–778.

Kirino, O., Rubin, B., and Casida, J. E. (1985). *N*-Phenyl-maleimides, -isomaleimides and -maleamic acids as selective herbicide antidotes. *Agric. Biol. Chem.* **49**, 267–268.

Komives, T., and Dutka, F. (1980). Mode of action of EPTC-antidotes on corn. *Cereal Res. Commun.* **8**, 627–633.

Komives, T., Marton, A. F., and Dutka, F. (1977). Effect of ortho substitution on the aminolysis of active esters in aprotic solvents. *Z. Naturforsch., B: Anorg. Chem., Org. Chem. 32B*, 1359–1360.

Komives, T., Balazs, M., Komives, A. V., and Dutka, F. (1985a). Effects of herbicide antidotes on reduced glutathione content and levels of cytochrome P-450 and glutathione *S*-transferase in *Zea mays* L. *In* "Cytochrome P-450: Biochemistry, Biophysics and Induction" (L. Vereczkey and K. Magyar, eds.), pp. 451-454. Akadémiai Kiadó, Budapest.

Komives, T., Komives, V. A., Balazs, M., and Dutka, F. (1985b). Role of glutathione-related enzymes in the mode of action of herbicide antidotes. *Proc. Br. Crop Prot. Conf.—Weeds.* **3**, 1155–1162.

Lambert, R., Kroneck, P. M. H., and Boger, P. (1984). Radical formation and peroxidative activity of phytotoxic diphenyl ethers. *Z. Naturforsch C: Biosci.* **39C**, 486–491.

Lamoureux, G. L., and Rusness, D. (1986). Tridiphane (2-[3,5-dichlorophenyl]-2-[2,2,2-trichloroethyl] oxirane), an atrazine synergist: Enzymatic conversion to a potent glutathione *S*-transferase inhibitor. *Pestic. Biochem. Physiol.* **26**, 323–342.

Lamoureux, G. L., Shimabukuro, R. H., Swanson, H. R., and Frear, D. S. (1970). Metabolism of 2-chloro-4-ethylamino-6-isopropyl-amino-*s*-triazine (atrazine) in excised sorghum leaf sections. *J. Agric. Food Chem. 18*, 81–86.

LaRossa, R. A., and Falco, S. C. (1984). Amino acid biosynthetic enzymes as targets of herbicide action. *Trends Biotechnol.* **2**, 158–161.

Larson, R. L., and Bussard, J. B. (1986). Microsomal flavonoid 3'-monooxygenase from maize seedlings. *Plant Physiol.* **80**, 483–486.

Law, M. Y., and Halliwell, B. (1986). Purification and properties of glutathione synthetase from spinach (*Spinacia oleracea*) leaves. *Plant Sci.* **43**, 185–191.

Lay, M.-M., and Casida, J. E. (1976). Dichloroacetamide antidotes enhance thiocarbamate sulfoxide detoxification by elevating corn root glutathione content and glutathione *S*-transferase activity. *Pestic. Biochem. Physiol.* **6**, 442–456.

Lay, M.-M., and Niland, A. M. (1985). Biochemical response of inbred and hybrid corn (*Zea mays*) to R-25788 and its distribution with EPTC in corn seedlings. *Pestic. Biochem. Physiol.* **23**, 131–140.

Leavitt, J. R. C., and Penner, D. (1979). *In vitro* conjugation of glutathione and other thiols with acetanilide herbicides and EPTC sulfoxide and the action of the herbicide antidote R-25788. *J. Agric. Food Chem.* **27**, 533–536.

Miaullis, J. B., Thomas, V. M., Gray, R. A., Murphy, J. J., and Hollingworth, R. M. (1978). Metabolism of R-25788 (*N,N*-diallyl-2,2-dichloroacetamide) in corn plants, rats and soil. *In* "Chemistry and Action of Herbicide Antidotes" (J. E. Casida and F. M. Pallos, eds.), pp. 109–131. Academic Press, New York.

Miller, N. E., and Halpert, J. (1986). Analogues of chloramphenicol as mechanism-based inactivators of rat liver cytochrome P-450: Modification of the propanediol side chain, the *p*-nitro-group, and the dichloromethyl moiety. *Mol. Pharmacol.* **29**, 391–398.

Mozer, T. J., Tiemeier, D. C., and Jaworski, E. G. (1983). Purification and characterization of corn glutathione *S*-transferase. *Biochemistry* **22**, 1068–1072.

Reed, D. J. (1985). Cellular defense mechanisms against reactive metabolites. *In* "Bioactivation of Foreign Compounds" (M. W. Anders, ed.), pp. 71–108. Academic Press, New York.

Reichhart, D., Salaun, J.-P., Benveniste, I., and Durst, F. (1979). Induction by manganese, ethanol, phenobarbital, and herbicides of microsomal cytochrome P-450 in higher plants. *Arch. Biochem. Biophys.* **196**, 301–303.

Rennenberg, H. (1982). Glutathione metabolism and possible biological roles in higher plants. *Phytochemistry* **21**, 2771–2781.

Rubin, B., and Casida, J. E. (1985). R-25788 effects on chlorsulfuron injury and acetohydroxyacid synthase activity. *Weed Sci.* **33**, 462–468.

Sedlak, J., and Lindsay, R. H. (1968). Estimation of total, protein-bound, and non-protein sulfhydryl groups in tissues with Ellman's reagent. *Anal. Biochem.* **25**, 192–205.

Shimabukuro, R. H. (1985). Detoxication of herbicides. *In* "Weed Physiology" (S. O. Duke, ed.), Vol. 2, pp. 215–240. CRC Press, Boca Raton, Florida.

Shimabukuro, R. H., Swanson, H. R., and Walsh, W. C. (1970). Glutathione conjugation: Atrazine detoxication in corn. *Plant Physiol.* **46**, 103–107.

Stephenson, G. R., and Ezra, G. (1985). Use of subtoxic herbicide pretreatments to improve crop tolerance to herbicides. *ACS Symp. Ser.* **276**, 69–84.

Stephenson, G. R., Ali, A., and Ashton, F. M. (1983). Influence of herbicides and antidotes on the glutathione levels of maize seedlings. *In* "Pesticide Chemistry: Human Welfare and the Environment" (J. Miyamoto and P. C. Kearney, eds.), pp. 219–224. Pergamon, Oxford.

Sweetser, P. B. (1985). Safening of sulfonylurea herbicides to cereal crops: Mode of herbicide antidote action. *Proc. Br. Crop Prot. Conf.—Weeds.* **3**, 1147–1154.

Taft, L. G. (1975). A study of the mechanism of action of the thiocarbamate protectant R-25788. *Diss. Abstr. Int. B* **36**, 5387.

West, C. A. (1980). Hydroxylases, monooxygenases, and cytochrome P-450. *In* "The Biochemistry of Plants" (D. D. Davies, ed.), Vol. 2, pp. 317–364. Academic Press, New York.

Wiegand, R. C., Shah, D. M., Mozer, T. J., Harding, E. I., Diaz-Collier, J., Saunders, C., Jaworski, E. G., and Tiemeier, D. C. (1986). Messenger RNA encoding a glutathione *S*-transferase responsible for herbicide tolerance in maize is induced in response to safener treatment. *Plant Mol. Biol.* **7**, 235–243.

Wilkinson, R. E. (1983). Gibberellin precursor biosynthesis inhibition by EPTC and reversal by R-25788. *Pestic. Biochem. Physiol.* **19**, 321–329.

Wilkinson, R. E. (1985). CDAA inhibition of kaurene oxidation in etiolated sorghum coleoptiles. *Pestic. Biochem. Physiol.* **23**, 19–23.

Zama, P., and Hatzios, K. K. (1986). Effects of CGA-92194 on the chemical reactivity of metolachlor with glutathione and metabolism of metolachlor in grain sorghum (*Sorghum bicolor*). *Weed Sci.* **34**, 834–841.

Chapter 6

Mechanisms of Action of Dichloroacetamide Safeners

Gail Ezra
PPG Industries, Inc.
P.O. Box 31
Barberton, Ohio 44203

Gerald R. Stephenson
Department of Environmental Biology
University of Guelph
Guelph, Ontario N1G 2W1, Canada

I. INTRODUCTION

In the past several years, attention has been focused on the potential of certain chemicals to confer tolerance to crops from herbicide injury. In some cases, these chemicals are structurally related to the herbicides themselves and have been termed antidotes, protectants, or safeners.

Herbicide safeners have been the subject of several comprehensive reviews (Pallos and Casida, 1978; Hatzios, 1983b; Stephenson and Ezra, 1983). The purpose of this chapter is to review the activities of one class of safeners, the chloroacetamides, particularly with regard to their involvement in enhanced herbicide metabolism. The correlation of observed physiological and biochemical responses and overall *in vivo* protection will be discussed.

The most extensively studied dichloroacetamide safener to date has been dichlormid (also known as R-25788 and DDCA), which was developed in the early 1970s by Stauffer Chemical Company to protect corn (*Zea mays* L.) from thiocarbamate herbicide injury (Pallos *et al.*, 1972, 1978) (Fig. 1). At that time,

Crop Safeners for Herbicides

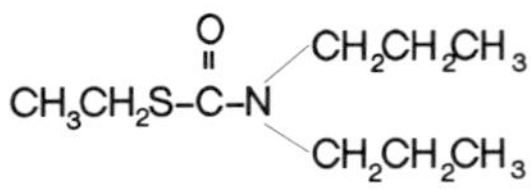

EPTC

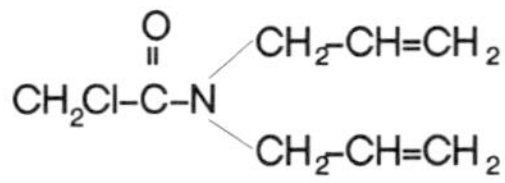

CDAA

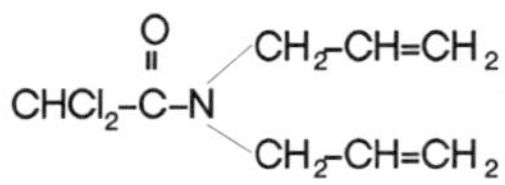

DICHLORMID

Fig. 1. Chemical structures of EPTC, CDAA, and dichlormid.

dichlormid was novel in that it protected corn, but not weeds, from herbicide injury. It could, therefore, be applied as a tank mixture together with the herbicide (Chang *et al.*, 1972).

Since the initial studies with thiocarbamate herbicides, dichlormid has been reported to protect corn from barban (Blair, 1978), perfluidone (Blair and Dean, 1976), and acetanilide herbicide injury (Leavitt and Penner, 1978a,b). Dichlormid also provices partial protection to corn from linuron (Chang *et al.*, 1973a), chlorsulfuron (Parker *et al.*, 1980), and most recently, imazaquin (Barrett and Olson, 1986) injury. In addition to corn, dichlormid partially protected grain sorghum [*Sorghum bicolor* (L.) Moench], wheat (*Triticum aestivum* L.), and rice (*Oryza sativa* L.) from several thiocarbamate and acetanilide herbicides (Parker and Dean, 1976; Miller *et al.*, 1978; Miller and Nalewaja, 1980); barley (*Hordeum vulgare* L.) from barban, chlorsulfuron, and perfluidone injury (Blair, 1978; Parker *et al.*, 1980); and sorghum from imazaquin injury (Barrett and Olson, 1986).

A new dichloroacetamide safener, BAS 145 138 (1-dichloro-acetyl-hexahydro-3,3,8α-trimethyl-pyrrolo-[1,2-α]-pyrimidin-6-(2*H*)-one) (Fig. 2), is now being developed by BASF to protect corn from injury by another acetanilide herbicide, metazachlor (Wuerzer *et al.*, 1983). Studies with BAS 145 138 may provide further insights into the mode of action of dichloroacetamide safeners.

Like dichlormid, BAS 145 138 is quite selective for corn. It did provide some protection to certain sorghum varieties (Elker and Stephenson, 1987), but did not

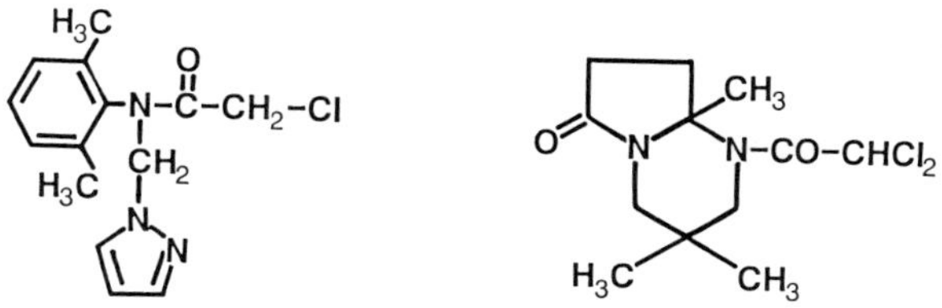

METAZACHLOR BAS 145 138

Fig. 2. Chemical structures of metazachlor and safener BAS 145 138.

protect proso millet (*Panicum miliaceum* L.) from metazachlor injury (Ezra and Stephenson, 1986). BAS 145 138 was also very effective in protecting corn from injury by other acetanilide herbicides such as alachlor and metolachlor, when combined with the herbicide at a ratio of 1 : 12 (safener to herbicide). However, it provided only partial protection to corn from thiocarbamate (EPTC) herbicide injury when applied preemergence (Ezra and Stephenson, 1986).

II. THEORIES ON THE MODE OF ACTION OF DICHLORMID

There has been, and still is, much controversy surrounding the mode of action of dichlormid. Early studies with dichlormid indicated that it did not act by inhibiting the uptake and/or translocation of EPTC in intact plants (Chang *et al.*, 1974). The question was therefore, whether dichlormid acted at a site of herbicide action or whether protection was the result of enhanced metabolism.

A. Effects of Dichlormid on Lipid and Gibberellic Acid Synthesis

Although EPTC was reported to inhibit RNA synthesis (Beste and Schreiber, 1972) and stimulate peroxidase (EC 1.11.1.7) production and lignin deposition (Harvey *et al.*, 1975), extensive research indicated that its primary mechanism of action involved inhibition of lipid and gibberellic synthesis (Still *et al.*, 1970; Wilkinson, 1978; Wilkinson and Ashley, 1979). Wilkinson and Smith (1975) showed that dichlormid prevented or reversed EPTC-induced inhibition of fatty acid synthesis in isolated spinach (*Spinacia oleracea* L.) chloroplasts. Wilkinson (1983) also showed that dichlormid prevented EPTC-induced inhibition of [^{14}C]mevalonate incorporation into kaurene in a cell-free extract of sorghum. Later, Wilkinson expanded his hypothesis to suggest that dichlormid prevented the inhibition of terpenoid biosynthesis caused by thiocarbamate and chloroacetanilide herbicides (see Chapter 10, this volume).

Leavitt and Penner (1979a), working with corn, showed that dichlormid prevented EPTC-induced inhibitions of epicuticular wax aggregation and proposed this as a major mechanism. The results of Görög *et al.* (1982) supported this hypothesis for the mechanism of action of dichlormid and other dichloroacetamide safeners.

B. Structure–Activity Relations

Noting the similarity in structure between EPTC and dichlormid (Fig. 1), Stephenson and co-workers (1978, 1979) proposed that dichlormid acted as a herbicidally inactive competitive inhibitor of EPTC at a site of toxic action in the plant. Using a soil-free bioassay, they showed that compounds (including dichlormid) with a structure similar to that of EPTC were active as safeners for the herbicide in corn. The most active safener for EPTC was the one most similar to it with respect to the N,N-disubstituted alkyl groups (Stephenson *et al.*, 1978, 1979), although dichlormid still had superior activity in the soil.

Stephenson *et al.* (1983) also observed that in two different series of structurally related chloroacetamide safeners for EPTC, safener activity was directly related to the elevation of glutathione (GSH) levels in corn roots. Furthermore, herbicides such as EPTC, alachlor, propachlor, and barban, which are all metabolized to GSH conjugates (Lamoureaux and Rusness, 1986a), also elevated GSH levels in corn seedlings. On the basis of these studies, they proposed that similar molecules may also act to elevate the substrates or enzymes required to detoxify the herbicide.

C. Effects of Dichlormid on Herbicide Metabolism

Detoxication of EPTC was shown to involve its sulfoxidation via a microsomal/NADPH system followed by conjugation to glutathione (Casida *et al.*, 1975). Casida *et al.* (1974) suggested that the sulfoxide formed was, paradoxically, both the herbicidally active compound and the first step in detoxication of the herbicide. Toxicity of the sulfoxide was attributed to its carbamoylation of physiologically important tissue thiols such as coenzyme A and GSH (Lay *et al.*, 1975). Conjugation of the sulfoxide to GSH was suggested to protect more sensitive thiol sites and render the herbicide inactive, the GSH–sulfoxide conjugate formed being subsequently broken down to other nontoxic, partially volatile components.

Lay and Casida (1976) demonstrated that corn pretreated with dichlormid had higher GSH levels and glutathione-*S*-transferase (GST, EC 2.5.1.18) activity, and metabolized EPTC sulfoxide faster than nonpretreated plants. From these observations, they proposed that dichlormid protection was the result of enhanced EPTC metabolism through elevations in GSH levels and GST activity as shown in Fig. 3.

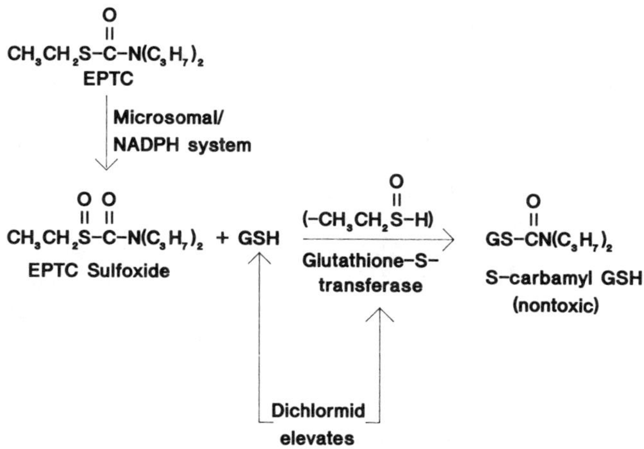

Fig. 3. Pathway for detoxication of EPTC and activity of dichlormid as proposed by Lay and Casida (1976).

Many questions regarding the mode of action of both EPTC and dichlormid were raised as a result of these and other studies. Was EPTC or EPTC sulfoxide the herbicidally active moiety? Was conjugation of EPTC sulfoxide to GSH enzymatic or nonenzymatic? How relevant were increases in GSH to overall protection? Did dichlormid protect corn from acetanilide and thiocarbamate herbicides by the same or different mechanisms?

Komives and Dutka (1980) confirmed the role of GSH in detoxication of EPTC, but questioned the role of EPTC sulfoxide in phytotoxicity. Using monooxygenase inhibitors, they demonstrated EPTC toxicity without conversion to the sulfoxide, suggesting that this conversion was not a prerequisite for phytotoxicity, but only for degradation. In addition, they showed that EPTC was indeed more phytotoxic than its corresponding sulfoxide, in contrast to the results of Casida *et al.* (1974).

Horváth and Pulay (1980) challenged the proposal that GSH was carbamoylated by EPTC sulfoxide and proposed EPTC sulfone as the true carbamoylating agent. Their work indicated that EPTC sulfoxide reacted with GSH only when EPTC sulfone was present. They suggested that two oxidation enzymes were responsible for oxidation of EPTC, but that the enzyme system converting sulfoxide to sulfone was destroyed when corn tissues were homogenized. In addition, they observed that both the thiol carbamoylating reactions and sulfone hydrolysis were much quicker than oxidation reactions to EPTC sulfone and reasoned that these observations had given rise to unsuccessful detection of EPTC sulfone in *in vivo* metabolism studies.

Carringer *et al.* (1978a) and Leavitt and Penner (1979b) confirmed EPTC sulfoxide conjugation to GSH, but could not detect GST activity associated with sulfoxide cleavage. On the other hand, elevation in GSH levels alone could not be sufficient to account for safening by dichlormid. Fedtke (1981) showed that a number of herbicides and plant growth regulators significantly increased GSH levels in corn and soybeans without protecting either from EPTC injury.

D. Cell Culture Studies with Dichlormid

Ezra and Gressel (1982) used corn cell cultures to study the rapid effects of EPTC and dichlormid and to follow the sequence of events which could contribute to protection. They found that dichlormid could competitively inhibit EPTC uptake within 30 min, whereas dissimilar molecules did not affect EPTC uptake (Ezra *et al.*, 1982). This inhibition was, however, greatest when dichlormid was given at concentrations higher than EPTC. Although this did not reflect the field ratios of 1:12 safener to herbicide, they argued that differential uptake of dichlormid and EPTC in intact plants may result in more equitable cellular distributions. In a later study, Lay and Niland (1985) found that hybrid corn seedlings absorbed dichlormid more readily from soil than EPTC, despite the fact that the initial soil concentration of dichlormid was 12-fold less than EPTC.

Work with corn cell cultures pointed to lipid synthesis as another early site of action of EPTC and dichlormid (Ezra and Gressel, 1982). Both compounds had rapid but opposite effects which could be observed 1–4 hr after treatment. Specifically, EPTC inhibited [^{14}C]acetate incorporation into fatty acids, inhibited the synthesis of both polar and neutral lipids, and altered the proportion of neutral lipids in the cell fraction. Dichlormid, on the other hand, stimulated [^{14}C]acetate incorporation, particularly into unsaturated fatty acids, and stimulated synthesis of neutral lipids without altering their proportion in the cell. Dichlormid did not affect polar lipid synthesis.

When given together with EPTC, dichlormid partially reversed EPTC inhibition of fatty acid and neutral lipid synthesis, but could not counteract the EPTC-induced alteration in proportions of the neutral lipid components, nor EPTC inhibition of the polar lipid fraction. From these data it seemed unlikely that dichlormid acted as a direct competitive inhibitor of EPTC at a site of lipid synthesis (Ezra *et al.*, 1983).

Elevations in GSH levels of cell cultures were observed on treatment with dichlormid, but this occurred only after 12 hr, and reached a maximum after 48 hr (Ezra and Gressel, 1982).

From these studies it was concluded that all the dichlormid effects, occurring at different times, contributed to overall safening. However, Ezra and Gressel (1982) did question the timing required for enhancement of GSH and GST activity, compared with the rate of degradation of EPTC. EPTC could be degraded to water-soluble products within 8 hr, whereas 1–2 days were needed for elevation of root GST activity, and up to 7 days for shoot GST activity (Lay and Casida, 1976; Ezra *et al.*, 1985a).

E. Effects of Dichlormid on Monooxygenase Enzymes

Another quite different mode of action of dichlormid has also been alluded to the elevation of, or interference with, the activity of monooxygenase (MFO, EC 1.14.14.1) enzymes. Leavitt and Penner (1979b) suggested that sulfoxidation of EPTC, thought to be mediated by monooxygenase enzymes, was a rate-limiting step in EPTC degradation, whereas conjugation to GSH was not. They proposed that dichlormid acted to enhance sulfoxidation, presumably by stimulating the activity of the monooxygenases.

The involvement of monooxygenase enzymes in dichlormid mode of action has been investigated only indirectly, through the use of monooxygenase inhibitors and insecticide synergists such as piperonyl butoxide and SKF-525A. Using these inhibitors, Dutka and Komives (1983) synergized the activity of EPTC, presumably by preventing sulfoxidation and hence detoxication. Hatzios (1981, 1983a) also demonstrated synergism of EPTC by the antioxidant propyl gallate and the herbicide tebuthiuron, even in the presence of dichlormid. Sweetser (1985) suggested that enhanced chlorsulfuron metabolism, which occurred in corn following dichlormid pretreatment, was also the result of increased monooxygenase activity.

Recently, Fedtke and Trebst (1987), reviewing the mechanism of action of herbicide safeners, speculated further on the involvement of the monooxygenase enzymes in the safening response. They agreed that enhanced monooxygenase activity could account for increased sulfonylurea herbicide metabolism following safener pretreatment, but doubted that it was involved in enhanced metabolism of thiocarbamate and acetanilide herbicides. Rather, they suggested that these herbicides acted by inhibiting monooxygenase activity, and that dichlormid treatment resulted in a shift in monooxygenase herbicide target sites leading to higher tolerance after safener application.

Direct studies on monooxygenase enzyme involvement in dichlormid mode of action have not been reported, possibly because of the difficulties involved in isolating, stabilizing and assaying these enzymes. More work will be needed to confirm their involvement in dichlormid action, and represents a challenging area for future research.

F. Other Effects of Dichlormid

In addition to stimulating GST activity, dichlormid has been reported to increase the activity of other enzymes. Carringer *et al.* (1978b) reported the *in vitro* stimulation of glutathione synthase (EC 6.3.2.3), while Adams *et al.* (1983) reported increases in the specific activity of ATP sulfurylase (EC 2.7.7.4), which is the first enzyme involved in sulfate assimilation. Recently, Rubin and Casida (1985) reported dichlormid stimulation of acetohydroxyacid synthase (AHAS; EC 4.1.3.18). They suggested that elevation in AHAS activity following dichlormid treatment may be sufficient to counteract chlorsulfuron inhibition of this enzyme. Barrett and Olson (1986), working on protection of corn from imazaquin injury, also reported a reversal of AHAS inhibition *in vivo*, but suggested that variation in enzyme activity did not necessarily correlate to corn hybrid sensitivity to imazaquin. Rather, they attributed protection from imazaquin injury to enhanced metabolism of imazaquin in safener-treated plants.

III. SUBTOXIC HERBICIDE PRETREATMENTS

Once the structure of dichlormid was released, it was evident that not only did it resemble EPTC but it was, in fact, a close structural analog of the chloroacetamide herbicide CDAA (Fig. 1). Chang *et al.* (1973b) were able to show that CDAA could also protect corn from EPTC injury, although it was not as effective as dichlormid.

The concept of herbicide selectivity and crop tolerance is relative, depending in part on such factors as rate of herbicide application and plant age. Now, the line demarking herbicide and safener was becoming blurred. Here was a chemical, CDAA, that could be a herbicide or protectant, depending on the crop in which it was used and the rate at which it was applied. The question arose as to how these effects could be occurring. Were low levels of chemicals elevating the pathways for metabolism? This had already been proposed by Stephenson *et al.* (1983) as a result of their structure–activity relationship studies. If so, could low levels of herbicides induce the metabolism pathways for their own metabolism? Jachetta and Radosevich (1981) had demonstrated that prior treatment of corn with atrazine resulted in increased rates of atrazine metabolism, although they did not show protection at the whole plant level.

To study this further, we pretreated corn with a number of different herbicides, including acetanilide and thiocarbamate herbicides known to be degraded by enzymatic conjugation to GSH (Lamoureux and Rusness, 1986a). Alachlor, propachlor, and CDAA all elevated GSH levels (Stephenson *et al.*, 1983; Ezra *et al.*, 1985a), but only CDAA elevated GST levels as well (Ezra *et al.*, 1985a). In addition, only CDAA pretreatment protected corn from injury by higher CDAA levels. We interpreted the data by suggesting that CDAA protection was the result of enhanced metabolism of the herbicide through elevation of the metabolic pathway. The lack of protection from alachlor could be explained by suggesting that enzymatic, as opposed to nonenzymatic, conjugation to GSH is much more crucial to alachlor metabolism, so that enhancement of GSH levels alone without an increase in GST activity would not be sufficient to provide protection.

The importance of enzymatic GSH conjugation can in part be corroborated with the use of the herbicide synergist tridiphane. Tridiphane is a herbicide, but in this context it was used at a rate which did not in itself result in crop injury. One mode of action of tridiphane was shown to involve inhibition of GST

(Lamoureux and Rusness, 1986b). In growth room experiments summarized in Table I, corn was much more susceptible to tridiphane in combination with alachlor than in combination with CDAA (Ezra *et al.*, 1985b). Evidently, high levels of nonenzymatic conjugation of CDAA to GSH that are known to occur may be sufficient to protect corn from injury when in the presence of a GST inhibitor. Nonenzymatic conjugation of alachlor is not as great; thus inhibition of the enzymatic pathway for degradation may be expected to result in injury.

From these data, we may infer that elevation of a metabolic pathway by low rates of a herbicide could result in crop protection. This is, however, an oversimplification. The thiocarbamate herbicide EPTC was effective in elevating GSH levels and enhancing GST activity in corn. However, unlike CDAA, EPTC pretreatment was not effective in protecting corn from later, higher rates of the herbicide (Stephenson and Ezra, 1985). In looking for reasons for the discrepancy, it became clear that the extent of enhancement of GSH and GST was similar for both herbicides, so it could not be the reason for the difference in response. However, conversion of EPTC to EPTC sulfoxide, as mentioned earlier, may be a rate limiting step in the detoxication pathway. If this pathway is not affected by EPTC pretreatments, enhanced GST activity alone would not be enough to result in increased metabolism.

IV. STUDIES WITH THE SAFENER BAS 145 138

BAS 145 138 is being developed as a safener for the acetanilide herbicide metazachlor. To determine whether metazachlor was metabolized in corn in a manner similar to the other acetanilide herbicides, i.e., via conjugation to GSH, excised roots and shoots were incubated with [^{14}C]metazachlor, and metabolism to water-soluble products was followed over a period of 1–4 hr. Both corn roots and shoots metabolized metazachlor very rapidly, most of the chemical being converted to the GSH conjugate within the first couple of hours. Verification of the conjugation product was by thin-layer chromatography (TLC),

Table I Effect of Tridiphane on Alachlor and CDAA Injury to Corn

Treatment	Rate (kg/ha)	Percentage inhibition of shoot height[a]	
		CDAA	Alachlor
Herbicide	2.0	0	9
Tridiphane	0.05	4	4
Herbicide + Tridiphane	2.0 + 0.05	5	36

[a] Height of control shoots was 3.5 cm.

high-performance liquid chromatography (HPLC), and fast atom bombardment (FAB) mass spectral comparison of the metabolite with synthetic metazachlor–GSH conjugate (Ezra *et al.*, 1986).

When pretreated with BAS 145 138, corn responded in a manner similar to that with dichlormid. Levels as low as 10^{-8} *M* could elevate root GST activity following a 2-day pretreatment, but like dichlormid, shoots required a much longer induction time, as much as 6 days, for increased activity (Table II). When pretreated for 7 days with 5 μ*M* BAS 145 138, corn shoots metabolized metazachlor at a faster rate than untreated shoots. Increased metabolism following a 2-day pretreatment of corn roots with 5 μ*M* BAS 145 138 was not quite so pronounced, but did occur.

In similar experiments, corn plants were treated with 5 μ*M* dichlormid for 2 days (roots) and 7 days (shoots) and then labeled with [^{14}C]metazachlor. As mentioned earlier, corn is known to respond to dichlormid with increases in GSH and GST. We found that dichlormid pretreatments enhanced metazachlor metabolism in both roots and shoots to an extent comparable to the BAS 145 138.

It may be concluded from these data that BAS 145 138 and dichlormid protected corn from metazachlor injury by enhancing metazachlor metabolism through elevations in GST activity. This may indeed be one mode of action of the safeners. However, the extent of dichlormid protection from metazachlor injury at the whole plant level remains to be resolved.

V. DISCUSSION AND CONCLUDING REMARKS

Since the first report of Lay and Casida (1976), the "enhanced metabolism" theory of safener action through elevated GSH levels and GST activity has gained increasing popularity. Thus, sorghum protection from metolachlor injury has been attributed to enhanced metabolism of the herbicide following dichlormid pretreatment (Fuerst and Gronwald, 1986). Furthermore, enhanced metab-

Table II Effects of Safener BAS 145 138 Pretreatment (1 μ*M*) on GST Activity in Corn

Treatment time (days)	GST activity (treated/control)[a]	
	Root	Shoot
1	1.36	1.00
2	1.96	1.16
6	2.05	1.33

[a] Control values were 1245 nmol/min/mg protein for roots and 347 nmol/min/mg protein for shoots.

olism as a primary mode of safener action has been used to explain the activity of other safeners unrelated to dichlormid, such as flurazole (Mozer *et al.*, 1983), cyometrinil, and oxabetrinil (CGA-92194) (Fuerst and Gronwald, 1986).

Part of the attraction to this theory must lie in the fact that thiocarbamates such as EPTC, and acetanilides such as propachlor, alachlor, CDAA, metolachlor (Lamoureux and Rusness, 1986a), acetochlor (Breaux, 1986), and most recently metazachlor (Ezra *et al.*, 1986), have all been shown to be metabolized initially by conjugation to GSH, although this conjugation is not always enzymatic (Lamoureux and Rusness, 1986a). In addition, the sulfonylurea herbicide chlorimuron-ethyl has also recently been shown to conjugate to homoglutathione in soybeans [*Glycine max* (L.) Merr.] (Brown and Ray, 1986).

Is there any evidence, however, that enhanced metabolism through increased GSH levels and GST activity may not always lead to enhanced protection? Can corn be protected by dichlormid without measured increases in GSH and GST?

Lay and Niland (1985) examined the responses of several inbred and hybrid corn lines to dichlormid. In general, the inbred corn lines were better protected from EPTC injury than the hybrid lines, although both responded to dichlormid by increasing GSH and, where measured, GST activity. Some of these data have been summarized in Table III.

In the examples given above, it is clear that while the intensity of the response to dichlormid differs, the actual amounts of GSH and GST in some of the corn lines are very similar, both before and after dichlormid treatment. It is interesting to note that there is not always a good correlation between elevation of GSH and GST and protection. In one case, the inbred line 'MO 17 Ht' did not show any increase in GSH levels following dichlormid treatment, and yet was protected

Table III Effects of Dichlormid on GSH Levels and GST Activity of Various Inbred and Hybrid Lines[a]

Corn line	GSH (μmol/g fresh wt)		Increase in GSH (%)	GST (% sulfoxide cleavage)		Increase in GST (%)	Protection by dichlormid (%)
	Dichlormid (ppm)			Dichlormid (ppm)			
	0	30		0	30		
Inbred							
'A 619 Ht'	0.681	1.037	52	16.6	30.4	81	100
'Mo 17 Ht'	1.008	1.098	9	—	—	—	100
'OH 551'	0.721	1.438	99	—	—	—	23
Hybrid							
'DeKalb XL25A'	0.628	1.245	98	22.6	33.4	50	0
'DeKalb XL55A'	0.409	0.899	120	18.5	35.0	89	11
'DeKalb XL72B'	0.618	0.976	58	12.4	33.4	169	87

[a] Adapted from Lay and Niland (1985).

from EPTC injury. Conversely, the hybrid lines 'DeKalb XL25A' and 'DeKalb XL55A' both responded well to dichlormid with increases in GSH and GST, but were not protected from EPTC injury.

In conclusion, while the role of elevated metabolism in safener action seems assured, other activities of dichloroacetamides that could afford protection should be investigated further. Those corn varieties in which protection can be separated from increased metabolism through GSH and GST would seem to be a good place to start.

References

Adams, C. A., Blee, E., and Casida, J. E. (1983). Dichloroacetamide herbicide antidotes enhance sulfate metabolism in corn roots. *Pestic. Biochem. Physiol.* **19**, 350–360.

Barrett, M., and Olson, G. L. (1986). Protection from imazaquin injury: Studies on the mode of action of herbicide protectants. *IUPAC Int. Cong. Pestic. Chem., 6th,* Abstr. No. 3D-14.

Beste, C. E., and Schreiber, M. M. (1972). RNA synthesis as the basis for EPTC and 2,4-D antagonism. *Weed Sci.* **20**, 8–11.

Blair, A. M. (1978). Interactions between barban and protectants on maize, oats and barley. *Weed Res.* **18**, 77–81.

Blair, A. M., and Dean, M. L. (1976). Improvement in selectivity of perfluidone against *Roetboellia exaltata* in maize with herbicide protectants. *Weed Res.* **16**, 47–52.

Breaux, E. J. (1986). Identification of the initial metabolites of acetochlor in corn and sorghum seedlings. *J. Agric. Food Chem.* **34**, 884–888.

Brown, H. M., and Ray, T. B. (1986). Biochemical basis for soybean basis to DPX-F6025 (Classic) herbicide. *Abstr. Weed Sci. Soc. Am.* **26**, 86.

Carringer, R. D., Rieck, C. E., and Bush, L. P. (1978a). Metabolism of EPTC in corn (*Zea mays*). *Weed Sci.* **26**, 157–160.

Carringer, R. D., Rieck, C. E., and Bush, L. P. (1978b). Effect of R-25788 on EPTC metabolism in corn (*Zea mays*). *Weed Sci.* **26**, 167–171.

Casida, J. E., Gray, R. A., and Tilles, H. (1974). Thiocarbamate sulfoxides: Potent selective and biodegradable herbicides. *Science* **184**, 573–574.

Casida, J. E., Kimmel, E. C., Ohkawa, H., and Ohkawa, R. (1975). Sulfoxidation of thiocarbamate herbicides and metabolism of thiocarbamate sulfoxides in living mice and liver enzyme systems. *Pestic. Biochem. Physiol.* **5**, 1–11.

Chang, F. Y., Bandeen, J. D., and Stephenson, G. R. (1972). A selective antidote for prevention of EPTC injury in corn. *Can. J. Plant Sci.* **52**, 707–714.

Chang, F. Y., Bandeen, J. D., and Stephenson, G. R. (1973a). *N,N*-diallyl-2,2-dichloroacetamide as an antidote for EPTC and other herbicides in corn. *Weed Res.* **13**, 399–406.

Chang, F. Y., Stephenson, G. R., and Bandeen, J. D. (1973b). Comparative effects of three EPTC antidotes. *Weed Sci.* **21**, 292–295.

Chang, F. Y., Stephenson, G. R., and Bandeen, J. D. (1974). Effects of *N,N*-diallyl-2,2-dichloroacetamide on ethyl *N,N*-di-*n*-propylthiocarbamate uptake and metabolism by corn seedlings. *J. Agric. Food Chem.* **22**, 245–248.

Dutka, F., and Komives, T. (1983). On the mode of action of EPTC and its antidotes. *In* "Pesticide Chemistry: Human Welfare and the Environment" (J. Miyamoto and P. C. Kearney, eds.), Vol. 3, pp. 213–218. Pergamon, Oxford.

Elker, S., and Stephenson, G. R. (1987). Comparative action of three herbicide antidotes in corn and sorghum. *Abstr. Weed Sci. Soc. Am.* **27**, 67–68.

Ezra, G., and Gressel, J. (1982). Rapid effects of a thiocarbamate herbicide and its dichloroacetamide protectant on macromolecular syntheses and glutathione levels in maize cell cultures. *Pestic. Biochem. Physiol.* **17**, 48–58.

Ezra, G., and Stephenson, G. R. (1986). The physiology of metazachlor and safener 145 138 interactions in corn. *IUPAC Int. Cong. Pestic. Chem., 6th* Abstr. No. 3D-12.

Ezra, G., Krochmal, E., and Gressel, J. (1982). Competition between a thiocarbamate herbicide and herbicide protectants at the level of uptake into maize cells in culture. *Pestic. Biochem. Physiol.* **18**, 107–112.

Ezra, G., Gressel, J., and Flowers, H. M. (1983). Effects of the herbicide EPTC and the protectant DDCA on incorporation and distribution of [2-^{14}]-acetate into major fractions of maize cell suspension cultures. *Pestic. Biochem. Physiol.* **19**, 225–234.

Ezra, G., Rusness, D. G., Lamoureux, G. L., and Stephenson, G. R. (1985a). The effect of CDAA (*N,N*-diallyl-2-chloroacetamide) pretreatment on subsequent CDAA injury in corn (*Zea mays* L.). *Pestic. Biochem. Physiol.* **23**, 108–115.

Ezra, G., Dekker, J. H., and Stephenson, G. R. (1985b). Tridiphane as a synergist for herbicides in corn (*Zea mays* L.) and proso millet (*Panicum milliaceum* L.). *Weed Sci.* **33**, 287–290.

Ezra, G., Stephenson, G. R., and Lamoureux, G. L. (1986). Metabolism of [^{14}C]metazachlor and action of safener 145 138 in corn. *Abstr. Weed Sci. Soc. Am.* **26**, 73.

Fedtke, C. (1981). Wirkung von herbiziden und herbizid antidots auf den glutathion-gehalt in maiz und soja bohnenwuzeln. *Z. Pflanzenkr. (Pflanzenpathol.) Pflanzenschutz, Sonderh.* **9**, 141–146.

Fedtke, C., and Trebst, A. (1987). Advances in understanding herbicide modes of actions. *In* "Pesticide Science and Biotechnology" (R. Greenhalgh and T. R. Roberts, eds.), pp. 161–168. Blackwell, Oxford.

Fuerst, E. P., and Gronwald, J. W. (1986). Induction of rapid metabolism of metolachlor in sorghum (*Sorghum bicolor*) shoots by CGA-92194 and other antidotes. *Weed Sci.* **34**, 354–361.

Görög, K., Muschinek, G. Y., Mustardy, L. A., and Faludi-Daniel, A. (1982). Comparative studies of safeners for the prevention of EPTC injury in maize. *Weed Res.* **22**, 27–33.

Harvey, B. M. R., Chang, F. Y., and Fletcher, R. A. (1975). Relationship between *S*-ethyl dipropylthiocarbamate injury and peroxidase activity in corn seedlings. *Can. J. Bot.* **53**, 225–230.

Hatzios, K. K. (1981). Synergistic interactions of tebuthiuron with EPTC + R-25788 and butylate + R-25788 in corn (*Zea mays*). *Weed Sci.* **29**, 601–604.

Hatzios, K. K. (1983a). Interactions of the herbicides EPTC and EPTC + R-25788 with ozone or antioxidants in corn. *J. Agric. Food Chem.* **31**, 1187–1191.

Hatzios, K. K. (1983b). Herbicide antidotes: Development, chemistry and mode of action. *Adv. Agron.* **36**, 265–316.

Horváth, L., and Pulay, Á. (1980). Metabolism of EPTC in germinating corn: Sulfone as the true carbamoylating agent. *Pestic. Biochem. Physiol.* **14**, 267–270.

Jachetta, J. J., and Radosevich, S. R. (1981). Enhanced degradation of atrazine by corn (*Zea mays*). *Weed Sci.* **29**, 37–44.

Komives, T., and Dutka, F. (1980). On the mode of action of EPTC and its protectants on corn. *Cereal Res. Commun.* **8**, 627–633.

Lamoureux, G. L., and Rusness, D. G. (1986a). Tridiphane [2-(3,5-dichlorophenyl)-2-(2,2,2-trichloroethyl)oxirane], an atrazine synergist: Enzymatic conversion to a potent glutathione-*S*-transferase inhibitor. *Pestic. Biochem. Physiol.* **26**, 323–342.

Lamoureux, G. L., and Rusness, D. G. (1986b). Xenobiotic conjugation in higher plants. *ACS Symp. Ser.* **229**, 61–105.

Lay, M.-M., and Casida, J. E. (1976). Dichloroacetamide antidotes enhance thiocarbamate sulfoxide detoxification by elevating corn root glutathione content and glutathione-*S*-transferase activity. *Pestic. Biochem. Physiol.* **6**, 422–456.

Lay, M.-M., Hubbell, J. P., and Casida, J. E. (1975). Dichloroacetamide antidotes for thiocarbamate herbicides: Mode of action. *Science* **189**, 287–289.

Lay, M.-M., and Niland, A. M. (1985). Biochemical response of inbred and hybrid corn (*Zea mays* L.) to R-25788 and its distribution with EPTC in corn seedlings. *Pestic. Biochem. Physiol.* **23**, 131–140.

Leavitt, J. R. C., and Penner, D. (1978a). Potential antidotes for acetanilide herbicide injury to corn (*Zea mays*). *Weed Res.* **18**, 281–286.

Leavitt, J. R. C., and Penner, D. (1978b). Protection of corn (*Zea mays*) from acetanilide herbicide injury with the antidote R-25788. *Weed Sci.* **26**, 653–659.

Leavitt, J. R. C., and Penner, D. (1979a). Prevention of EPTC-induced epicuticular wax aggregation in corn (*Zea mays*) with R-25788. *Weed Sci.* **27**, 47–50.

Leavitt, J. R. C., and Penner, D. (1979b). *In vitro* conjugation of glutathione and other thiols with acetanilide herbicides and EPTC sulfoxide and the action of the herbicide antidote R-25788. *J. Agric. Food Chem.* **27**, 533–536.

Miller, S. D., and Nalewaja, J. D. (1980). Herbicide antidotes with triallate. *Agron. J.* **72**, 662–664.

Miller, S. D., Nalewaja, J. D., and Pudelko, J. (1978). Effect of herbicide antidotes on barban. *Weed Sci.* **26**, 116–118.

Mozer, T. J., Tiemeier, D. C., and Jaworski, E. G. (1983). Purification and characterization of corn glutathione-*S*-transferase. *Biochemistry 22*, 1068–1072.

Pallos, F. M., and Casida, J. E., eds. (1978). "Chemistry and Action of Herbicide Antidotes." Academic Press, New York.

Pallos, F. M., Brokke, M. E., and Arneklev, D. R. (1972). Belgian Patent 782, 120.

Pallos, F. M., Gray, R. A., Arneklev, D. R., and Brokke, M. E. (1978). Antidotes protect corn from thiocarbamate injury. *In* "Chemistry and Action of Herbicide Antidotes" (F. M. Pallos and J. E. Casida, eds.), pp. 15–20. Academic Press, New York.

Parker, C., and Dean, M. L. (1976). Control of wild rice in rice. *Pestic. Sci.* **7**, 403–416.

Parker, C., Richardson, W. G., and West, T. M. (1980). Potential for expanding the selectivity of DPX 4189 by use of herbicide safeners. *Proc. Br. Crop Prot. Conf.—Weeds* **1**, 15–21.

Rubin, B., and Casida, J. E. (1985). R-25788 effects on chlorsulfuron injury and acetohydroxyacid synthase activity. *Weed Sci.* **33**, 462–468.

Stephenson, G. R., and Ezra, G. (1983). Herbicide antidotes: A new era in selective chemical weed control. *In* "Plant Growth Regulating Chemicals" (L. G. Nickell, ed.), Vol. 2 pp. 193–212. CRC Press, Boca Raton, Florida.

Stephenson, G. R., and Ezra, G. (1985). Use of subtoxic herbicide pretreatments to improve crop tolerance to herbicides. *ACS Symp. Ser.* **276**, 69–84.

Stephenson, G. R., Bunce, N. J., Makowski, R. I., and Curry, J. C. (1978). Structure–activity relationships for S-ethyl-dipropyl thiocarbamate (EPTC) antidotes in corn. *J. Agric. Food Chem.* **26**, 137–140.

Stephenson, G. R., Bunce, N. J., Makowski, R. I., Bergsma, M. D., and Curry, J. C. (1979). Structure–activity relationships for antidotes to thiocarbamate herbicides in corn. *J. Agric. Food Chem.* **27**, 543–547.

Stephenson, G. R., Ali, A., and Ashton, F. M. (1983). Influence of herbicides and antidotes on the glutathione levels of maize seedlings. *In* "Pesticide Chemistry: Human Welfare and the Environment" (J. Miyamoto and P. C. Kearney, eds.), Vol. 3, pp. 219–224. Pergamon, Oxford.

Still, G. G., Davis, D. G., and Zander, G. L. (1970). Plant epicuticular lipids: Alterations by herbicidal carbamates. *Plant Physiol.* **46**, 307–314.

Sweetser, P. B. (1985). Safening of sulfonylurea herbicides in cereal crops: Mode of herbicide antidote action. *Proc. Br. Crop Prot. Conf.—Weeds.* **3**, 1147–1154.

Wilkinson, R. E. (1978). Physiological response of lipid components to thiocarbamates and antidotes. *In* "Chemistry and Action of Herbicide Antidotes" (F. M. Pallos and J. E. Casida, eds.), pp. 85–108. Academic Press, New York.

Wilkinson, R. E. (1983). EPTC inhibition of gibberellin precursor biosynthesis and reversal of the inhibition by *N,N*-diallyl-2,2-dichloroacetamide. *In* "Pesticide Chemistry: Human Welfare and the Environment" (J. Miyamoto and P. C. Kearney, eds.), Vol. 3, pp. 233–236. Pergamon, Oxford.

Wilkinson, R. E., and Ashley, D. (1979). EPTC induced modification of gibberellin biosynthesis. *Weed Sci.* **27**, 270–274.

Wilkinson, R. E., and Smith, A. E. (1975). Reversal of EPTC-induced fatty acid synthesis inhibition. *Weed Sci.* **23**, 90–92.

Wuerzer, B., Nuyken, W., and Rohr, W. (1983). Reliability of safeners for metazachlor under varying precipitation levels. *Aspects Appl. Biol.* **4**, 345–354.

Chapter 7

Mechanisms of Action of Thiazole Safeners

E. J. Breaux,* M. A. Hoobler, J. E. Patanella, and G. A. Leyes
Monsanto Agricultural Company
St. Louis, Missouri 63167

I. INTRODUCTION

The physiological basis for the safening effect of flurazole (Fig. 1), the active ingredient in Screen seed protectant, is poorly understood at present. Flurazole effectively protects corn (*Zea mays* L.) and grain sorghum [*Sorghum bicolor* (L.) Moench] seedlings from the phytotoxic effects of the chloroacetanilide herbicides acetochlor and alachlor (Fig. 1).

Several possible mechanisms which have been proposed (Hatzios, 1983a) to explain how herbicide safeners work are: (1) the safeners interfere with the herbicide "target site"; (2) the safeners reduce herbicide uptake; and (3) the safeners enhance the metabolic detoxication of the herbicide. Recently, it has been proposed that crop safeners for chloroacetanilide herbicides may function by enhancing oxidative bioactivation of these herbicides to reactive intermediates (metabolite X, Fig. 2). These intermediates are then rapidly detoxified by the safened seedlings to the nonphytotoxic glutathione (GSH) conjugate (Hatzios,

* Present address: Rhone-Poulenc Agrochimie, 14–20 rue Pierre Baizet, 69009 Lyon, France.

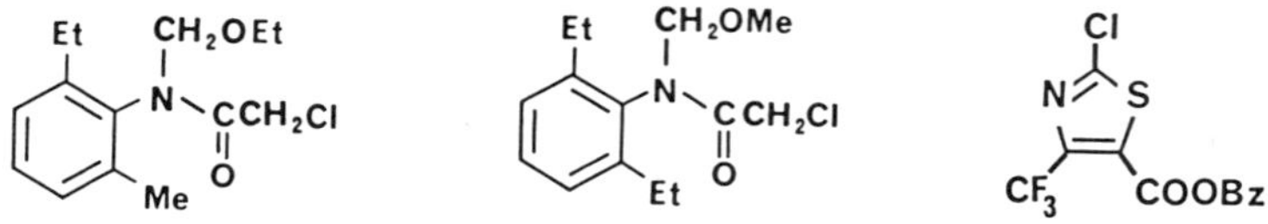

ACETOCHLOR ALACHLOR FLURAZOLE

Fig. 1. Chemical structures of the herbicides acetochlor and alachlor and of the safener flurazole.

1983b). One aspect of safener physiology that has been overlooked thus far is the metabolic fate of safeners by the treated crop seedlings.

The overall objectives of the present study, therefore, were to: (1) determine the initial metabolism of the herbicide acetochlor in corn and grain sorghum seedlings, attempting to trap and identify any bioactivated metabolic intermediates preceding the conjugation of this herbicide with GSH; (2) examine the effect of the safener flurazole on acetochlor metabolism by corn and sorghum seedlings; (3) examine the effect of flurazole on the GSH detoxication system of corn and sorghum seedlings; and (4) investigate the initial fate of radiolabeled flurazole in safened corn and sorghum seedlings.

II. RESULTS AND DISCUSSION

A. The Initial Fate of Acetochlor in Flurazole-Safened Seedlings

As noted previously, flurazole effectively safens corn seedlings from acetochlor phytotoxicity (Brinker *et al.*, 1982). To determine whether flurazole had an effect on the detoxication of acetochlor in safened corn seedlings, isotopically labeled herbicide ([U-^{14}C]phenyl; specific activity 9.8 mCi/mmol) was applied to the shoots of 3- to 5-day-old etiolated corn seedlings safened or unsafened with flurazole and the uptake and metabolism of acetochlor were monitored with time (Fig. 3). The metabolites of acetochlor formed in both safened and unsafened corn seedlings were identified by comparison to synthetic standards and to previously identified metabolites (Breaux, 1986). In both cases, the first detectable metabolites were chromatographically similar to the previously identified GSH and cysteine conjugates (Fig. 3, peaks a and b, respectively) of radiolabeled acetochlor (Breaux, 1986).

A second method used to determine if the metabolism of ^{14}C-labeled acetochlor followed the same detoxication pathway in safened and unsafened corn seedlings

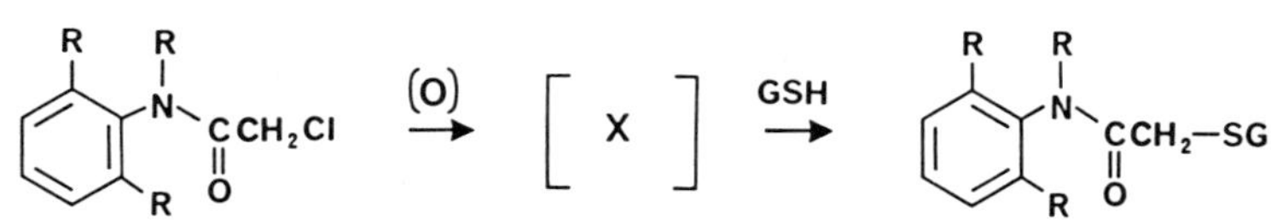

CHLOROACETANILIDE

Fig. 2. Proposed metabolic bioactivation of chloroacetanilide herbicides induced by herbicide safeners.

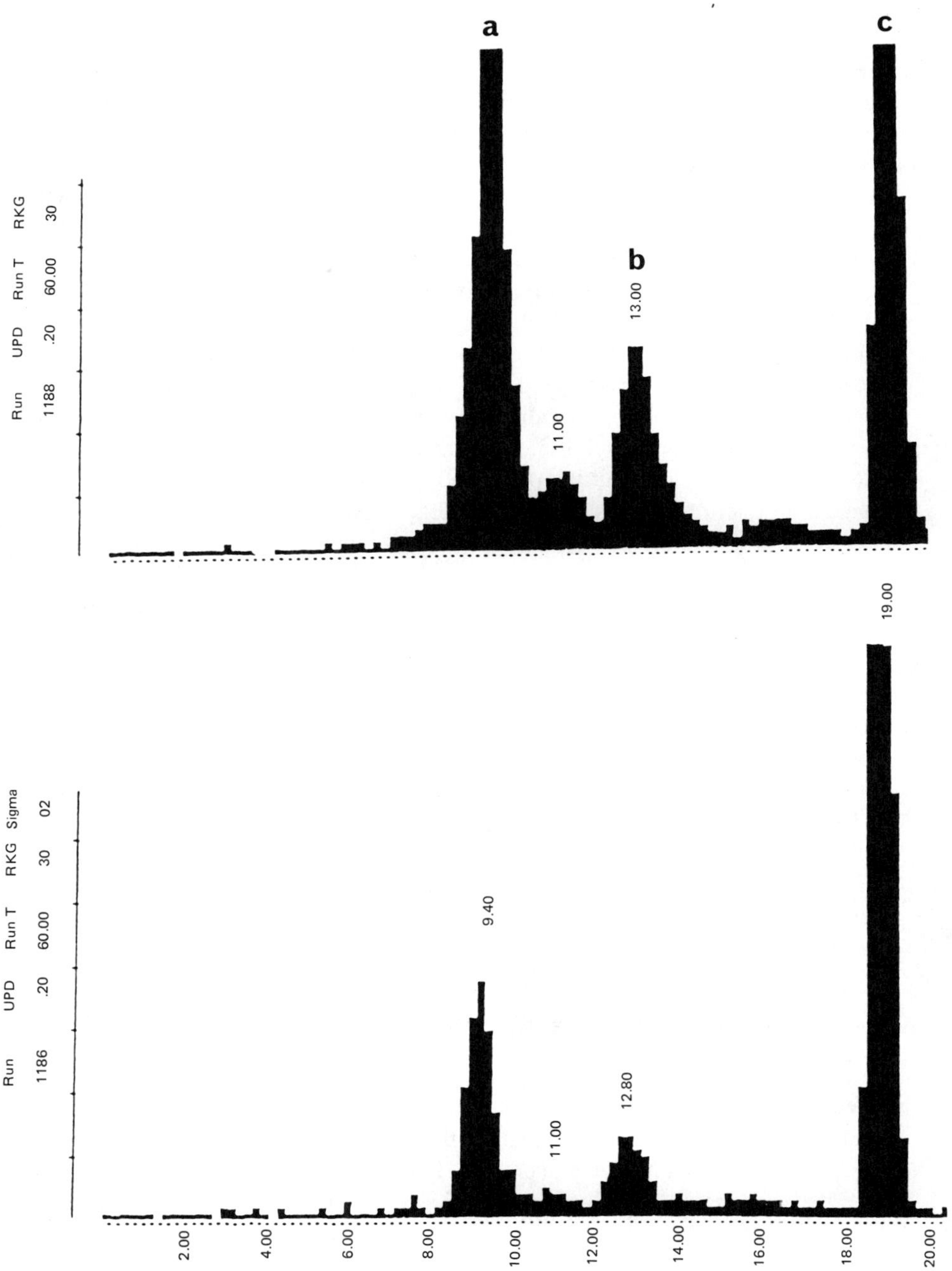

Fig. 3. HPLC analysis of acetochlor metabolites in safened (top chromatogram) and unsafened (bottom chromatogram) corn seedlings. Peaks: a = GSH conjugate, 9.4 min; b = cysteine conjugate, 12.8 min; c = acetochlor, 19.0 min. HPLC conditions: column, Waters microbondapak C-18, 7.8 mm × 30 cm. Eluents: A, 60:40 acetonitrile:water containing 1% acetic acid. B, 80:20 acetonitrile:water containing 1% acetic acid. Gradient conditions: A to B in 10 min, gradient 10 (Waters 650 programmer). Flow rate = 2 ml/min. Detector: Packard RAM 7500 radioactivity detector.

was a Raney nickel desulfurization procedure. This procedure selectively hydrogenates thioether conjugates to desulfurized derivatives (Diesperger and Sandermann, 1979). In the case of chloroacetanilide metabolites, studies with synthetic standards revealed that the GSH, cysteine, and other initial thioether conjugates were all converted to the corresponding acetanilide when using this catalytic reduction method (Fig. 4). Raney nickel desulfurization is ideally suited for HPLC analysis as can be seen in Fig. 5. The polar thioether conjugates such as the GSH conjugate have relatively short retention times, whereas the common reduction product formed (Fig. 5) is relatively nonpolar and has a much longer retention time upon reverse-phase HPLC analysis. Both safened and unsafened corn and grain sorghum seedlings were analyzed using this procedure. In each case, the same nonpolar desulfurization derivative of the GSH conjugate of acetochlor was formed as the major product (Fig. 5). Although these types of experimental results do not definitively rule out the presence of any bioactivated metabolic intermediates, they seriously question their existence. Unlike the case of thiocarbamate herbicides in which there is evidence for bioactivation (Lay *et al.*, 1975), there is no plausible chemical analogy for the chloroacetanilide herbicides. Therefore, the proposed bioactivation of chloroacetanilides awaits the trapping and unambiguous identification of plausible reactive intermediates.

Although the safener did not affect the type of metabolites formed, flurazole was found to have an effect on the rate of metabolism in the seed-treated seedlings. For example, it was found that the safener doubled the amount of metabolized acetochlor in etiolated 'Pioneer 3320' corn seedlings after 4 hr (Table I). A similar result was obtained with a second cultivar of corn ('Pioneer 3780'). No difference in acetochlor uptake between safened and unsafened corn seedlings was observed, however (data not shown). These results are in agreement with those reported recently by Fuerst and Gronwald (1986). These researchers found that seed safeners caused an increase in the rate of chloroacetanilide metabolism by treated sorghum seedlings.

B. The Effect of Flurazole on GSH Levels of Corn and Sorghum Seedlings

We recently reported new analytical techniques to identify (Breaux *et al.*, 1988) and quantify (Breaux *et al.*, 1987) plant thiols such as GSH and homoglutathione (hGSH; glutamyl-cysteinyl-β-alanine). These methods were based on the selective derivatization of the thiols with radioactive maleimide such as ^{14}C-labeled

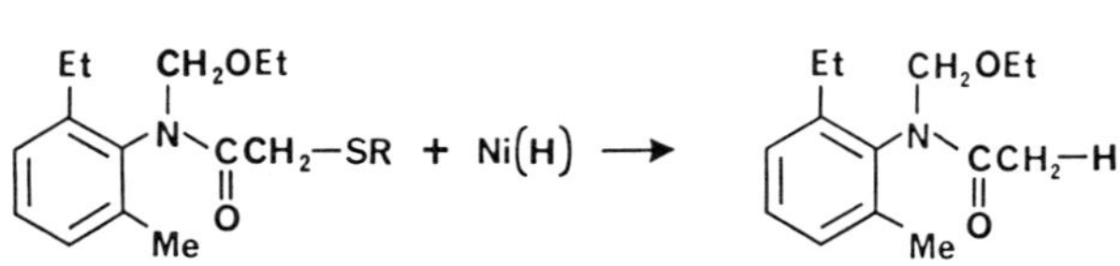

Fig. 4. Reaction scheme illustrating the Raney nickel desulfurization procedure with thioether conjugates of acetochlor.

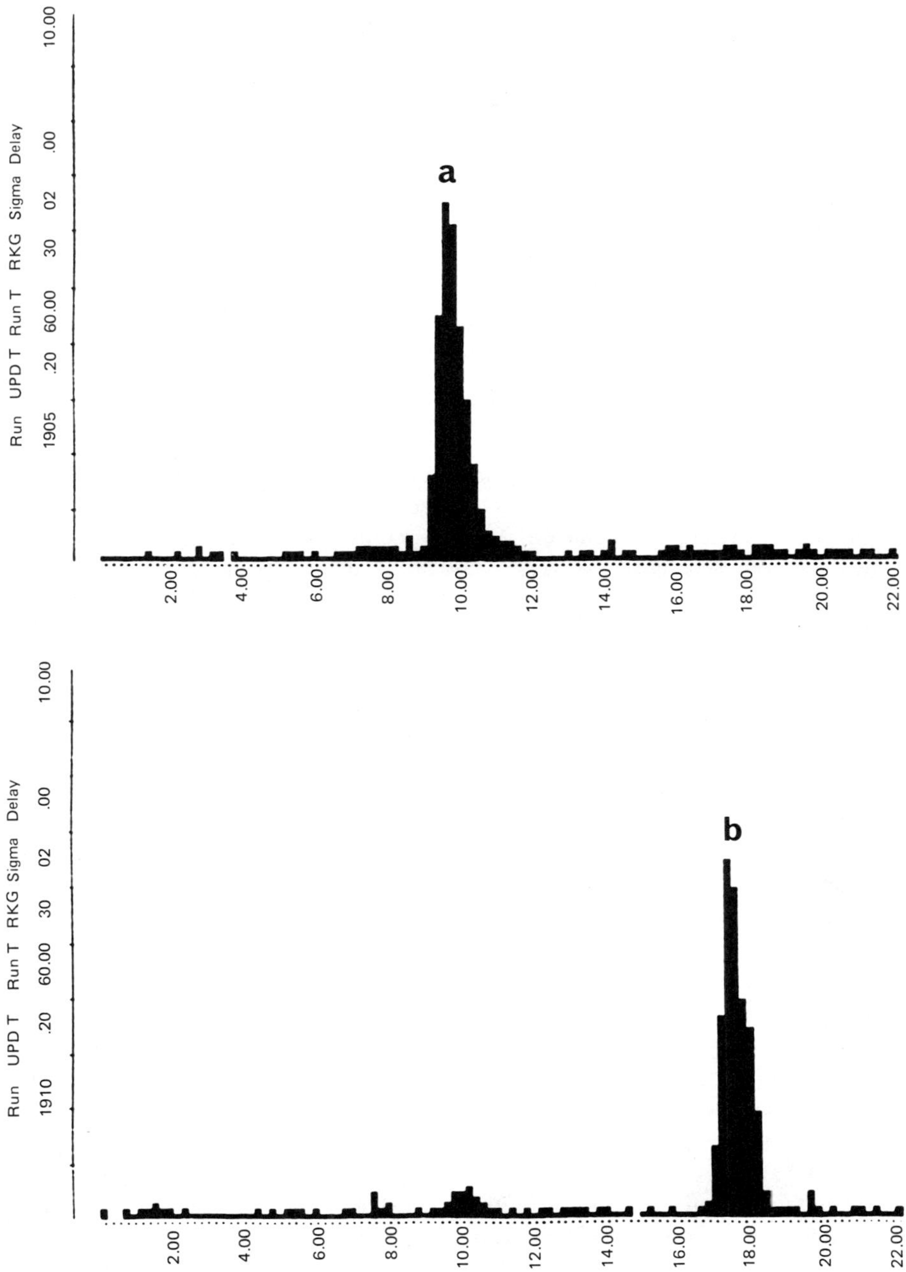

Fig. 5. HPLC analysis of a mixture of initial metabolites of acetochlor in flurazole-safened corn seedlings before (top chromatogram) and after (bottom chromatogram) the application of the Raney nickel desulfurization procedure. Peaks: a = metabolite mixture, 9.8 min. b = acetochlor reduction product, 17.7 min. HPLC conditions: column, Waters micro-bondapak C-18, 7.8 mm × 30 cm. Eluents: A, water; B, acetonitrile. Gradient: A to B in 30 min, linear gradient (curve 6). Flow rate = 3ml/min. Detector: Packard RAM 7500 radioactivity detector.

Table I The Effects of Flurazole on Acetochlor Metabolism in Safened Corn Seedlings

Corn cultivar	Treatment[a]	Acetochlor metabolized after 4 hr [b,c] (%)
'Pioneer 3320'	Safened	60.2 ± 3.7
	Unsafened	30.4 ± 5.7
'Pioneer 3780'	Safened	61.8 ± 4.9
	Unsafened	39.8 ± 3.2

[a] Safened seedlings were treated with 0.5% flurazole by seed weight.
[b] Based upon HPLC analysis of shoot extracts.
[c] The differences between the means for the safened and unsafened seedlings were significant at the 0.01 level.

N-ethylmaleimide (NEM) and *N-para*-bromophenylmaleimide (BPM). The procedure was designed as an HPLC method for use with commercial chromatographic detectors for radioactivity. The isotopically labeled thiol maleimide adducts were separated by reverse phase HPLC and quantified with the aid of the radiochemical detector. The advantages of this procedure are sensitivity and selectivity. The problems due to co-chromatography of UV absorbing and fluorescing natural plant products are avoided. As can be seen in Fig. 6, the NEM adduct of GSH is easily separated from the other thiol maleimide in extracts isolated from corn and wheat (*Triticum aestivum* L.) tissues.

It has been reported that safeners such as dichlormid enhance the level of biochemical detoxication activity in safened plants (Lay and Casida, 1976). Recently, Mozer and co-workers (1983) showed that flurazole caused an increase in glutathione-*S*-transferase (GST; EC 2.5.1.18) activity in corn seedlings. Flurazole has also been reported to enhance the levels of GSH, the other important component of the metabolic detoxication of chloroacetanilide herbicides, through GSH conjugation in grain sorghum roots (Rubin *et al.*, 1985). Since these herbicides are reported to be active in the shoot of grass plants (Narsaiah and Harvey, 1977), we employed the analytical methods described above to determine the effect of flurazole on the GSH levels of corn and grain sorghum. Seedlings safened or unsafened with flurazole were grown in the dark for 3–5 days. At harvest, the shoots were excised and assayed using the NEM method, which is selective for GSH determination. The major thiol in both safened and unsafened corn and grain sorghum seedling shoots was GSH. The quantitative results are shown in Table II. Shoots from safened grain sorghum seedlings were found to contain approximately 50% more GSH than shoots from unsafened seedlings. Similarly, shoots from safened corn seedlings contained approximately 25% more GSH than shoots from unsafened seedlings. These results coupled with the previously reported results on GST activity enhancement (Mozer *et al.*, 1983) indicate that flurazole functions as a safener by enhancing the biochemical detoxication capacity of corn and grain sorghum seedlings.

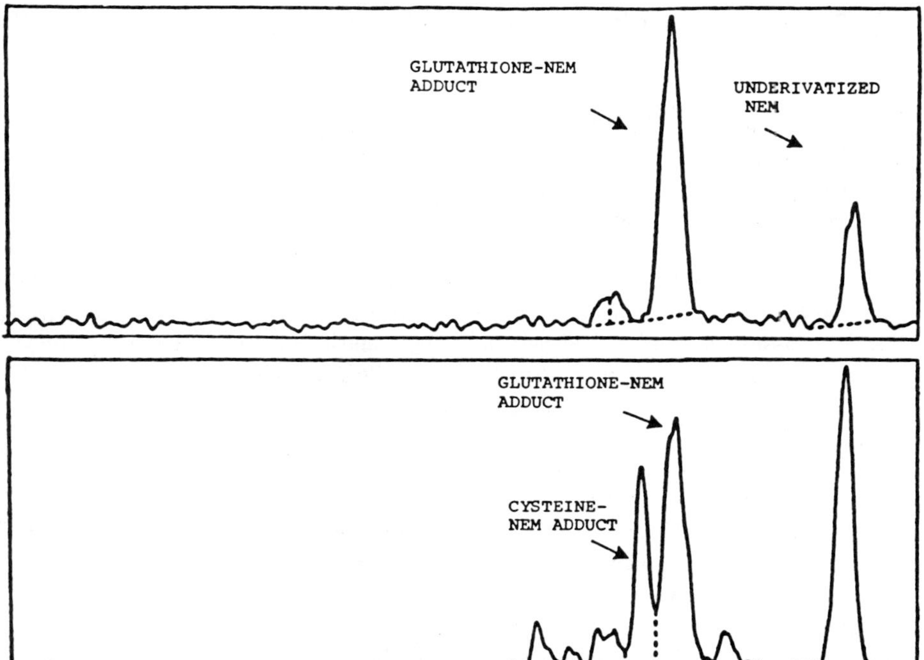

Fig. 6. HPLC analysis of ^{14}C-labeled *N*-ethylmaleimide (NEM) derivatives of plant thiols from corn (top chromatogram) and wheat (bottom chromatogram) extracts. Peaks: NEM–GSH conjugate, 7.6 min; NEM–cysteine conjugate, 7.3 min; underivatized NEM, 9.2 min. HPLC conditions: column, Altex ultrasphere C-18, 10 mm × 25 cm. Eluents: A, 1% acetic acid; B, acetonitrile. Gradient conditions: A to B in 10 min, linear gradient (curve 6). Flow rate = 4 ml/min. Detector = Berthold Model 503 radioactivity detector.

Table II The Effect of Flurazole on GSH Content of Corn and Sorghum Shoots

Species	Treatment[a]	GSH level[b] (μ/g GSH/g fresh wt)
Corn	Safened	370.2 ± 23.7
('Pioneer 3320')	Unsafened	293.8 ± 30.8
Grain sorghum	Safened	161.4 ± 24.1
('DeKalb D59E')	Unsafened	102.1 ± 11.4

[a] Safened seedlings were treated with 0.5% flurazole by seed weight.

[b] The differences between the means of safened and unsafened seedlings were significant at the 0.05 level.

C. Flurazole Uptake and Metabolism in Corn and Sorghum Seedlings

As noted above, the safener flurazole causes an increase in the rate of metabolic deactivation of the herbicide acetochlor. Furthermore, both GSH levels and GST activity in safened corn and grain sorghum seedlings are increased by this safener. The next question is: how does the safener enhance the level and activity of these detoxication systems? In an attempt to shed some light on this question, we initiated an investigation on the metabolic fate of isotopically labeled flurazole in corn and grain sorghum. Radiolabeled flurazole was applied directly to 3- to 5-day-old etiolated shoots and uptake and metabolism were monitored with time. This shoot application technique resulted in high uptake of the safener and allowed the exposure time to be accurately recorded. Approximately 90% of the applied safener was absorbed by the seedlings after 2 hr. Very little activity was translocated from the application zone. In all cases, the seeds and roots of treated seedlings contained less than 2% each of the shoot applied activity. Reverse phase HPLC analysis of corn and grain sorghum shoot extracts produced the chromatograms shown in Figs. 7 and 8, respectively. After 2 hr, three similar metabolites were formed in both corn and grain sorghum seedlings. No parent safener (retention time = 23.6 min) was present in either corn or grain sorghum extracts after 2 hr. The two most polar metabolites (peaks a and c in Figs. 7 and 8) were identified by chromatographic and mass spectral comparison to authentic standards. The least polar of these two metabolites was the carboxylic acid derivative of flurazole (peak c in Figs. 7 and 8; compound c in Fig. 9), whereas the most polar metabolite was identified as the 2-hydroxy-thiazolecarboxylate (peak a in Figs. 7 and 8; compound a in Fig. 9). The third major initial metabolite of flurazole (peak b in Figs. 7 and 8; compound b in Fig. 9) was chromatographically identical to the GSH conjugate of this safener. Negative and positive ion Fast Atom Bombardment (FAB) mass spectra of this metabolite were also identical to those obtained for an authentic sample. The negative ion FAB mass spectrum for this metabolite is shown in Fig. 10. It was subsequently found that the conjugation of flurazole with GSH is enzymatic and is catalyzed efficiently by GSTs. The initial metabolism of flurazole in corn and grain sorghum seedlings is outlined in Fig. 9. In both grass crops, the GSH conjugate of flurazole (metabolite b in Fig. 9) is one of the initially detected metabolites.

The finding that flurazole forms a GSH conjugate may provide the answer to the actual mechanism of the enhanced levels of GSH and GST activity in corn and grain sorghum seedlings treated with this safener. In the case of the increased level of GSH, the explanation may lie in the interference of the GSH conjugate of flurazole with the normal regulation of GSH biosynthesis in corn and grain-sorghum seedlings. The GSH biosynthesis pathway in plants has been reviewed recently by Rennenberg (1982). The coupling of glutamate and cysteine to form glutamyl-cysteine is catalyzed by the enzyme glutamyl-cysteine synthetase or

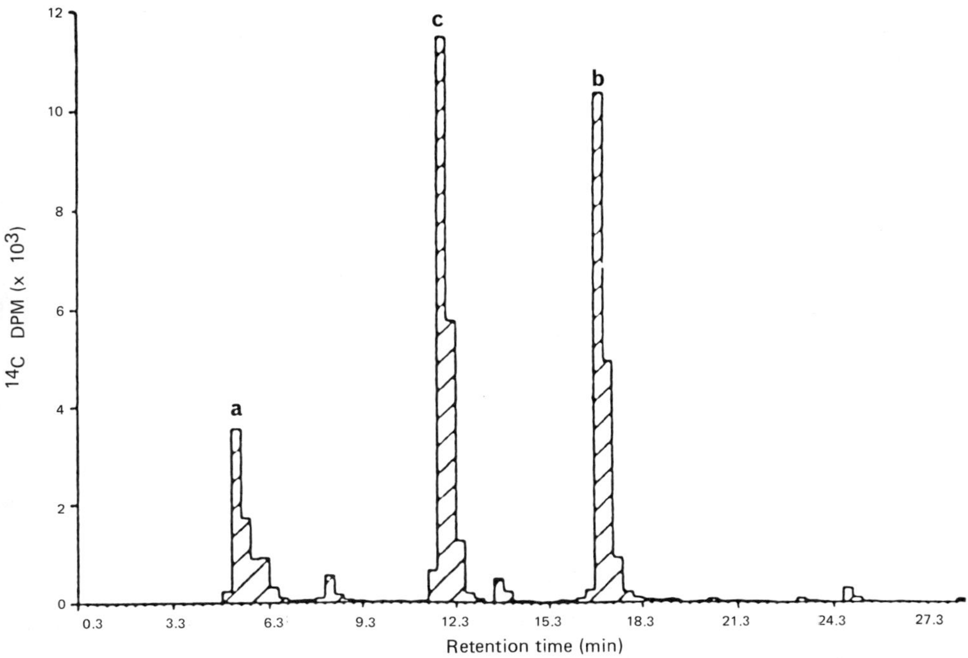

Fig. 7. HPLC analysis of ^{14}C-labeled metabolites of flurazole from corn seedlings, 2 hr after treatment with radiolabeled flurazole. Peaks: a = 2-hydroxy-thiazolecarboxylate, 5.3 min; b = GSH conjugate of flurazole, 12.0 min; c = carboxylic acid derivative of flurazole, 17.3 min. HPLC conditions: column, Waters microbondapak C-18, 7.8 mm × 30 cm. Eluents: A, 0.001 *M* aqueous potassium dihydrogen phosphate; B, acetonitrile. Gradient conditions: A to B in 30 min, linear gradient. Flow rate = 3 ml/min.

GSH synthetase I (EC 6.3.2.2). The second step, the coupling of glycine to glutamyl-cysteine, is catalyzed by GSH synthetase II (EC 6.3.2.3). The overall regulation of GSH biosynthesis is through feedback inhibition of glutamyl-cysteine synthetase by the end product of the pathway, GSH. It has recently been reported that in animal systems, the GSH conjugates of xenobiotics can bind to glutamyl-cysteine synthetase and override the feedback inhibition caused by GSH (Kondo *et al.*, 1984). It is possible that a similar mechanism may account for the flurazole-induced increase in corn and grain sorghum levels of GSH. In this case, the GSH conjugate of flurazole may bind to glutamyl-cysteine synthetase of corn and grain sorghum, thereby circumventing the feedback regulation of the GSH biosynthesis pathway.

There are several recent reports that other chemicals which conjugate with GSH also function as safeners. For example, Rubin and co-workers (1985) reported that substituted phenylmaleimides increase GSH levels in grain sorghum and act as alachlor safeners on this crop. Another study showed that the chloroacetamide allidochlor (also known as CDAA) can function as a safener

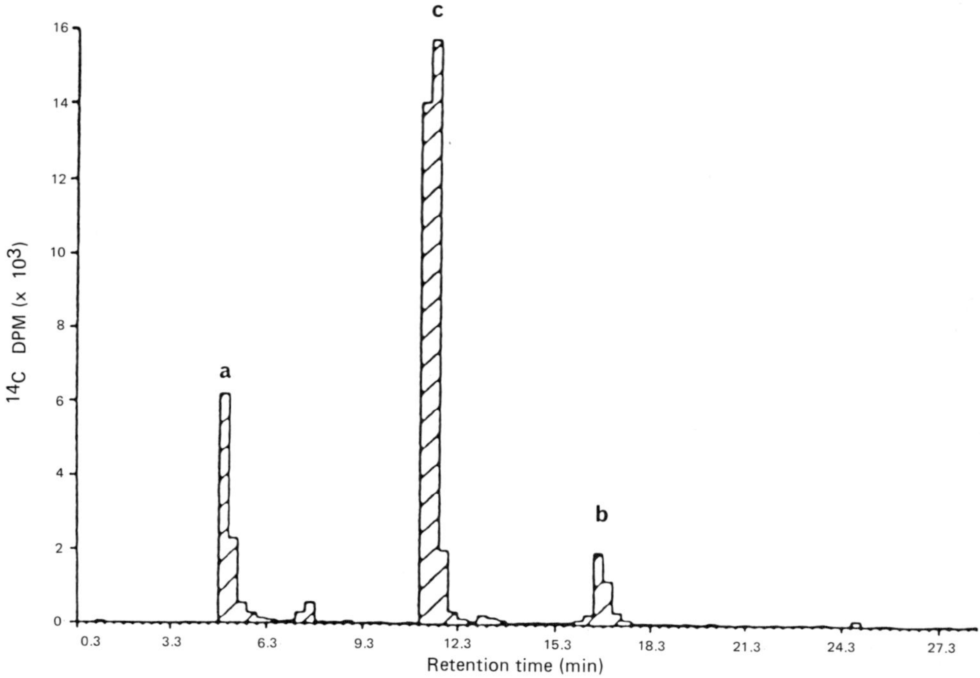

Fig. 8. HPLC analysis of ^{14}C-labeled metabolites of flurazole from grain sorghum seedlings, 2 hr after treatment with radiolabeled flurazole. Peaks and HPLC conditions same as in Fig. 7.

when used at low rates (Ezra *et al.*, 1985). Finally, both Fedtke (1981) and Stephenson *et al.* (1983) have reported additional examples of enhanced GSH levels following treatment with herbicides which can react with GSH. Although it has not been demonstrated *in vivo*, dichloroacetamide safeners such as dichlormid may also react with GSH since Martin *et al.* (1980) have shown that dichloroacetamide antibiotics form transient GSH conjugates *in vitro*. A GST was needed to catalyze the conjugation, however.

At present, flurazole is the only safener which has been found to conjugate with GSH in plants. However, all safeners presently in use contain an electrophilic site and have the potential to react with GSH. At present, it is not clear how the GSH conjugate of flurazole induces the biosynthesis of the GST enzymes, the conjugation catalysts. There are several examples, however, of the coordinate induction of coupled detoxication systems in animal systems by xenobiotics which can form GSH conjugates such as the allelochemical benzyl isothiocyanate (Sparnins, *et al.*, 1982).

III. CONCLUDING REMARKS

The effects of the seed safener flurazole on the metabolism of the chloroacetanilide herbicide acetochlor were examined and it was found that although the

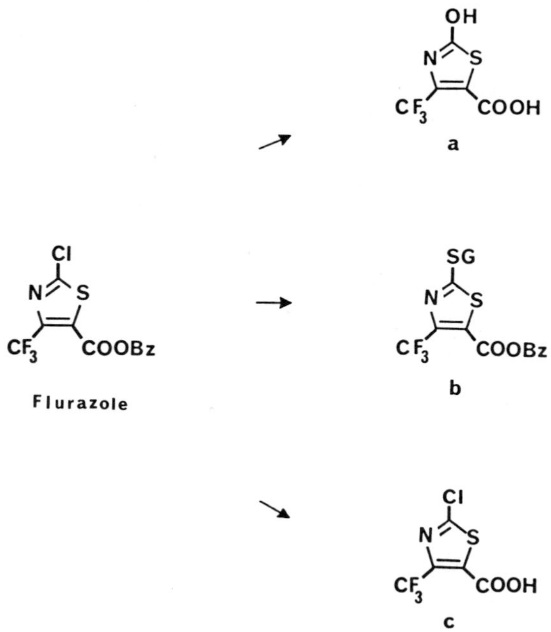

Fig. 9. Proposed initial metabolism of the safener flurazole in corn and sorghum seedlings. Metabolites: a = 2-hydroxy-thiazolecarboxylate; b = GSH conjugate; c = carboxylic acid derivative of flurazole.

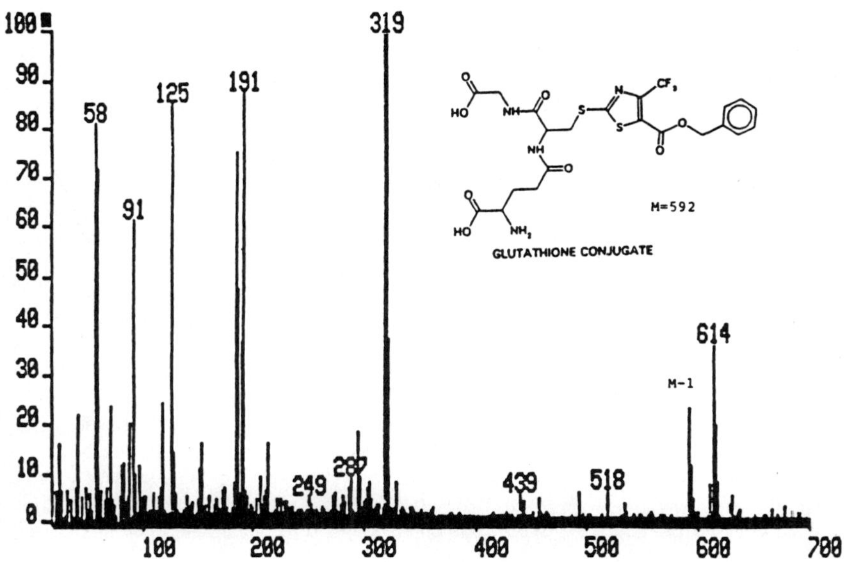

Fig. 10. Negative ion FAB mass spectrum of the GSH conjugate of flurazole identified in corn and grain sorghum seedlings.

safener did not detectably change the metabolic pathway responsible for detoxication of acetochlor, it did alter the rate of metabolism of acetochlor. Acetochlor was converted to GSH conjugates in both safened and unsafened corn and grain sorghum seedlings. The safener caused an increase in both GSH levels and the activity of GSTs which catalyze the conjugation of acetochlor with GSH. Studies on the metabolism of flurazole in corn and grain sorghum showed that flurazole was rapidly absorbed and metabolized by seedlings of these grass crops. No parent flurazole was detected 2 hr after treatment in corn or grain sorghum seedlings. The three initial metabolites of flurazole were isolated and identified. One of these metabolites was the GSH conjugate of this safener. Based on the results of these studies, it is proposed that the key to the puzzle of the mechanism of safener action may be the formation of its conjugate with GSH. The GSH conjugate of flurazole may cause the deregulation of GSH biosynthesis, leading to an increase in GSH levels and GST activity.

Acknowledgments

We would like to thank Dr. P. Feng for the synthesis of the GSH conjugate of flurazole. We are also indebted to Dr. H. Fujiwara and Dr. T. Solsten for the mass spectral analyses.

References

Breaux, E. J. (1986). Identification of the initial metabolites of acetochlor in corn and soybean seedlings. *J. Agric. Food Chem.* **34**, 884–888.

Breaux, E. J., Patanella, J. E., and Sanders, E. F. (1987). Chloroacetanilide herbicide selectivity: Analysis of glutathione and homoglutathione in tolerant, susceptible and safened seedlings. *J. Agric. Food Chem.* **35**, 474–478.

Breaux, E. J., Patanella, J. E., Sanders, E. F., and Fujiwara, H. (1988). Identification of biologically important thiols. *Biomed. Environ. Mass Spectrom.* **15**, 123–128.

Brinker, R. J., Schafer, D. E., Hakes, H. R., and Radke, R. O. (1982). Evaluation of MON-4606 as a seed safener for acetochlor and alachlor in grain sorghum. *Proc., South. Weed Sci. Soc.* **35**, 390.

Diesperger, H., and Sandermann, H. (1979). Soluble microsomal glutathione-*S*-transferase activities in pea (*Pisum sativum* L.) seedlings. *Planta 146*, 643–648.

Ezra, G., Rusness, D. G., Lamoureux, G. L., and Stephenson, G. R. (1985). The effect of CDAA (*N,N*-diallyl-2-chloroacetamide) pretreatments on subsequent CDAA injury to corn (*Zea mays* L.). *Pestic. Biochem. Physiol.* **23**, 108–115.

Fedtke, C. (1981). Action of herbicides and herbicide antidotes on glutathione content of corn and soybean roots. *Z. Pflanzenkr. (Pflanzenphysiol.) Pflanzenschutz, Sonderh.* **9**, 141–146.

Fuerst, E. P. and Gronwald, J. W. (1986). Induction of rapid metabolism of metolachlor in sorghum (*Sorghum bicolor*) shoots by CGA-92194 and other antidotes. *Weed Sci.* **34**, 354–361.

Hatzios, K. K. (1983a). Herbicide antidotes: Development, chemistry and mode of action. *Adv. Agron.* **36**, 265–316.

Hatzios, K. K. (1983b). Effects of CGA-43089 on responses of sorghum (*Sorghum bicolor*) to metolachlor combined with ozone or antioxidants. *Weed Sci.* **31**, 280–284.

Kondo, T., Taniguchi, N., and Kawakami, Y. (1984). Significance of glutathione-*S*-transferases for glutathione metabolism in human erythrocytes. *Eur. J. Biochem.* **145**, 131–136.

Lay, M.-M., and Casida, J. E. (1976). Dichloroacetamide antidotes enhance thiocarbamate sulfoxide detoxification by elevating corn root glutathione content and glutathione-*S*-transferase activity. *Pestic. Biochem. Physiol.* **6**, 442–456.

Lay, M.-M., Hubbell, J. P., and Casida, J. E. (1975). Dichloroacetamide antidotes for thiocarbamate herbicides: Mode of action. *Science* **189**, 287–289.

Martin, J. L., Gross, B. J., Morris, P., and Pohl, L. R. (1980). Mechanism of glutathione dependent dechlorination of chloramphenicol and thiophenicol by cytosol of rat liver. *Drug Metab. Dispos.* **8**, 371–375.

Mozer, T. J., Tiemeier, D. C., and Jaworski, E. G. (1983). Purification and characterization of corn glutathione-*S*-transferase. *Biochemistry* **22**, 1068–1072.

Narsaiah, D. B. and Harvey, G. R. (1977). Alachlor placement in the soil as related to phytotoxicity to maize (*Zea mays* L.) seedlings. *Weed Res.* **17**, 163–168.

Rennenberg, H. (1982). Glutathione metabolism and possible biological roles in higher plants. *Phytochemistry* **21**, 2771–2781.

Rubin, B., Kirino, O., and Casida, J. E. (1985). Chemistry and action of *N*-phenylmaleamic acids and their progenitors as selective herbicide antidotes. *J. Agric. Food Chem.* **33**, 489–494.

Sparnins, V. L., Venegas, P. L., and Wattenberg, L. W. (1982). Glutathione-*S*-transferase activity enhancement by compounds inhibiting chemical carcinogenesis and by dietary constituent. *JNCI, J. Natl. Cancer Inst.* **68**, 493–496.

Stephenson, G. R., Ali, A., and Ashton, F. M. (1983). Influence of herbicides and herbicide antidotes on the glutathione levels of maize seedlings. *In* "Pesticide Chemistry: Human Welfare and the Environment" (J. Miyamoto and P. C. Kearney, eds.), Vol. 3, pp. 219–224. Pergamon, Oxford.

Chapter 8

Differential Effects of Oxabetrinil and Fenclorim against Metolachlor and Pretilachlor Injury on Various Grasses

Edith Ebert and Hans R. Gerber
Ciba-Geigy Ltd., Agricultural Division
CH-4002 Basle, Switzerland

I. INTRODUCTION

Metolachlor and pretilachlor are two acetanilide herbicides which have been developed for selective use in different crops. Metolachor has been introduced in maize (*Zea mays* L.) and soybeans [*Glycine max* (L.) Merr.] to control grasses (Gerber *et al.*, 1974). Pretilachlor is selective in transplanted rice (*Oryza sativa* L.) for the control of barnyard grass [*Echinochloa crus-galli* (L.) Beauv.] and sedges (Murakami and Ebner, 1984).

The introduction of crop safeners for these herbicides has broadened their range of application. The addition of oxabetrinil (also known as CGA-92194) as seed dressing allows the use of metolachlor in grain sorghum [*Sorghum bicolor* (L.) Moench] to control closely related grasses like barnyard grass (Ellis *et al.*, 1980). Oxabetrinil is a follow-up product of the safener cyometrinil (known also as CGA-43089) (Rufener *et al.*, 1982).

Pretilachlor, on the other hand, is fully tolerated in water-sown rice when the rice safener fenclorim is added to the herbicide in a tank mixture. Besides sedges and other weeds, barnyard grass can be selectively controlled (Rufener and

Quandranti, 1983). Both safeners develop their best antagonistic activity in the respective crop-herbicide–safener combination. The safeners do not antagonize the weed spectrum controlled by both herbicides including grassy and broad-leaved weeds.

The aim of this study was to examine and compare the physiological and biochemical responses of selected crop and weed species exposed to combined treatments of the acetanilide herbicides metolachlor and pretilachlor and the safeners oxabetrinil and fenclorim.

II. EFFECT OF SAFENERS ON HERBICIDAL ACTIVITY

Selected plant species including grain sorghum and rice as crops and weeds associated with these crops were treated with seed dressings of the two safeners oxabetrinil and fenclorim. Seed treatment brings the safener in direct contact with the germinating plant tissues, thus ensuring the optimum possible uptake of the safener during germination and early growth phase. Metolachlor or pretilachlor were applied as preemergent treatment. The herbicides were given at field relevant application rates and produced substantial damage to all species tested. The presence of the safeners caused differential reduction of the phytotoxicity.

Seed coating was achieved by shaking the seeds in a flask with a wettable powder containing the safener in concentrations of 2, 1, or 0.5 g a.i. per kilogram of seeds. Within 10 days after treatment seeds were planted in 9-cm-diameter plastic pots filled with an alluvial soil with the following components: 38% clay, 17% silt, and 45% sand (4.4% organic matter). The pots were covered with 2 mm of fine soil and then sprayed with the herbicides at 2.0, 1.0, 0.5, 0.25, 0.125, 0.062 and 0.031 kg a.i./ha. For rice, 1 to 0.015 kg/ha was applied. The amount of water was 500 liters/ha. Two replications were performed. Greenhouse conditions were 25–27°C day, 20°C night, day length 14 h. Additional illumination was provided by sodium high pressure lamps. Light intensity was 200 $\mu Em^{-2}sec.^{-1}$ After 20 days the phytotoxicity was evaluated by a visual rating corresponding to a linear percentage rating scale: 100 = no growth, 0 = as untreated.

The phytotoxicity in the presence of different herbicide–safener combinations is summarized in Table I, which represents the results of selected conditions, optimized for herbicide–safener combinations in grain sorghum and rice.

No safening effect was observed on seed-dressed barnyard grass, millet foxtail [*Setaria italica* (L.) Beauv.], yellow foxtail [*Setaria glauca* (L.) Beauv.], large crabgrass [*Digitaria sanguinalis* (L.) Scop.], and also the two dicotyledonous weeds, white mustard (*Sinapis alba* L.) and velvetleaf (*Abutilon theophrasti* Medik.). A complete safening effect was achieved only with the combinations grain sorghum–metolachlor–oxabetrinil and rice–pretilachlor–fenclorim. Slight to very pronounced safening effects were observed with these herbicide–safener

Table I Protection[a] of Selected Plant Species against Metolachlor and Pretilachlor Injury with Seed Dressings of Oxabetrinil and Fenclorim

	Metolachlor (0.25 kg/ha)		Pretilachlor (1.0 kg/ha)	
Plant species	**Oxabetrinil**	**Fenclorim**	**Oxabetrinil**	**Fenclorim**
Sorghum ('Funk G623')	+ + + +	+	+ + +	+ +
Rice ('Starbonet')[b]	+	+ + +	+ +	+ + + +
Shattercane	+ + +	0	+ + +	0
Johnsongrass	+ +	0	+ +	0
Proso millet	+	0	+	0
Alexandergrass	+ +	0	+	0
Barnyard grass	0	0	0	0
Millet foxtail	0	0	0	0
Yellow foxtail	0	0	0	0
Large crabgrass	0	0	0	0
White mustard	0	0	0	0
Velvetleaf	0	0	0	0

[a]Safening rating scale: + + + + = complete safening effect, equal to herbicide-untreated control; + + + = pronounced safening effect, up to 75% reduction in phytotoxicity; + + = moderate safening effect, up to 50% reduction in phytotoxicity; + = slight safening effect, up to 25% reduction in phytotoxicity.

[b]Herbicide dosages for rice were 0.03 kg/ha for metolachlor and 0.5 kg/ha for pretilachlor.

combinations on the weeds shattercane [*Sorghum bicolor* (L.) Moench], johnsongrass [*Sorghum halepense* (L.) Pers.], proso millet (*Panicum miliaceum* L.), and alexandergrass [*Brachiaria plantaginea* (Link) A.S. Hitchc.] (Table I).

The seed treatment of weeds with safeners must be considered an artificial system which is irrelevant to the field situation. However, it offers the possibility to compare the mode of action of the safeners in a variety of plant species.

The safening efficiency depends not only on the plant species, but also on the given herbicide concentration. Dose–response curves in Figs. 1 and 2 show the safening effect of oxabetrinil and fenclorim in six plant species at varying concentrations of metolachlor and pretilachlor. Application of the safener shifts the sigmoidal dose–response curves to higher herbicide rates. From these dose–response curves the inherent efficiency of the safening action can be evaluated.

A safening index was used as a quantitative term defined by the ratio of the herbicide concentration giving 50% inhibition in the presence of the safener (H + S) divided by the herbicide concentration giving 50% phytotoxicity in the absence of the safener (H).

$$\text{Safening Index (SI)} = \frac{50\% \text{ phytotoxicity (H + S)}}{50\% \text{ phytotoxicity (H)}}$$

High SI values indicate an efficient safening effect and values close to one represent no effect.

The SI values derived from the dose–response curves of Figs. 1 and 2 are summarized in Table II. The high SI values for the two crops grain sorghum and

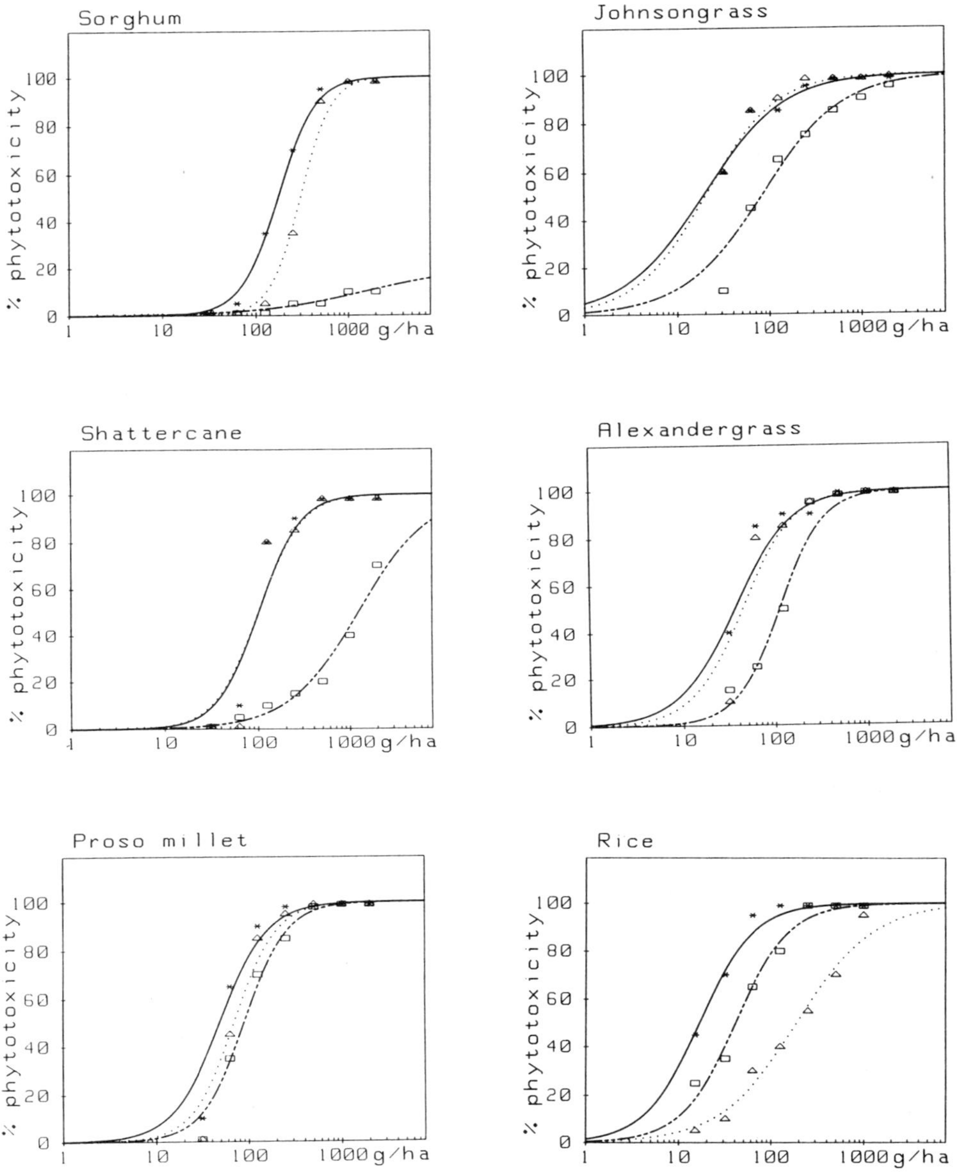

Fig. 1. Dose–response curves of six grass species to combinations of metolachlor with the safeners oxabetrinil and fenclorim. Herbicide without safener, *——*; Herbicide with fenclorim, △– – –△; Herbicide with oxabetrinil, □– -- –□. Phytotoxicity rating scale: 100 = no growth, 0 = as untreated.

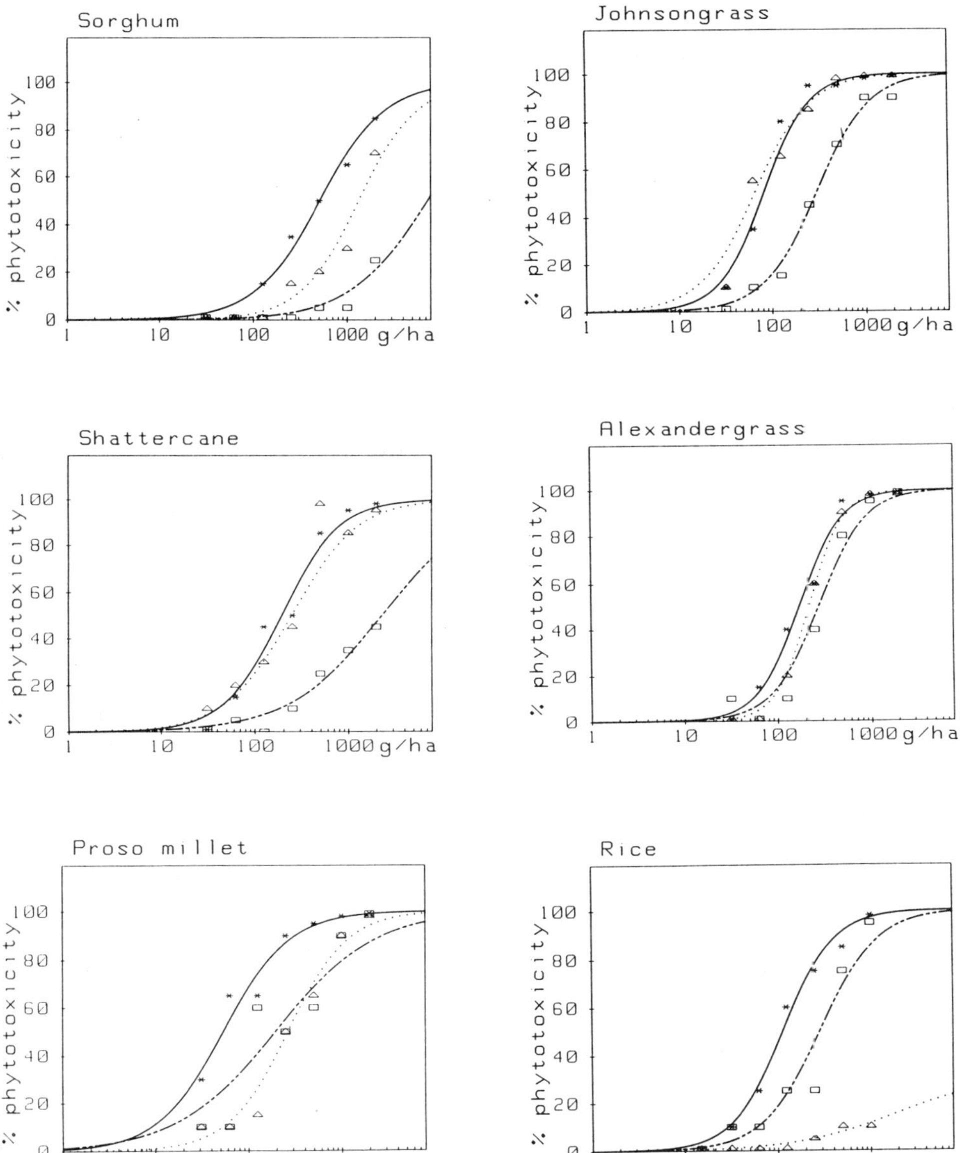

Fig. 2. Dose–response curves of six grass species to combinations of pretilachlor with the safeners oxabetrinil and fenclorim. Legend same as in Fig. 1.

Table II Safening Index[a] of Combined Treatments of the Herbicides Metolachlor and Pretilachlor and the Safeners Oxabetrinil and Fenclorim on Six Grass Species

	Metolachlor (0.25 kg/ha)		Pretilachlor (1.0 kg/ha)	
Plant species	**Oxabetrinil**	**Fenclorim**	**Oxabetrinil**	**Fenclorim**
Sorghum ('Funk G623')	>>20.0	1.7	>15.0	2.6
Rice ('Starbonnet')[b]	2.5	11.3	2.5	>>20.0
Shattercane	12.3	1.0	11.9	1.3
Johnsongrass	4.3	1.1	3.7	0.8
Proso millet	1.8	1.4	3.6	4.9
Alexandergrass	3.1	1.2	1.6	1.3

[a]Safening index = (50% phytotoxicity (H + S)) / (50% phytotoxicity (H)).
[b]Herbicide rates for rice were 0.03 kg/ha for metolachlor and 0.5 kg/ha for pretilachlor.

rice confirm that with the corresponding safener–herbicide combination a maximum of protection can be achieved. The low SI values for the weeds indicate that the weeds, contrary to the crops, may lack the physiological systems involved in the protective action of the safeners against injury from these herbicides.

III. PHYSIOLOGICAL INTERACTIONS OF ACETANILIDE HERBICIDES AND THEIR SAFENERS IN PLANTS

The mechanism of the herbicidal action of acetanilide injury in plants is not well understood. It seems to be associated with membrane integrity and/or lipid biosynthesis (Ebert, 1980; Fedtke, 1982). Specific physiological effects of the safeners oxabetrinil and fenclorim on plants have not been reported.

Physiological effects of acetanilide herbicides, particularly metolachlor, in the presence or absence of the sorghum safener, cyometrinil, have been reported at macroscopic and histological levels (Ebert, 1980, 1982) as well as correlations between morphological effects and ethylene formation (Paradies *et al.*, 1981). The effects of the acetanilide herbicides and their safeners on wax biosynthesis are well documented.

Epicuticular wax formation on leaves of corn is inhibited by EPTC (Görög *et al.*, 1982) and that of sorghum by metolachlor (Ebert, 1982). Protected plants show no depletion of epicuticular waxes. The same is seen in scanning electron micrographs (SEM) of rice leaves which have been treated with pretilachlor and fenclorim (Fig. 3).

Barnyard grass, which is not protected by either oxabetrinil or fenclorim, shows depletion of the waxes due to metolachlor and pretilachlor treatments (Fig. 4). Seed dressing with the safeners did not restore wax biosynthesis. Figure 4 shows leaves of barnyard grass plants (lower surface of the first leaf) which have

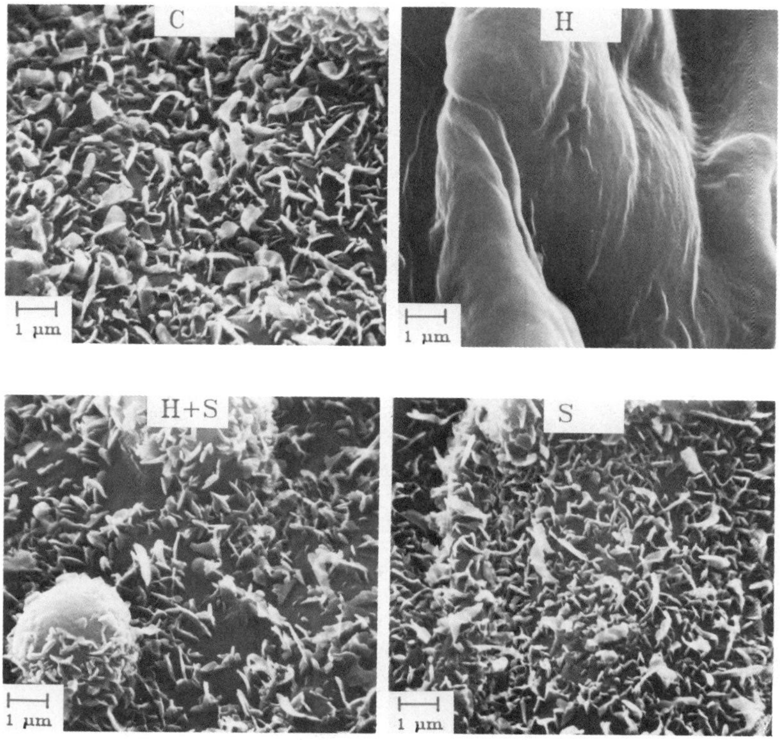

Fig. 3. Scanning electron micrographs (SEM) of the lower surface of rice cv. 'Starbonnet' second leaves of 4-day old plants (×3000). (C) Leaf from control plant; (H) leaf from plant treated with pretilachlor; (H + S) leaf from plant treated with pretilachlor + fenclorim; (S) leaf from plant treated with fenclorim.

been treated with pretilachlor and/or fenclorim. Rice and barnyard grass plants have been treated and evaluated by SEM, as reported by Ebert (1982).

The reduction of the epicuticular waxes on the leaf surfaces of treated plants increases the leaf uptake of metolachlor. In safener-treated plants, leaf and shoot uptake of metolachlor is reduced. This may have an influence on the safening action of the antidotes in sorghum, under certain conditions (Ebert, 1982; Ketchersid *et al.*, 1982).

IV. INFLUENCE OF SAFENERS ON HERBICIDE UPTAKE

Most of the physiological studies have been done with cyometrinil in relation to metolachlor injury in sorghum. Cyometrinil was replaced by oxabetrinil, which belongs to the same chemical class, because of better performance in the field (Rufener *et al.*, 1982).

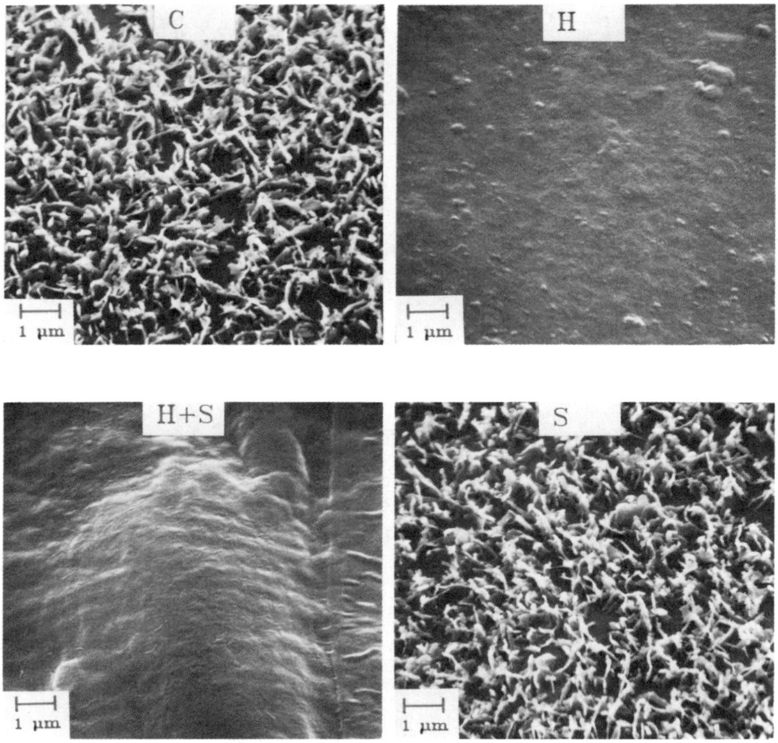

Fig. 4. Scanning electron micrographs (SEM) of the lower surface of *Echinochloa crus-galli* first leaves of 4-day old plants (×3000). (C) Leaf from control plant; (H) leaf from plant treated with pretilachlor; (H + S) leaf from plant treated with pretilachlor + fenclorim; (S) leaf from plant treated with fenclorim.

In various grasses the predominant site of uptake of metolachlor was found to be the shoot (Gerber *et al.*, 1974). It has also been shown that under field conditions the preferred site of uptake for cyometrinil is the coleoptile of sorghum seedlings (Nyffeler *et al.*, 1980). Since the safener penetrates the seed coat into the interior of the seed (Laanio, 1980), a seed dressing is recommended as the method of safener application. Treating sorghum seeds with oxabetrinil increased shoot tolerance to metolachlor more than 10-fold, but root tolerance was not increased significantly (Fuerst and Gronwald, 1986).

The developing coleoptile and shoot of sorghum and other grasses seem to be most sensitive to injury by metolachlor. Protection of these plant parts seems to play a major role in the mode of action of the safeners.

In soil of different composition and different moisture content treated with metolachlor, sorghum plants grown from safened seeds were healthier than plants grown from unsafened seeds. The amount of metolachlor—as measured by gas chromatographic (GC) analysis—in coleoptiles of unsafened plants was higher than that in safened plants. Thus, cyometrinil either reduced the uptake of metolachlor or facilitated its degradation. At the same time, no cyometrinil was found by GC analysis in any of the sorghum coleoptiles examined (Ketchersid *et al.*, 1982).

The great variability of the crops and weed species with respect to their susceptibility to metolachlor and pretilachlor injury (Figs. 1 and 2) as well as the different antidotal effects of oxabetrinil and fenclorim, ranging from positive effects to no effects (see Table I), may result from one or more of the following reasons: (1) Independently of the safeners' action, different plant species vary in the amount of metolachlor and pretilachlor taken up. Higher amounts would increase phytotoxicity due to the herbicides, and lower amounts would reduce it. (2) The two safeners influence uptake of herbicides. (3) The two safeners do not influence the uptake of the herbicides but they differentially influence the rate of detoxication of the herbicides in the various plants. These possibilities were examined in selected crops and weed species.

Seeds of sorghum, shattercane, barnyard grass, and rice dressed with either oxabetrinil or fenclorim were germinated in vermiculite. Three to four days after germination, the roots of the plants were immersed in glass tubes with tap water which contained [^{14}C]metolachlor or [^{14}C]pretilachlor. After 48 hr, the plants were harvested and the total radioactivity was determined. There was no significant difference in fresh weight caused by the treatments except for the highest concentration of metolachlor alone, which gave a 20% reduction in grain sorghum and rice.

Figures 5 and 6 show clearly that on a fresh weight basis, different amounts of both herbicides are taken up by various plant species, as measured by the incorporation of ^{14}C-label into the plant material.

The lowest amount of ^{14}C-labeled metolachlor per gram of fresh weight is found in grain sorghum. Rice and the weeds shattercane and barnyard grass absorbed higher amounts of metolachlor (Fig.5).

The uptake of pretilachlor into rice, grain sorghum, shattercane, and barnyard grass is shown in Fig. 6. These plant species vary in their ability to absorb this herbicide. Grain sorghum again absorbed the lowest amount of pretilachlor. Rice and shattercane absorbed twice as much whereas barnyard grass absorbed three times as much pretilachlor as did grain sorghum (Fig. 6).

Only a minor influence in ^{14}C uptake after application of metolachlor or pretilachlor between safened and unsafened seedlings was observed (Fig. 7). There is a tendency of higher herbicide uptake by safened plants, especially at concentrations of metolachlor and pretilachlor which are not phytotoxic. This was clearly seen in rice and barnyard grass (Fig. 8). This tendency was already observed in rice (cv. 'IR 54') which was treated with pretilachlor and fenclorim

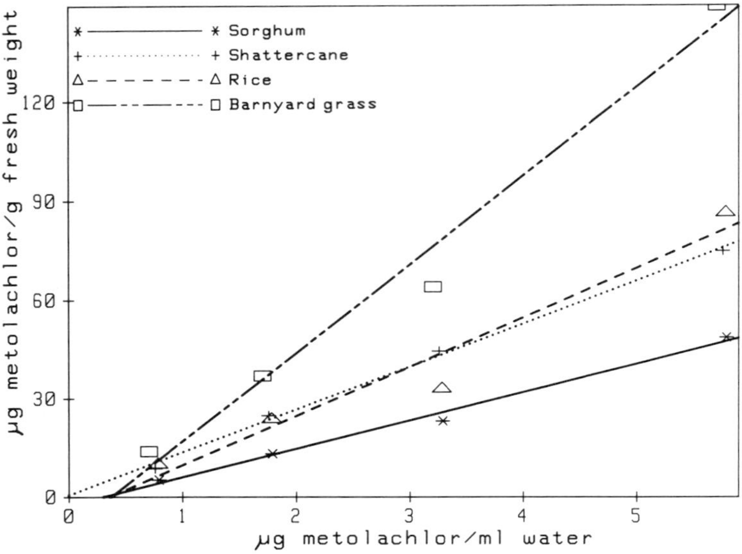

Fig. 5. Uptake of metolachlor by sorghum, shattercane, rice and barnyard grass 48 hr after treatment. Experimental conditions: protected and unprotected 3- to 4-day old plants were exposed to [^{14}C]metolachlor for 48 hr. Metolachlor and its degradation products were extracted in methanol/H_2O (80%). The total radioactivity was determined and calculated per gram fresh weight. The results were calculated from two different experiments with 20 plants each (*Echinochloa crus-galli*, 30 plants).

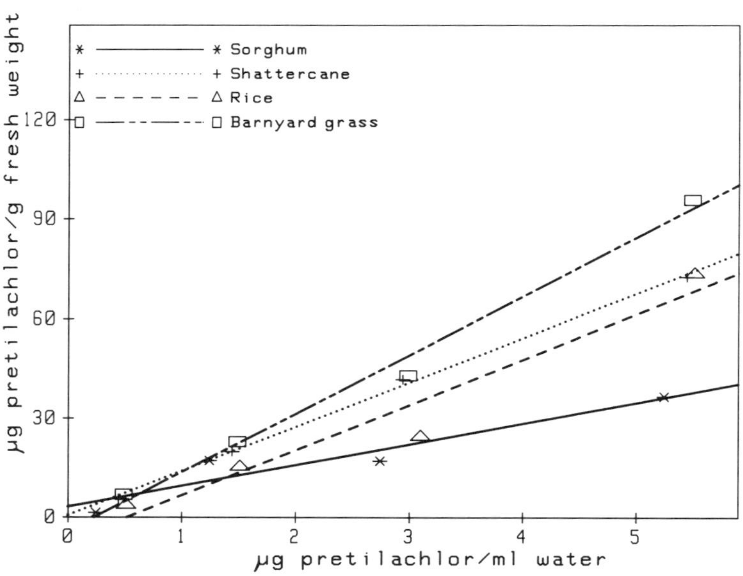

Fig. 6. Uptake of pretilachlor by sorghum, shattercane, rice and barnyard grass 48 hr after treatment. Experimental conditions same as in Fig. 5.

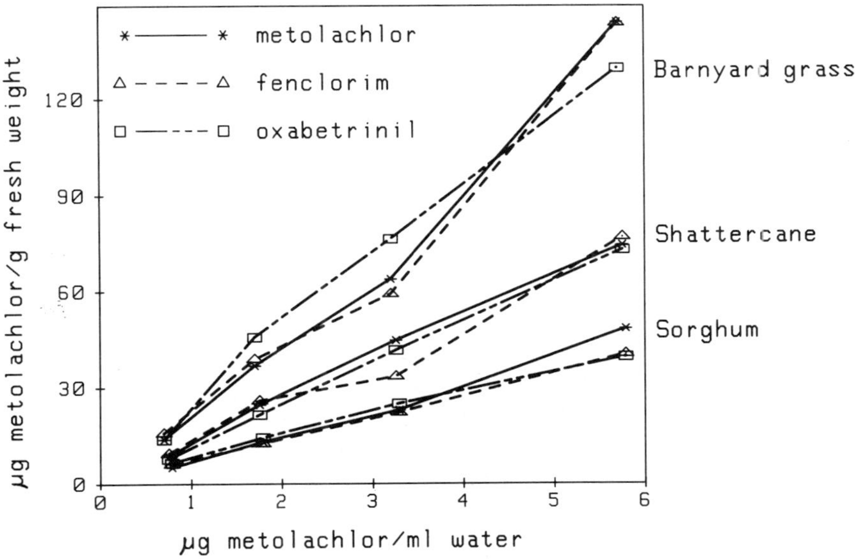

Fig. 7. Influence of oxabetrinil and fenclorim on the uptake of metolachlor by grain sorghum, shattercane, and barnyard grass, 48 hr after treatment. Experimental conditions same as in Fig. 5.

(Stingelin, 1986). There was also no difference between the sorghum safener oxabetrinil and the rice safener fenclorim with respect to their influence on herbicide uptake (Figs. 7 and 8).

From these studies it is concluded that even closely related plant species such as grain sorghum and shattercane differ greatly in their ability to absorb the herbicides metolachlor and pretilachlor. The two safeners had only a minor influence on the uptake of these herbicides by all plants examined in this study. Differences in herbicide uptake between safened and unsafened plants are substantially smaller than differences in herbicide uptake observed among plant species.

V. INFLUENCE OF OXABETRINIL AND FENCLORIM ON THE DEPLETION RATE OF [^{14}C]METOLACHLOR IN FOUR GRASSES

It was further interesting to investigate whether the differential response of the various grasses to protection against metolachlor injury by oxabetrinil and fenclorim may be based on the differential detoxication of this herbicide.

The total uptake of metolachlor into the four safened and unsafened grasses, namely grain sorghum, shattercane, rice, and barnyard grass, is shown in Figs.

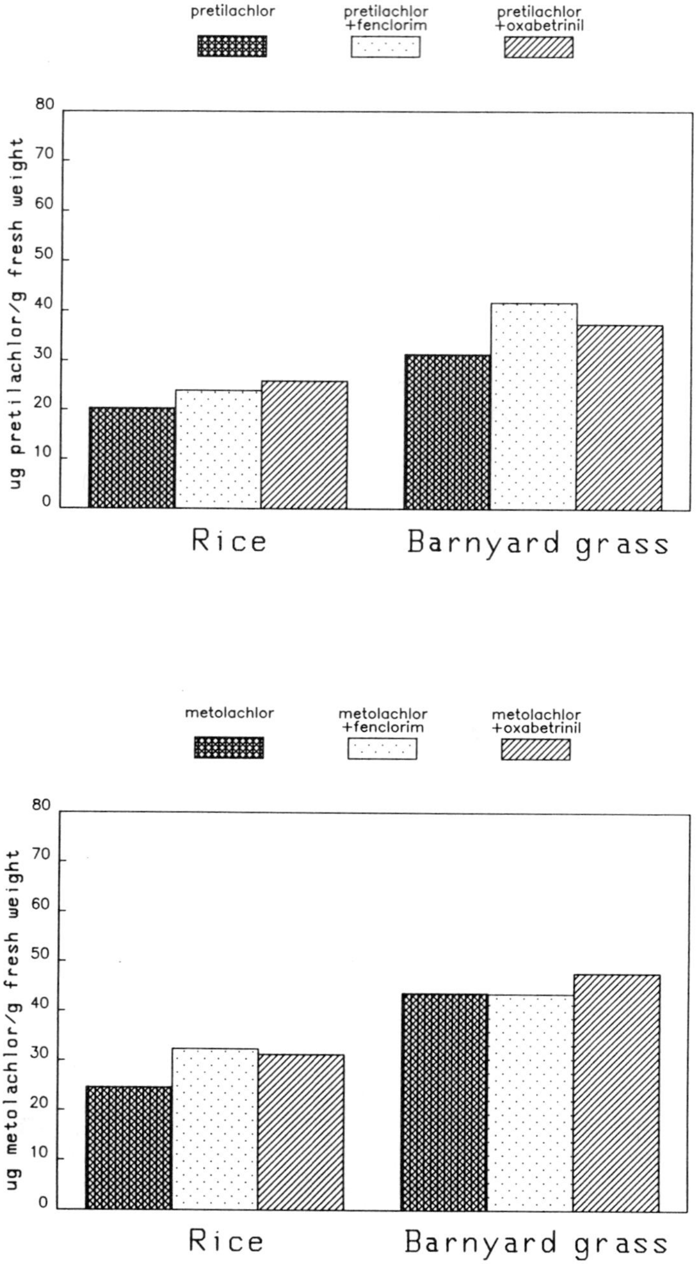

Fig. 8. Influence of oxabetrinil and fenclorim on the uptake of metolachlor and pretilachlor by rice and barnyard grass, 48 hr after treatment. Experimental conditions same as in Fig. 5. Pretilachlor concentration used was 2 μg/ml H_2O.

7 and 8. From this experiment, the relative amounts of the parent herbicide (metolachlor) and its metabolites in roots and shoots of safened and unsafened-grasses were determined. Metolachlor was separated from its degradation products by HPLC with a water–acetonitrile gradient. The total radioactivity of roots and shoots is taken as 100%. The actual recovery of [^{14}C]metolachlor into roots and shoots was 50% for grain sorghum, 40% for shattercane, 30% for rice, and 20% for barnyard grass. Figure 9 presents a survey of the depletion rate of metolachlor in roots and shoots of the four grasses after an incubation time of 48 hr. In unsafened grain sorghum, unmetabolized metolachlor is found in the roots, none in the shoots. Seed dressing with oxabetrinil increased the degradation of metolachlor in the roots. Fenclorim activated the depletion of metolachlor in the roots to a lower degree than oxabetrinil. The shoots are free of the unmetabolized herbicide.

In shattercane the depletion rate of metolachlor was similar to that of grain sorghum. Earlier, it was shown that shattercane takes up more metolachlor per gram fresh weight than grain sorghum (Fig. 5). Therefore the persistence of metolachlor in this species is higher than in grain sorghum. Oxabetrinil and fenclorim influence the depletion of metolachlor in shattercane to a similar degree as in grain sorghum. The levels of parent metolachlor in safener-treated plants are decreased in the roots compared to those of unsafened plants.

In rice plants, metolachlor is degraded more slowly than in sorghum. Oxabetrinil does not influence essentially the degradation of metolachlor. Fenclorim, however, strongly increases the degradation of metolachor in the roots (Fig. 9).

The degradation of metolachlor in the roots and shoots of barnyard grass is slower than in grain sorghum. Interestingly, neither oxabetrinil nor fenclorim increase degradation of metolachlor as it is seen under natural conditions in the unsafened control plants.

In summary, it can be concluded that in grasses investigated in this study, a major factor of the safener action against metolachor injury lies in their capability to increase metabolism of the herbicide as previously reported for pretilachlor in rice (cv. 'IR 54') and fenclorim (Stingelin, 1986).

VI. DISCUSSION AND CONCLUDING REMARKS

The differential responses of grasses toward protection by safeners against metolachlor or pretilachlor injury may result from several reasons: (1) The uptake of metolachlor or pretilachlor per unit of fresh weight varies from species to species. Although the same amounts of safeners per seed weight have been applied, the two safeners shift the sigmoidal dose–response curves individually for each species to higher herbicide rates. (2) In undamaged plants, the safeners

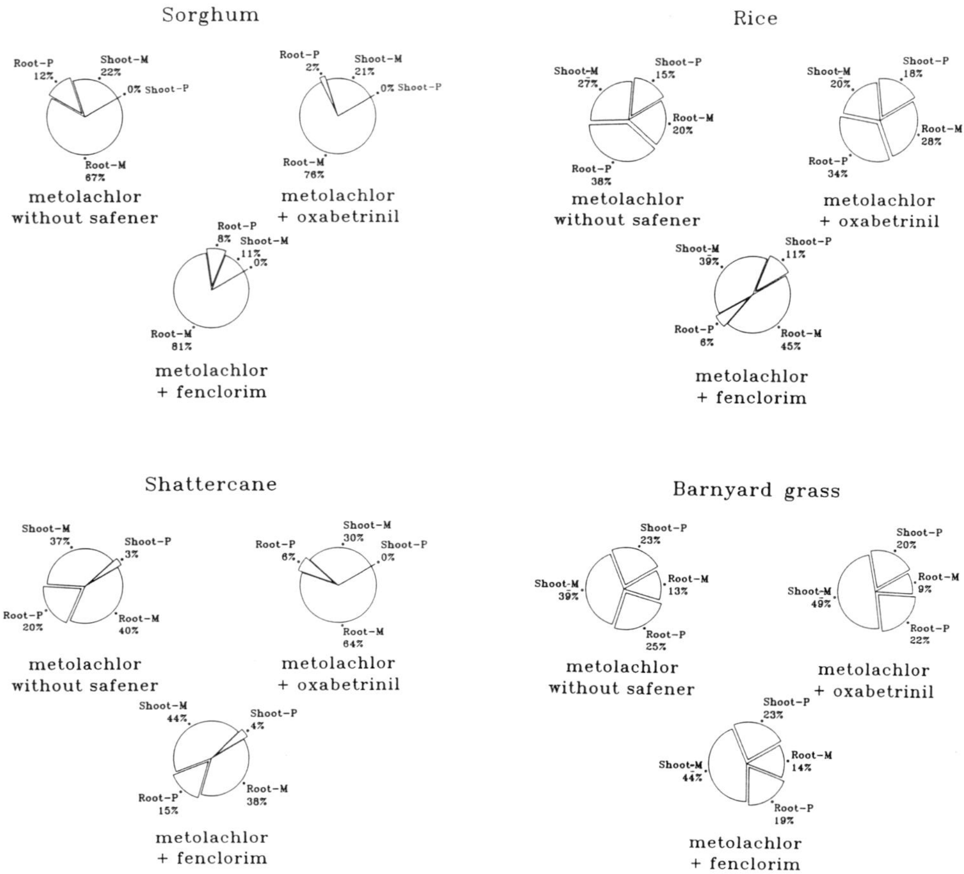

Fig. 9. Relative percentage distribution of [^{14}C]metolachlor and its degradation products in roots and shoots of sorghum, shattercane, rice, and barnyard grass. Root-P = percentage of metolachlor in roots; Root-M = percentage of metolachlor metabolites in roots; Shoot-P = percentage of metolachlor in shoots; Shoot-M = percentage of metolachlor metabolites in shoots. Experimental conditions: safened and unsafened plants (3–4 days old) were exposed to [^{14}C]metolachlor for 48 hr. Metolachlor and its metabolites were extracted with aqueous methanol (80%, v/v) and separated via HPLC with a water:acetonitrile gradient. Data presented are averages of two experiments with 20 plants each (barnyard grass experiments had 30 plants each).

had only a minor influence on the total uptake of the herbicides, as measured by ^{14}C uptake. However, the safeners change the degradation rate of the herbicides. As a consequence, oxabetrinil as well as fenclorim reduces the amount of the parent herbicide in the root and shoot tissue of those grasses which are susceptible to them.

In intact plants, the elongation rates of single leaves were taken to demonstrate the inhibitory effect of the herbicides and the safening effect of the two safeners. The chemicals were applied to the roots in a nutrient solution. In grain sorghum the combined application of cyometrinil and metolachlor prevented the growth reduction caused by metolachlor (Christ, 1981). In rice fenclorim prevented the growth reduction caused by pretilachlor (Christ, 1985). In these studies it was shown that in sorghum the safener must be applied at the same time as metolachlor to prevent the damage. A delayed application of the safener had no effect. In rice, however, it was possible to treat with fenclorim 1 day after the application of pretilachlor and still achieve a full safening effect on this plant. From these data it is concluded that the mechanism of action of these two safeners (oxabetrinil and fenclorim) may be different.

Effects of acetanilide herbicides and their safeners have been associated with wax biosynthesis, which is inhibited by metolachlor or pretilachlor. Oxabetrinil prevents metolachlor from inhibiting synthesis of the dominating C-28 to C-30 chain length constituents of the sorghum epicuticular waxes (Ebert and Ramsteiner, 1984). In his review, Fedtke (1985) compared the enzymatic pathways of long chained fatty acids and thiolcarbamate as well as chloroacetanilide metabolism. He speculated that oxidative enzyme systems such as mixed function oxygenases (MFO; EC 1.14.14.1) which are involved in wax biosynthesis, may also lead to an increased oxygenation and thereby metabolism of these herbicides. This hypothesis is supported by the findings of Hatzios (1983a,b), who showed that piperonyl butoxide, an inhibitor of MFO, synergized the activity of EPTC and metolachlor on corn and grain sorghum, respectively.

It has recently been shown that oxabetrinil influences the metabolism of metolachlor in sorghum, which proceeds via glutathione conjugation (Blattmann *et al.*, 1986). Fuerst and Gronwald (1986) as well as Zama and Hatzios (1986) showed an increased conjugation rate of metolachlor in grain sorghum in the presence of oxabetrinil. The same effect was observed in the case of pretilachlor and fenclorim in rice (Stingelin, 1986).

Furthermore, it has been shown by Zama and Hatzios (1986) that addition of tridiphane, an inhibitor of the enzyme glutathione-*S*-transferase (GST; EC 2.5.1.18), reduced the formation of GS–metolachlor in untreated and oxabetrinil-treated seedlings. They suggested that oxabetrinil may stimulate the spontaneous and enzymatic conjugation of metolachlor with GSH and in this way protect grain sorghum against injury by metolachlor.

It has also been shown that a mechanism of safener action against alachlor injury to corn may be attributed, at least partly, to the ability of cyometrinil to stimulate the activity or induce the *de novo* synthesis of GSTs that conjugate alachlor to GSH (Mozer *et al.*, 1983). Wiegand *et al.* (1986) have proven the existence of different GST species in corn. GST I conjugates GSH to alachlor and atrazine. GST II, which is induced by dichlormid, conjugates GSH to alachlor. Furthermore, Wiegand *et al.* (1986) have shown that increases in GST enzymes occur at the transcriptional level. The GST isozymes, active on alachlor degradation in corn, are reported to be different from the isozyme conjugating EPTC with GSH (Lay and Casida, 1976).

Similar investigations for weeds such as shattercane, barnyard grass, and other grasses shown in Table I are not available. Since GST enzyme levels and activities may vary in different plant families and species, it is envisaged that different safener structures would exhibit a high degree of specificity for selected plant species.

Acknowledgments

The SEM micrographs were prepared by Mrs. Ch. Brücher. The technical assistance of Mr. J. Gaudin and Mrs. Ch. Gottschling is greatly appreciated. The stimulating discussion with Drs. K. Pfister, H.C. Steinrücken, and L. Streit is also gratefully appreciated.

References

Blattmann, P., Gross, D., Kriemler, H. P., and Ramsteiner, K. (1986). Identification of thio-lactic acid type conjugates as major degradation products in the glutathione dependent metabolism of the α-chloroacetamide herbicides metolachlor (CGA 24705), dimetachlor (CGA 17020) and pretilachlor (CGA 26423) in plants. *IUPAC Int. Congr. Pestic. Chem., 6th*, Abstr. No. 7A-02.

Christ, R. A. (1981). The effect of CGA 43089 as a safener of metolachlor in sorghum [*Sorghum bicolor*] (recordings of elongation rates on single sorghum leaves). *Weed Res.* **21**, 1–8.

Christ, R. A. (1985). Effect of CGA 123407 as a safener for pretilachlor in rice (*Oryza sativa* L.). Recordings of elongation rates of single rice leavs. *Weed Res.* **25**, 193–200.

Ebert, E. (1980). Herbicidal effects of metolachlor (2-chloro-*N*-[2-ethyl-6-methylphenyl]-*N*-[2-methoxy-1-methyl-ethyl] acetamide) at the cellular level in sorghum. *Pestic. Biochem. Physiol.* **13**, 227–236.

Ebert, E. (1982). The role of waxes in the uptake of metolachlor into sorghum in relation to the protectant CGA 43089. *Weed Res.* **22**, 305–311.

Ebert, E., and Ramsteiner, K. (1984). Influence of metolachlor and the metolachlor protectant CGA 43089 on the biosynthesis of epicuticular waxes on the primary leaves of *Sorghum bicolor* Moench. *Weed Res.* **24**, 383–389.

Ellis, J. F., Peek, J. W., Boehle, J.,Jr., and Müller, G. (1980). Effectiveness of a new safener for protection of sorghum (*Sorghum bicolor*) from metolachlor injury. *Weed Sci.* **28**, 1–5.

Fedtke, C. (1982). "Biochemistry and Physiology of Herbicide Action." Springer-Verlag, Berlin and New York.

Fedtke, C. (1985). Einsatzmöglichkeiten und Wirkungsweisen von Herbizid-Antidots. *Z. Pflanzenkr. Pflanzenschutz.* **92**, 654–664.

Fuerst, E. P., and Gronwald, J. W. (1986). Induction of rapid metabolism of metolachlor in sorghum *(Sorghum bicolor)* shoots by CGA 92194 and other antidotes. *Weed Sci.* **34**, 354–361.

Gerber, H. R., Müller, G., and Ebner, L. (1974). CGA 24705, a new grasskiller herbicide. *Proc. Br. Weed Control Conf.* **12**, 787–794.

Görög, K., Muschinek, G. Y., Mustardy, L. A., and Faludi-Daniel, A. (1982). Comparative studies of safeners for the prevention of EPTC injury in maize. *Weed Res.* **22**, 27–33.

Hatzios, K. K. (1983a). Effects of CGA 43089 on responses of sorghum *(Sorghum bicolor)* to metolachlor combined with ozone and antioxidants. *Weed Sci.* **31**, 280–284.

Hatzios, K. K. (1983b). Herbicide antidotes: Development, chemistry, and mode of action. *Adv. Agron.* **36**, 265–316.

Ketchersid, M. L., Vietor, D. M., and Merkle, M. G. (1982). CGA 43089 effects on metolachlor uptake and membrane permeability in grain sorghum *(Sorghum bicolor)*. *J. Plant Growth Regul.* **1**, 285–294.

Laanio, T. (1980). "Biochemistry," Biochem. Rep. 1/80. Agricultural Division, Ciba-Geigy, Basle, Switzerland.

Lay, M.-M., and Casida J. E. (1976). Dichloroacetamide antidotes enhance thiocarbamate sulfoxide detoxification by elevating corn root glutathione content and glutathione content and glutathione-*S*-transferase activty. *Pestic. Biochem. Physiol.* **6**, 442–456.

Mozer, T. J., Tiemeier, D. C., and Jaworski, E. G. (1983). Purification and characterization of corn glutathione-*S*-transferase. *Biochemistry* **22**, 1068–72.

Murakami, S., and Ebner, L. (1983). Pretilachlor—A new selective herbicide for transplanted rice in Japan. *Proc. Asian-Pac. Weed Sci. Soc., Conf., 9th, 1983*, pp. 532–538.

Nyffeler, A., Gerber, H. R., and Hensley, J. R. (1980). Laboratory studies on the behaviour of the herbicide safener CGA 43089. *Weed Sci.* **28**, 6–10.

Paradies, I., Ebert, E., and Elstner, E. F. (1981). Metolachlor (2-chloro-*N*-[2-ethyl-6-methylphenyl]-*N*-[2-methoxy-1-methylethyl)acetamide) and the metolachlor safener CGA 43089 [α-cyano-methoximino)-benzacetonitrile] in sorghum seedlings: Correlations between morphological effects and ethylene formation. *Pestic. Biochem. Physiol.* **15**, 209–212.

Rufener, J., and Quadranti, M. (1983). Early weed control in wet sown rice: The role of the safener CGA 123407. *Proc. Int. Congr. Plant Prot. 9th, pp. 332–338.*

Rufener, J., Nyffeler, A., and Peek, J. W. (1982). CGA 92194, a new safener to protect sorghum from injurious effects of metolachlor. *Proc. Br. Crop Prot. Conf.—Weeds* **1**,461–467.

Stingelin, J. (1986). "Biochemistry," Biochem. Rep. 1/86. Agricultural Division, Ciba-Geigy, Basle, Switzerland.

Wiegand, R. C., Shah, D. M. Mozer, T. J., Harding, E. I., Diaz-Collier, J., Sanders, C., Jaworski, E. G., and Tiemeier, D. C. (1986). Messenger RNA encoding a glutathione-*S*-transferase responsible for herbicide tolerance in maize is induced in response to safener treatment. *Plant Mol. Biol.* **7**, 235–243.

Zama, P., and Hatzios, K. K. (1986). Effects of CGA 92194 on the chemical reactivity of metolachlor with glutathione and metabolism of metolachlor in grain sorghum (*Sorghum bicolor*). *Weed Sci.* **34**, 834–841.

Chapter 9

Protection of Grass Crops from Sulfonylurea and Imidazolinone Toxicity

Michael Barrett
Department of Agronomy
University of Kentucky
Lexington, Kentucky 40546

I. INTRODUCTION

The sulfonylurea and imidazolinone herbicides are two recently introduced families of herbicide chemistry which represent significant additions to the arsenal of weed management tools. Representatives of these herbicides (Fig. 1) are used in soybean [*Glycine max* (L.) Merr.], wheat (*Triticum aestivum* L.), barley (*Hordeum vulgare* L.), rice (*Oryza sativa* L.), and noncrop weed management systems. These herbicides are highly active at low application rates (Anonymous, 1985a,b, 1986a,b,c) and, in many cases, are relatively persistent in soils (Anonymous, 1986a,b,c; Mills *et al.*, 1987; Witt *et al.*, 1985).

A number of safeners have shown potential in protecting grass crops from toxicity due to these herbicides. This opens new opportunities for understanding the mechanism of herbicide safening as both the site of action and selectivity mechanisms have been established for these herbicides. Thus, it is possible to study safener effects on both the biochemical target for the herbicides and selectivity mechanisms as a basis for the herbicide protection.

Crop Safeners for Herbicides

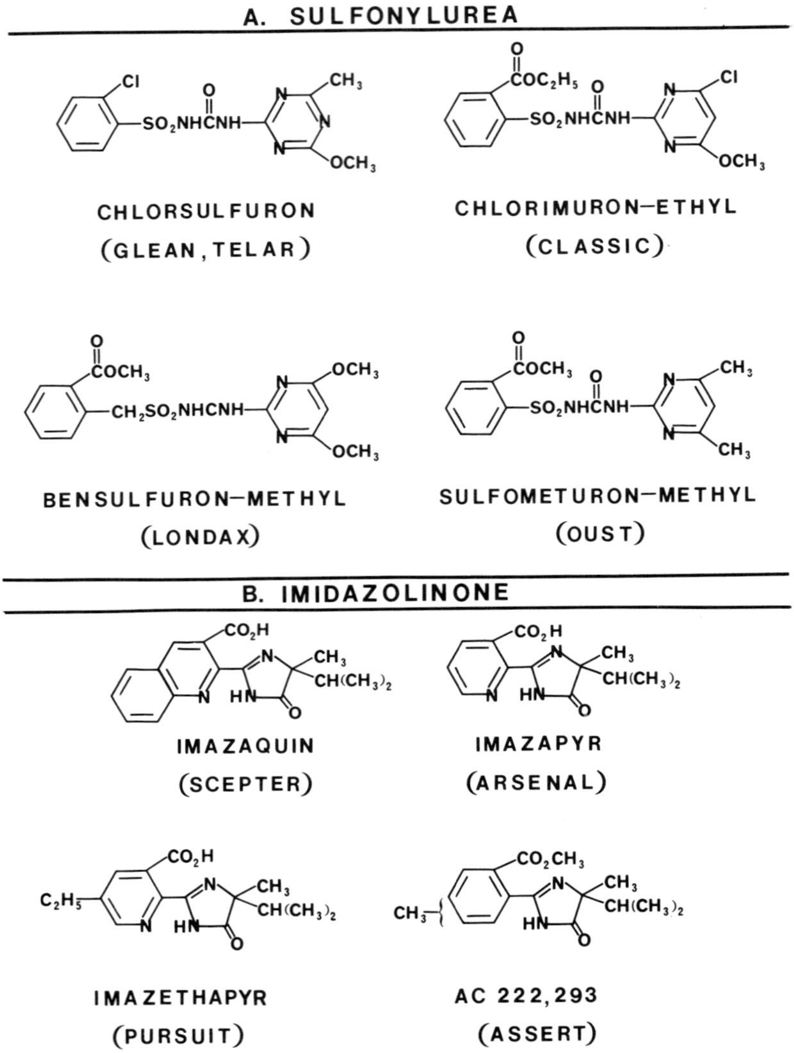

Fig. 1. Chemical structures, common names or code numbers, and trade names (if any) for representative sulfonylurea and imidazolinone herbicides.

This chapter reviews the efficacy of selected chemical safeners as protectants of grass crops against injury from sulfonylurea and imidazolinone herbicides. Our present knowledge of the mode of action, metabolism, and selectivity of these herbicidal classes and the potential involvement of these processes in the protective action of safeners is also reviewed.

II. MECHANISM OF ACTION OF SULFONYLUREA AND IMIDAZOLINONE HERBICIDES

A. Sulfonylureas

Initial studies on the sulfonylurea mechanism of action (Ray, 1982a) indicated that chlorsulfuron did not inhibit cell elongation induced by indoleacetic acid (IAA), cytokinin-promoted cell expansion, or gibberellic acid induced cell elongation. In addition, neither mitochondrial activity (O_2 uptake) nor photosynthesis (O_2 evolution and CO_2 fixation) was affected by chlorsulfuron.

Cell division was identified as the most sensitive plant process for chlorsulfuron inhibition (Ray, 1982a). Chlorsulfuron (1 ppm) reduced the mitotic index 87% in broad bean (*Vicia faba* L.) root tips. Analysis of the frequency distribution of the mitotic stages indicated that the inhibition of cell division was not due to an arrested mitotic stage. Rost (1984) confirmed the sensitivity of the cell division process to chlorsulfuron and noted a major blockage in G_2 and a minor blockage at G_1.

Deoxyribonucleic acid (DNA) synthesis (thymidine incorporation) and ribonucleic acid (RNA) synthesis (uridine incorporation) were inhibited by chlorsulfuron treatment (Ray, 1982a). Chlorsulfuron (1 ppm) inhibited DNA synthesis 60% within 2 hr of treatment of corn (*Zea mays* L.) roots. Lower chlorsulfuron levels, 0.01 and 0.10 ppm, inhibited DNA synthesis 40 and 80%, respectively, 6 hr after treatment. RNA synthesis was also inhibited 43% within 12 hr of chlorsulfuron treatment of pea (*Pisum sativum* L.) roots. Protein synthesis was not affected.

Subsequent experiments demonstrated that chlorsulfuron did not directly inhibit DNA synthesis. Chlorsulfuron did not affect the activity of DNA polymerase (EC 2.7.7.7) isolated from *Micrococcus luteus* or corn (Ray, 1982b). Also, thymidine kinase (EC 2.7.1.21) activity in crude extracts of roots of barley or corn was not inhibited by chlorsulfuron. Limited availability of purine and/or pyrimidine bases could also decrease DNA synthesis. However, addition of deoxynucleotides to the solution bathing corn roots did not overcome the inhibition of DNA synthesis or growth due to chlorsulfuron treatment. Rost (1984) hypothesized that the chlorsulfuron disruption of cell division was due to inhibition of "cell cycle specific RNA synthesis."

Experiments with bacterial cultures provided the first evidence for the actual biochemical lesion caused by sulfonylurea treatment. Inhibition of *Salmonella typhimurium* growth on media containing L-valine and treated with sulfometuron-methyl was reversed by addition of L-isoleucine (LaRossa and Schloss, 1984). The bacterial enzyme acetolactate synthase (ALS; EC 4.1.3.18), also called acetohydroxyacid synthase (AHAS), was identified as the sulfonylurea site of action in bacteria.

ALS catalyzes the first step in the biosynthesis of valine, leucine, and isoleucine (Goodwin and Mercer, 1983). Subsequently, it was found that ALS in plants is also inhibited by sulfonylureas (Chaleff and Mauvais, 1984; Ray, 1984). The inhibition of pea root growth following chlorsulfuron treatment could be overcome by additions of valine and isoleucine to the root medium (Ray, 1984) as shown in Table I. ALS isolated from a variety of plant species, both sulfonylurea-tolerant and susceptible, was inhibited (Table II) by all tested sulfonylurea herbicides (Ray, 1984). Sulfonylurea concentrations reducing ALS activity 50% (I_{50}) ranged from approximately 3 to 100 n*M*.

Additional evidence for ALS as the primary site of sulfonylurea activity has come from studies of sulfonylurea-resistant bacteria, yeast and plant mutants (Chaleff and Ray, 1984; Falco and Dumas, 1985; LaRossa and Schloss, 1984). In many of these mutants, the presence of an altered ALS form which is less sensitive to inhibition by sulfonylureas is the basis for the developed resistance.

B. Imidazolinones

Imidazolinone herbicides were also shown to inhibit ALS activity (Shaner *et al.*, 1984). This is despite any obvious structural similarity between the imidazolinones and the sulfonylureas. The herbicide concentrations required for ALS inhibition by the imidazolinones are considerably higher than those for the

Table I Reversal of Chlorsulfuron Inhibition of Excised Pea Root Growth by Amino Acid Supplements[a]

	Average net root length (mm)	
Supplement[b]	**Control**	**28 n*M* Chlorsulfuron**
No addition	52.8 ± 1.1	10.3 ± 0.3
Leu,Ile,Ala,Val	48.8 ± 1.8	50.2 ± 1.0
Leu,Ala,Val	49.2 ± 1.6	29.3 ± 0.5
Leu,Ile,Ala	44.1 ± 1.2	11.1 ± 0.5
Leu,Ile,Val	42.9 ± 1.0	47.4 ± 0.8
Ile,Ala,Val	45.0 ± 1.1	34.1 ± 1.2
Leu,Ile	24.7 ± 2.3	11.0 ± 0.5
Leu,Val	50.9 ± 1.0	18.8 ± 0.5
Leu,Ala	50.9 ± 1.0	10.7 ± 0.3
Ile,Val	44.8 ± 0.8	42.0 ± 1.2
Ile,Ala	50.9 ± 0.9	10.7 ± 0.4
Val,Ala	44.6 ± 1.1	20.2 ± 1.3
Leu	26.1 ± 2.4	11.9 ± 0.5
Ala	54.1 ± 1.5	10.4 ± 0.5
Ile	34.8 ± 0.8	10.1 ± 0.3
Val	47.6 ± 1.1	20.9 ± 0.5

[a]From Ray (1984).
[b]The concentration of each amino acid was 100 μ*M*.

Table II Sulfonylurea I_{50} Values for Acetolactate Synthase Isolated from Various Plants[a]

	I_{50} (n*M*)			
Plant	**Chlorsulfuron**	**Sulfometuron-methyl**	**DPX5384**	**Chlorimuron-ethyl**
Pea	20.9	15.3	63.8	6 0
Wheat	21.8	12.9	13.6	5.1
Rice	24.0	18.1	16.3	5.8
Soybean	32.4	46.4	74.6	7.7
Wild mustard (*Brassica* sp.)	10.9	9.3	12.6	3.1
Barnyard grass (*Echinochloa crusgalli* (L.) Beauv.)	24.3	21.1	21.8	3.6
Morning glory (*Ipomea* sp.)	24.3	19.8	97.9	6.8

[a]From Ray (1986).

sulfonylureas. Reported ALS K_i values for the imidazolinones range from 1.7 μ*M* for AC 222, 164 to 12 μ*M* for imazapyr (Shaner *et al.*, 1984). Additions of valine, leucine, and isoleucine can partially reverse imidazolinone inhibition (Table III) of corn seedling growth (Anderson and Hibberd, 1985; Shaner and Reider, 1986). Corn seedlings supplied with the exogenous amino acids and treated with 15 μ*M* imazapyr produced 83% of the dry weight of control plants (Shaner and Reider, 1986). Plants treated with imazapyr without the amino acids produced 34% of the growth of the untreated plants. Additions of leucine, valine, and isoleucine to corn cell suspension cultures completely reversed the growth inhibition caused by 1 μ*M* imazapyr when compared to cultures with the amino

Table III Reversal of Imazapyr Inhibition of Corn Seedling Growth by Amino Acid Supplements[a]

	Growth[b]	
Treatment	**Dry weight (mg/plant)**	**Percentage of control**
Control	178	100
Imazapyr	75[c]	34.2[c]
Imazapyr + Val, Leu, Ile	147	82.6

[a]From Shaner and Reider (1986).

[b]Corn grown 14 days in sand culture watered daily with half-strength Hoagland's solution with or without 15 μ*M* imazapyr and with or without 1 m*M* amino acids. Shoot weights recorded 14 days after the start of the treatment.

[c]Significantly ($P = 0.001$) different from control.

acids alone (Anderson and Hibberd, 1985). However, the supplementary amino acids were slightly (12%) inhibitory themselves to the cultured cells. Additional evidence for the involvement of ALS inhibition in imidazolinone phytotoxicity was provided by the discovery of a corn mutant resistant to imidazolinones (Anderson and Georgson, 1986). The basis for the corn resistance was attributed to an altered ALS form that was insensitive to inhibition by imazapyr.

Imidazolinone herbicide effects on other plant physiological processes are similar to those of sulfonylureas. DNA synthesis in corn was inhibited 70–90% within 24 hr of treatment with 150 μM imazapyr (Shaner and Reider, 1986). The inhibition of DNA synthesis could be overcome by the addition of valine, leucine, and isoleucine. Neutral sugars were increased 39% within 24 hr after treatment with imazapyr. Respiration and RNA synthesis were inhibited slightly, 32 and 15%, respectively. Lipid and protein synthesis rates were not affected by imazapyr 24 hr after treatment. However, soluble protein levels were reduced 40% and free amino acid levels increased 32% within 24 hr after imazapyr treatment.

C. Cross-Sensistivity to the Imidazolinones and Sulfonylureas

The evidence supports ALS inhibition as the common site of action for both the imidazolinone and the sulfonylurea herbicides. The two groups have very different I_{50} values for ALS inhibition with imidazolinone concentrations being 10- to 100-fold higher than those of the sulfonylureas (Ray, 1986; Shaner *et al.*, 1984). One difficulty in interpreting the I_{50} values is that both herbicide families exhibit a slow, tight form of binding (Muhitch *et al.*, 1987; Ray, 1984). This makes I_{50} values determined shortly after ALS exposure to the herbicides an order of magnitude higher than I_{50} values determined after several hours of incubation. Another complication in comparing I_{50} values is the different protocols used to isolate the ALS in the separate studies with the two herbicide groups.

The disparity in chemical structure between the two herbicide groups (Fig. 1) suggests a different site/mode of ALS inhibition or a pleiotropic binding site (Fedtke and Trebst, 1987). Schloss (1984) proposed that sulfonylureas interfere with binding of the second ketoacid molecule to an ALS–FAD–hydroxyethyl-thiamine pyrophosphate complex. This was based on evidence for sulfonylurea induced alterations of enzyme-bound FAD. The remote analogy of the imidazolinone structure to valine and *N*-phthalyl-L-valine, both of which allosterically control plant ALS activity, suggested that imidazolinone action was directed toward allosteric control rather than the catalytic center (Fedtke and Trebst, 1987). Determination of whether the two herbicides inhibit ALS activity through the same or differing mechanisms will require additional experimentation. However, it is important to note that an altered ALS form conferring resistance in corn to imidazolinones is also less sensitive to inhibition by sulfonylureas

(Anderson and Hibberd, 1985). The "resistance factor" (defined as the ratio of the I_{50} value for the mutant ALS versus the wild type ALS) was 3000 for the imidazolinone but only 30 for the sulfonylureas (Fedtke and Trebst, 1987).

III. SELECTIVITY MECHANISMS FOR SULFONYLUREA AND IMIDAZOLINONE HERBICIDES

A. Sulfonylureas

1. Herbicide Metabolism

The primary bases for the range in tolerance to chlorsulfuron among plant species are quantitative and qualitative differences in metabolism of the herbicide to nontoxic forms. Although uptake and translocation differences exist between tolerant and sensitive plants, the magnitude of the differences was not considered great enough to account for the large (100- to 4000-fold) difference in chlorsulfuron toxicity between these species (Sweetser *et al.*, 1982).

Sweetser *et al.* (1982) established that chlorsulfuron-resistant monocotyledonous plants such as wheat, barley, wild oats (*Avena fatua* L.), and annual bluegrass (*Poa annua* L.) metabolized >90% of the chlorsulfuron in leaves within 24 hr of treatment. Sensitive dicotyledonous plants such as cotton (*Gossypium hirsutum* L.), soybean, mustard (*Brassica* sp.), and sugar beet (*Beta vulgaris* L.) had 80% or more unmetabolized chlorsulfuron in the leaves within the same period. The proposed pathway of chlorsulfuron metabolism in the tolerant grass plants includes hydroxylation at the 5 position of the phenyl ring of the herbicide followed by conjugation with a carbohydrate (Fig. 2). Tolerant plants contained 40 to 60% of the absorbed chlorsulfuron radioactivity present as

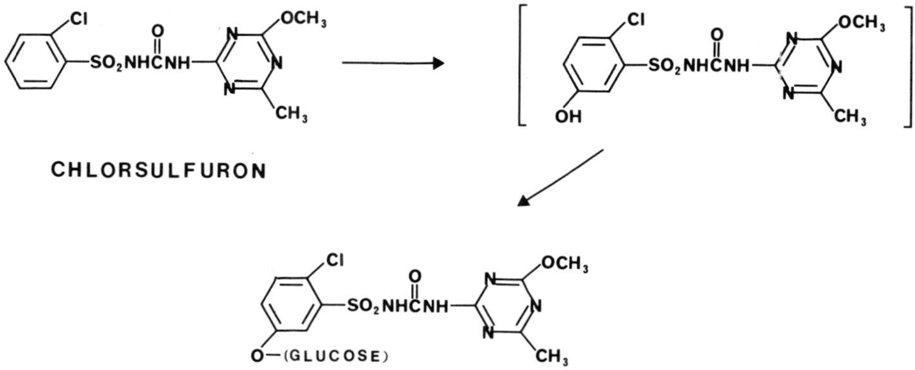

Fig. 2. Proposed pathway for chlorsulfuron metabolism in tolerant grasses. (From Sweetser *et al.*, 1982.)

an *O*-glucoside after herbicide application. Sensitive broadleaf plants formed little (<5%) or no *O*-glucoside of chlorsulfuron.

Variations in the phytotoxicity of chlorsulfuron among broadleaf species were also attributed to differential detoxication of the herbicide. Hageman and Behrens (1984) observed an approximate 18,000-fold difference in the chlorsulfuron rate required to reduce growth in sensitive velvetleaf (*Abutilon threophrasti* Medik.) versus tolerant eastern black nightshade (*Solanum ptycanthum* Dun.). There were few significant differences in chlorsulfuron retention on the plant foliage, absorption, or translocation between the two species. Eastern black nightshade was much more efficient in converting absorbed chlorsulfuron to metabolites. Over a 3 day period, eastern black nightshade metabolized 81% of the chlorsulfuron, whereas velvetleaf converted only 7% of the herbicide to metabolites.

Hutchinson *et al.* (1984) proposed an alternative pathway for chlorsulfuron metabolism in tolerant broadleaves as compared to that found in monocots (Sweetser *et al.*, 1982). Chlorsulfuron was first hydroxylated at the 6-methyl position of the triazine ring to form 2-chloro-*N*-([4-(hydroxymethyl)-6-methoxy-1,3,5-triazin-2-yl]aminocarbonyl)benzenesulfonamide (Fig. 3). This hydroxylated metabolite is then conjugated to a sugar. Hydroxylated chlorsulfuron was much less toxic than the parent herbicide.

Differential metabolism was also reported to account for the tolerance of soybean to chlorimuron-ethyl (Claus, 1987).

2. Altered ALS Forms

While plant tolerance to sulfonylureas is based on detoxication of the parent herbicide, resistance in plants, bacteria, and yeast is associated with altered ALS forms (Chaleff and Ray, 1984; Falco and Dumas, 1985; LaRossa and Schloss,

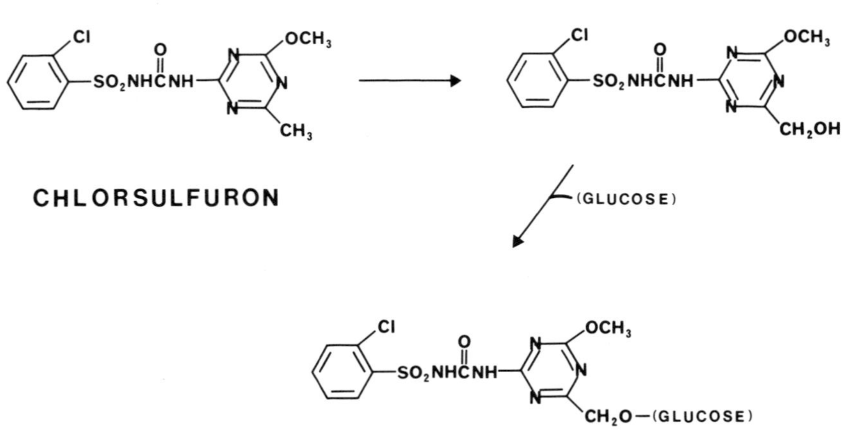

Fig. 3. Proposed pathway for chlorsulfuron metabolism in tolerant broadleaves. (From Hutchinson *et al.*, 1984.)

1984). The altered ALS is less sensitive to sulfonylurea inhibition. A single nucleotide change in the gene coding for the ALS enzyme is responsible for the observed resistance. The mutation for resistance in the gene from *Escherichia coli* (Yadav *et al.* 1986) substituted valine for alanine in ALS while the change in the yeast gene was substitution of a serine for proline in the ALS, (Falco and Dumas, 1985).

B. Imidazolinones

1. Metabolism

Imidazolinone tolerance in plants is also associated with differential metabolism of the parent herbicide to nontoxic products. However, the metabolic pathways detoxifying imazaquin are not as well documented as for chlorsulfuron. Shaner and Robson (1985) found imazaquin half-lives of 3, 12, and 30 days for tolerant soybean, moderately tolerant velvetleaf, and susceptible cocklebur (*Xanthium strumarium* L.), respectively (Table IV). Increased tolerance of velvetleaf at the four leaf velvetleaf stage compared to the cotyledon stage was attributed to a 72% decrease in imazaquin absorption and a 62% decrease in the imazaquin half-life in the older plants compared to younger plants. The selectivity among grass species for the imidazolinone herbicide AC 222, 293 (isomeric mixture of 6-4-isopropyl-4-methyl-5-oxo-2-imidazolin-2-yl]-*meta*-toluate and methyl 2-[4-isopropyl-4-methyl-5-oxo-2-imidazolin-2-yl]-*para*-toluate) was attributed to metabolism of the parent ester to herbicidally active free acids only in the sensitive grass species (Brown *et al.*, 1987). Tolerant wheat and corn contained less that 1% of the extracted radioactivity as the AC 222, 293 free acid 2 weeks after herbicide treatment while sensitive wild oat had 7.8% present as the more herbicidally active meta-isomer of AC 222, 293 acid. All three species detoxified AC 222, 293 by oxidation of the aryl methyl group to an alcohol which was subsequently conjugated with glucose. The primary difference between the three species was the efficient activation in wild oats of AC 222, 293 to the free acid.

Table IV Metabolism of Foliar-Applied Imazaquin 3 Days after Application[a]

Species	Plant part	Percentage of imazaquin remaining	Imazaquin half-life[b] (days)
Soybean	Treated leaf	38.3	3
Cocklebur	Treated leaf	99.9	30
Velvetleaf	Treated leaf	88.7	12
LSD (0.05)		10.3	2

[a]From Shaner and Robson (1985).
[b]Half-life was determined in excised tissue.

2. Alteration of the Target Site

Tolerance to imidazolinone herbicides appears to be primarily based on differential detoxication systems among plant species. Resistance to imidazolinone injury due to an altered ALS form in corn was also demonstrated (Anderson and Georgson, 1986). ALS from imidazolinone-resistant corn was 3000 times less sensitive to herbicide inhibition than ALS from normal corn.

IV. PROTECTION FROM HERBICIDE TOXICITY

A. Sulfonylureas

1. Protectant Activity

Parker *et al.* (1980) were the first to demonstrate a reduction in chlorsulfuron activity on grass species following protectant treatment (Table V). As with other herbicides (Hatzios, 1983), successful protection was achieved only with monocotyledonous species and not with dicot species (Parker *et al.*, 1980; Rubin and Casida, 1985; Sweetser, 1985). Seed dressing of corn, wheat, rice, or sorghum [*Sorghum bicolor* (L.) Moench] with naphthalic anhydride (NA) at 0.5% (w/w) increased the tolerance of all species to chlorsulfuron (Tables V and VI). Chlorsulfuron at 5 g/ha reduced corn shoot and root growth 50–60%, but following seed dressing with NA a chlorsulfuron rate of 40 g/ha was needed to reduce growth to the same extent. An approximate chlorsulfuron "safening factor" of four- to eightfold was observed for NA treatment of corn, sorghum, and rice. Wheat and barley were also protected by NA treatment, although barley was very tolerant of chlorsulfuron under the experimental conditions employed.

Table V Protection of Corn from Chlorsulfuron Injury with Naphthalic Anhydride (NA) or Dichlormid[a]

	Protectant		
Chlorsulfuron rate[b] (g/ha)	**None**	**NA (0.5%w/w)**	**Dichlormid (0.5%,w/w)**
	Shoot dry weight (% of control)		
0	100	78	80
5	15	80	70
10	10	70	45
20	5	50	20
Standard error 8%			

[a]From Parker *et al.* (1980).

[b]Corn ('Julia') seed, treated or untreated, was planted 2 cm deep in a sandy loam soil and chlorsulfuron was applied to the soil surface. Plants were harvested approximately 1 month after planting.

Table VI Protection of Sorghum and Rice from Chlorsulfuron Injury by Seed Treatment with Naphthalic Anhydride (NA, 0.5%(w/w))[a]

Chlorsulfuron rate (g/ha)	Sorghum[b]		Rice	
	No protectant	NA	No protectant	NA
	Shoot dry weight (%)			
0	100	100	100	85
5	62	90	30	70
10	30	70	10	65
20	15	59	5	57
40	10	48	2	40

Standard error sorghum = 8
Standard error rice = 7

[a]From Parker *et al.* (1980).
[b]Sorghum or rice seeds, treated or untreated, were planted 2 cm deep in a sandy loam soil and chlorsulfuron was applied to the soil surface. Plants were harvested 4–6 weeks after planting.

The safening factor for chlorsulfuron on wheat and barley was two- to fourfold, but the chlorsulfuron rates used ranged from 100 to 400 g/ha.

Dichlormid was less effective than NA in protecting the species studied by Parker *et al.* (1980) from chlorsulfuron damage. Whereas NA gave an eightfold or greater increase in the chlorsulfuron rate tolerated by corn, dichlormid seed dressing (0.5% w/w) or soil-applied (1 kg/ha) increased the tolerance two- to fourfold (Table V). Dichlormid was also less effective than NA when wheat or barley were used as the test species.

Hatzios (1984) confirmed the activity of NA in protecting corn from soil-applied chlorsulfuron. Both a carbamothioate-tolerant('DeKalb XL72AA') and a carbamothioate-sensitive ('DeKalb XL67') corn hybrid could be protected from chlorsulfuron injury with 0.5 or 1.0% (w/w) seed treatment with NA. The protection of the carbamothioate-tolerant hybrid was greater than that of the carbamothioate-sensitive hybrid at the low (30 and 60 g/ha) chlorsulfuron rates. The carbamothioate-sensitive hybrid was, however, more severely injured by chlorsulfuron. Only NA, of the protectants tested, was effective in decreasing the injury caused by postemergence chlorsulfuron applications to corn 10–15 cm tall.

Similar to the results of Parker *et al.* (1980), dichlormid was not very effective in protecting corn from chlorsulfuron (Hatzios, 1984). Tests with the two oxime protectants, cyometrinil and CGA-92194, produced chlorsulfuron protection levels intermediate between NA and dichlormid (Hatzios, 1984). A carbamothioate-tolerant corn hybrid was more effectively protected from chlorsulfuron injury than the carbamothioate-sensitive hybrid by both oxime protectants.

O'Leary and Prendeville (1985) also observed decreased corn phytotoxicity from chlorsulfuron following seed treatment with NA (1% w/w or a 5 ppm (w/v) water suspension). NA was very effective in preventing both foliar and root injury when chlorsulfuron (0.06 ppm (w/v)) exposure was limited to the below ground corn shoot zone. However, when chlorsulfuron was placed in the corn root zone, the foliage was significantly less protected than with only shoot exposure and the roots were not protected at all. Corn roots were demonstrated to be more sensitive to injury from soil residues of sulfonylureas than were shoot tissues (Groves and Foster, 1985; Mueller *et al.*, 1985; Witt *et al.*, 1985).

Protection of corn against injury from foliar-applied chlorsulfuron could be achieved by NA treatment of seeds or application of NA (0.7 kg/ha) as a spray to the foliage 1 week prior to chlorsulfuron application. NA application 1 week prior to chlorsulfuron treatment or a simultaneous application of NA plus chlorsulfuron to the foliage of plants previously receiving NA seed dressing significantly increased the protection over seed dressing alone. This was true when seed dressing was done as a NA water suspension, but not with a dry application of NA to the seeds.

The treatment of corn foliage with NA either 1 week prior or simultaneously with chlorsulfuron also protected the corn roots from chlorsulfuron toxicity. Simultaneous application of NA with chlorsulfuron tended to be less effective for protection of the roots than the application of NA 1 week ahead of chlorsulfuron.

In addition to work with chlorsulfuron, corn was also protected from metsulfuron-methyl by seed dressing with NA (0.5% w/w) or to a lesser extent by M-32988 (Mersie and Foy, 1984).

2. Mechanisms of Crop Protection against Sulfonylureas

While there is acceptance of the phenomenon of chemical protection of grass crops from chlorsulfuron injury, there is less agreement on the protective mechanisms involved. Studies have focused on protectant effects on the target site (ALS), detoxication of the herbicide, and/or protectant/auxin interactions as the bases for the safening of grass crops against sulfonylurea herbicides.

Rubin and Casida (1985) showed protection of corn from chlorsulfuron toxicity with dichlormid (1.0 kg/ha) applied as a soil drench treatment 5 to 7 days after planting and followed by a chlorsulfuron soil drench treatment 2 days later. A safening factor of greater than 4 was observed with the four hybrids used. Flurazole and CGA-92194 were less effective than the dichlormid under the conditions of this experiment. The level of protection offered by dichlormid was greater than that observed by Parker *et al.* (1980) and the relative safening by dichlormid versus CGA-92194 was reversed from the work of Hatzios (1984). The studies of Rubin and Casida (1985) also differ from previous work (Ray, 1984) in that exogenous amino acids (leucine, valine, isoleucine) only slightly reversed chlorsulfuron toxicity to roots of corn ('DeKalb XL25A'). The amino acids also did not significantly change chlorsulfuron toxicity to roots of pea or 'FR619' corn. The combination of dichlormid plus amino acids appeared to be additive in protecting 'XL25A' but dichlormid alone or in combination with the amino acids was ineffective in protecting 'FR619' corn or pea.

While dichlormid did not affect ALS activity *in vitro*, with or without chlorsulfuron present, Rubin and Casida (1985) detected a 30 and 24% increase in extractable ALS from 'XL25A' roots and shoots following dichlormid treatment, respectively. They hypothesized that elevation of ALS activity may contribute to the protectant action of dichlormid. However, the ALS increase appeared minor in comparison to the degree of the safening.

Where Rubin and Casida (1985) concentrated on protectant effects on ALS activity, Sweetser (1985) focused on alterations in chlorsulfuron, metsulfuron-methyl, and sulfometuron-methyl detoxication due to protectant treatment. Naphthalic acid, dichlormid, or cyometrinil treatment decreased the chlorsulfuron and metsulfuron-methyl half-lives more than 50% in corn (Table VII). Generally, naphthalic acid was slightly more effective than dichlormid and cyometrinil. The most effective stimulation of sulfonylurea metabolism in wheat occurred following treatment with naphthalic acid or cyometrinil. Dichlormid decreased the half-life of sulfometuron-methyl in wheat but had little effect on the rate of chlorsulfuron or metsulfuron-methyl metabolism in this species.

Feeding naphthalic acid (20 ppm) into cut leaves of corn rapidly produced a reduction in chlorsulfuron half-life. Maximum reduction in the chlorsulfuron half-life was achieved within 4 hr of protectant treatment. The effect of naphthalic acid treatment could be inhibited by inclusion of cyclohexamide in that treatment solution. Interestingly, cyclohexamide alone could reduce the persistence of chlorsulfuron in the corn leaf tissue.

Although no data were presented on the metabolic fate of chlorsulfuron in corn tissue, Sweetser (1985) stated that the metabolism of chlorsulfuron in corn included the pathways of tolerant broadleaf plants (Hutchinson *et al.*, 1984) and tolerant grasses (Sweetser *et al.*, 1982). Hydroxylation of chlorsulfuron at the

Table VII Effect of Protectant Treatment on Sulfonylurea Half-life in Various Species[a]

Plant species	Sulfonylurea herbicide	Relative tolerance[c]	Protectant[b]: None	Naphthalic acid	Dichlormid	Cyometrinil
			Half-life (hr)			
Corn	Chlorsulfuron	S-I	3.5	1.0	1.1	1.6
Wheat	Chlorsulfuron	T	1.8	0.8	1.8	0.7
Soybean	Chlorsulfuron	S	>50	>50		
Brassica kaber	Chlorsulfuron	S	>50	>50		
Corn	Metsulfuron-methyl	T	5.5	1.4	1.6	1.6
Wheat	Metsulfuron-methyl	S-I	2.0	1.0	1.8	0.8
Wheat	Sulfometuron-methyl	I	17.3	7.9	13.3	9.6

[a]*From Sweetser (1985).*

[b]Antidotes were applied to leaves of the species 16 hr prior to excising the leaves, placing the cut ends into herbicide solutions, and following metabolism of the absorbed herbicide over time.

[c]Relative tolerance to postemergence herbicide application. T = tolerant, I = intermediate, and S = sensitive.

6-methyl position of the triazine ring and hydroxylation at the 5-position of the phenyl ring was stated to occur in the corn. The hydroxylated triazine ring metabolite of chlorsulfuron was still phytotoxic, but less than the parent chlorsulfuron (Hutchinson *et al.*, 1984). The toxicity of the initial hydroxylated triazine ring metabolite of chlorsulfuron in corn, in comparison to the presence of only the detoxified chlorsulfuron product in wheat, was proposed to account for the sensitivity of corn to the herbicide despite a short half-life for chlorsulfuron in the corn tissue. Thus, although the half-life of chlorsulfuron in corn following treatment with naphthalic acid was actually less than the half-life of chlorsulfuron in wheat (Table VII), equivalent tolerance for the herbicide between the two species could not be expected.

Sweetser *et al.* (1982) further suggested, based on the metabolic pathways of chlorsulfuron detoxication, that a cytochrome P-450 system was involved in the process. The protectants were proposed to act through an induction of the P-450 system which could be blocked by cycloheximide.

Subsequent work by Frear *et al.* (1987) stands in contrast to the experimental findings of Sweetser (1985). Although naphthalic acid was an effective protectant for preventing corn injury from chlorsulfuron, the effects of naphthalic acid treatment on chlorsulfuron metabolism and ALS levels observed by Frear *et al.* (1987) do not agree with the results of Sweetser (1985) or Rubin and Casida (1985). The only chlorsulfuron detoxication pathway in corn detected by Frear *et al.* (1987) was the same as previously shown for wheat (Sweetser *et al.*, 1982). Thus, by this proposal, the initial chlorsulfuron metabolites would be nontoxic in both corn and wheat. Similar chlorsulfuron half-lives in the two species would be assumed to result in similar levels of tolerance to chlorsulfuron. However, as corn is more sensitive than wheat to chlorsulfuron, Frear *et al.* (1987) suggested that differential metabolism may not be the determinant factor in the variation in chlorsulfuron sensitivity between the two species.

Naphthalic acid did not increase the rate of chlorsulfuron metabolism in corn in the experiments of Frear *et al.* (1987). The half-life of chlorsulfuron was between 8 and 12 hr in these studies compared to the 3.5 hr (Table VII) found by Sweetser (1985). Frear *et al.* (1987) applied chlorsulfuron to intact leaves of corn seedlings while Sweetser (1985) fed chlorsulfuron to detached corn leaves by placing the cut ends into the treatment solution. The time required for foliar penetration in the study of Frear *et al.* (1987), compared to the rapid chlorsulfuron uptake with the detached leaf, could contribute to the difference between the two studies. Also, Sweetser (1985) utilized a chlorsulfuron pulse (1-hour) treatment while Frear *et al.* (1987) employed a continuous (24-hr) chlorsulfuron exposure. Variations in experimental technique could contribute to the differences in chlorsulfuron persistence in corn seen in the two experiments. Other possibilities to explain the differences are plant growth conditions and the corn hybrids used in the two studies. Neither paper presents much detail on these areas. However, it is unlikely that these factors could have led to a difference in the chlorsulfuron detoxication pathways and the lack of naphthalic acid stimulation of the detoxication mechanism.

Frear *et al.* (1987) separated extractable ALS from corn shoots on Sephacryl S-400 and noted ALS activity associated with 7×10^5 to 2×10^6 K_d (Fraction I) and 2×10^5 to 4×10^5 K_d (Fraction II) fractions. Chlorsulfuron seed treatment (0.5 g/100 ml acetone/250 g seed for 2 min) reduced extractable ALS Fraction II activity 42%, but actually increased ALS Fraction I activity 14%. Naphthalic acid seed treatment had little effect on extractable ALS activities, although Fraction II activity was increased 14%. ALS increases from naphthalic acid treatment comparable to the 24% increase following dichlormid treatment (Rubin and Casida, 1985) were not observed.

Because naphthalic anhydride was converted into naphthalic acid in corn, and subsequent metabolism has not been detected (Riden and Asbell, 1975), Frear *et al.* investigated the activity of auxins in protecting corn against chlorsulfuron injury. Naphthalic acid, phenylacetic acid, and indoleacetic acid were all effective in protecting corn (Fig. 4) when corn shoots were treated with 10 μg of the acids 24 hr before chlorsulfuron application.

B. Imidazolinones

1. Protectant Activity

While several groups studied protectant activity and protectant mode of action for sulfonylurea herbicides, this section on protection from imidazolinone injury is based on the work of Barrett and Olson (1986).

Attempts to protect grass crops from imazaquin injury grew out of field observations that grain sorghum was safened from imazaquin toxicity by prior seed treatment with CGA-92194. Aside from the potential to employ imazaquin as a herbicide in otherwise susceptible grass crops, concern with potential imazaquin soil residues damaging corn or sorghum grown in rotation with soybean (Anonymous, 1986c; Mills *et al.*, 1987) provides a need to increase the tolerance of grass crops to low levels of imazaquin in the soil. The soil concentrations of imazaquin following soybean would be too low to affect significant weed control but would be high enough to potentially damage the corn or sorghum crop.

Greenhouse experiments verified the activity of a number of herbicide protectants for increasing the tolerance of corn or sorghum to imazaquin applications. NA seed treatment (1% w/w) was generally the most effective protectant for increasing corn and sorghum tolerance to soil applied imazaquin (Figs. 5 and 6). Based on the imazaquin rates needed to reduce growth 50% (GR_{50}), the imazaquin safening factors for NA were approximately 8 for corn and 9 for sorghum. However, it should be noted that the imazaquin GR_{50} rates for the

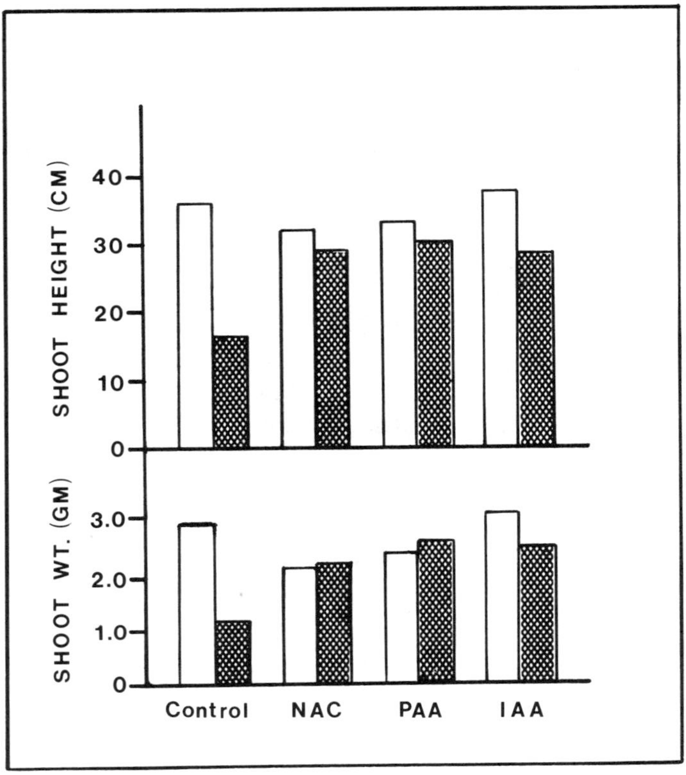

Fig. 4. Auxin protection of corn from chlorsulfuron injury. Six-day old corn seedling shoot tissue was treated with 10 μg of NAC, PAA, or IAA for 24 hr prior to application of 5 μg chlorsulfuron. The shoot measurements were made 7 days after chlorsulfuron treatment. NAC = β-naphthaleneacetic acid, IAA = indoleacetic acid, PAA = phenylacetic acid. (From Frear *et al.*, 1987).

safened corn or sorghum are still below or close to a theoretical level of imazaquin in soil following an application for weed control purposes. Imazaquin is applied at 140 g/ha (Anonymous, 1986c). Protection of corn and sorghum from soil-applied rates of imazaquin for weed control does not appear feasible.

CGA-92194 (0.1% w/w seed dressing), flurazole (0.03% w/w seed dressing), and dichlormid (0.56 kg/ha soil treatment) also increased corn and sorghum tolerance to soil applications of imazaquin (Figs. 5 and 6). CGA-92194, flurazole and dichlormid provided safening factors of 3, 2, and 2, respectively, with corn. Sorghum safening factors of approximately 4, 6, and 2 were found for the three protectants, respectively.

The protection of corn and sorghum was much greater with postemergence application of imazaquin compared to preemergence imazaquin treatment (Figs. 7 and 8). Little corn injury was seen from the imazaquin treatment where seed

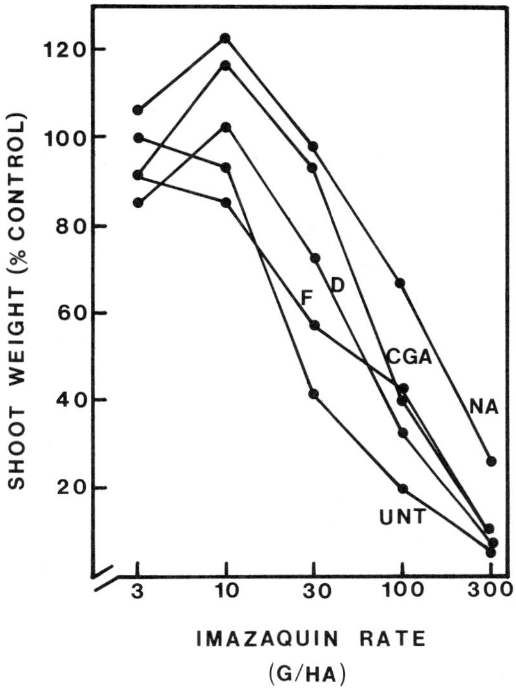

Fig. 5. Protection of corn from preemergence imazaquin application. Corn seed was treated with naphthalic anhydride (1% w/w), CGA-92194 (0.1% w/w), flurazole (0.03% w/w), or not treated and planted into silt loam soil. Dichlormid treatment (0.56 kg/ha) was incorporated into the soil prior to planting. Fresh weights of shoots were taken 2 weeks after planting. NA = naphthalic anhydride, CGA = CGA-92194, F = flurazole, D = dichlormid, UNT = untreated control.

dressing of NA was used. Dichlormid was the least effective of the protectants and gave safening factors of only 2 and 1 from postemergence imazaquin toxicity to corn or sorghum, respectively. NA, CGA-92194, and flurazole provided safening factors of 4, 4, and 3, respectively, with corn, and 5, 4, and 5, respectively, for sorghum. The poor fit of the regression equation used to calculate a GR_{50} value for the NA treatment ($r^2 = 0.44$) plus the lack of significant growth reduction of NA-treated corn plants due to postemergence imazaquin application makes the NA safening estimate for corn suspect.

The data from the postemergence applications of imazaquin to protected corn and sorghum may indicate increased protection if a delay occurs between protectant treatment and imazaquin treatment. The experiments of Parker *et al.* (1980) suggest a similar phenomenon for protection of corn from chlorsulfuron injury. The difference in the protection levels between soil-applied and foliar-applied imazaquin may also be related to the sensitivity of the roots compared to foliage to imazaquin (Mills *et al.*, 1987). Shoot toxicity from

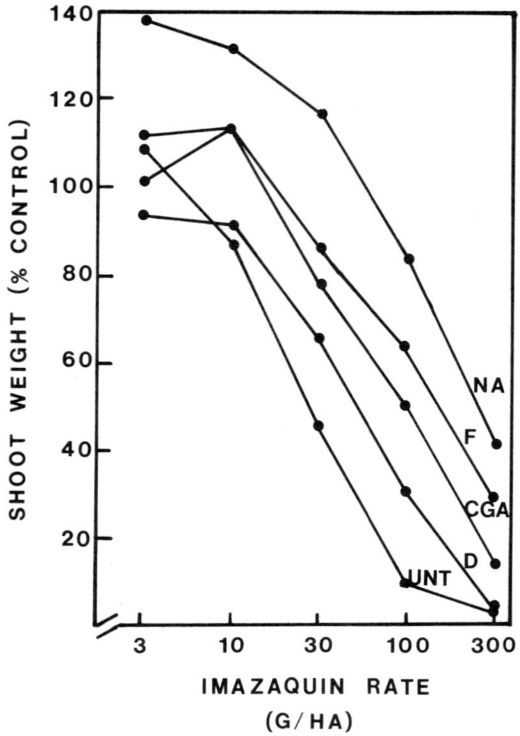

Fig. 6. Protection of sorghum from preemergence imazaquin application. Sorghum seed was treated with naphthalic anhydride (1% w/w), CGA-92194 (0.1% w/w), flurazole (0.03% w/w), or not treated and planted into silt loam soil. Dichlormid treatment (0.56 kg/ha) was incorporated into the soil prior to planting. Shoot fresh weights were taken 2 weeks after planting. NA = naphthalic anhydride, CGA = CGA-92194, F = flurazole, D = dichlormid, UNT = untreated control.

imazaquin soil applications is a combination of the indirect effects of root growth inhibition plus the direct effect of imazaquin translocated from the roots to the shoot. Imazaquin was shown to translocate from the roots to the shoots of broadleaf plant species (Shaner and Robson, 1985). A postemergence application would allow time for development of a functional corn root system prior to the imazaquin treatment and the contribution of root injury to shoot growth inhibition might be minimized under these conditions. The ability to protect the foliage from direct imazaquin toxicity would be more evident with postemergence treatment.

The overall impression is that NA gave the most effective and consistent protection of corn and sorghum from imazaquin damage. Flurazole and CGA-92194 also protected corn and sorghum, whereas dichlormid was the least effective protectant for imazaquin in these crops.

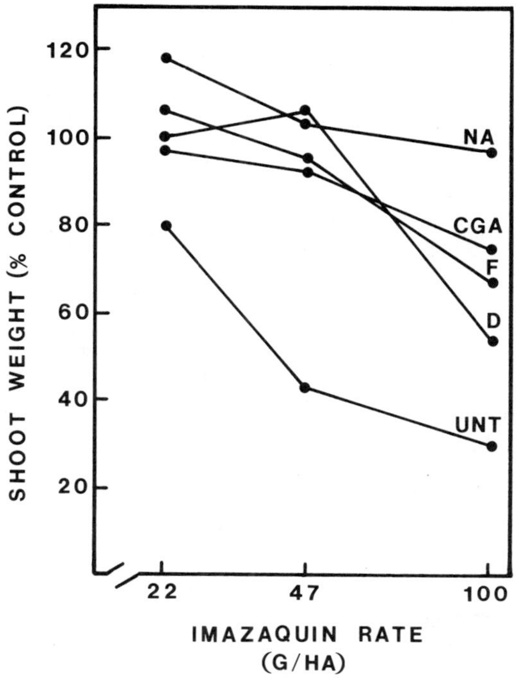

Fig. 7. Protection of corn from imazaquin application at the spike stage of corn growth. Corn seed treated with naphthalic anhydride (1% w/w), CGA-92194 (0.1% w/w), flurazole (0.03% w/w), or untreated was planted in a silt loam soil. Dichlormid (0.56 kg/ha) was incorporated into the soil prior to planting. Imazaquin was applied at the spike stage of corn growth. Shoot fresh weights were taken 2 weeks after imazaquin application. NA = naphthalic anhydride, CGA = CGA-92194, F = flurazole, D = dichlormid, UNT = untreated control.

2. Mechanisms of Crop Protection against Imidazolinones

The observed protection of corn and sorghum from imazaquin toxicity could arise from one or more interactions of the applied safeners and the imazaquin mode of action. Hatzios (1983) suggested that protectants could interfere with the herbicide at the site of action. Rubin and Casida (1985) proposed that increased ALS activity following protectant treatment could partially offset chlorsulfuron phytotoxicity. The knowledge that imazaquin also inhibits ALS activity (Shaner *et al.*, 1984) made protectant interactions with this enzyme activity the initial focus of experiments to elucidate the mechanism of the protection. The opportunity to study the effect of herbicide safeners on an identified herbicide site of action is relatively unique.

The first experiments examined the effect of protectants on ALS levels in etiolated sorghum shoots. Use of etiolated plant material for ALS extractions is a common procedure (Ray, 1984; Rubin and Casida, 1985; Shaner *et al.*, 1984).

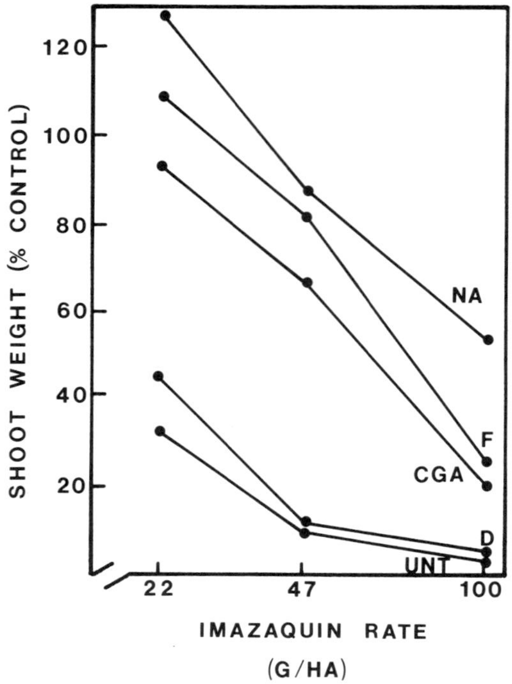

Fig. 8. Protection of sorghum from imazaquin application at the spike stage of sorghum growth. Sorghum seed was treated with naphthalic anhydride (1% w/w), CGA-92194 (0.1% w/w), flurazole (0.03% w/w), or not treated and planted into silt loam soil. Dichlormid treatment (0.56 kg/ha) was incorporated into the soil prior to planting. Shoot fresh weights were taken two weeks after planting. NA = naphthalic anhydride, CGA = CGA-92194, F = flurazole, D = dichlormid, UNT = untreated control.

Both NA and CGA-92194 appeared to slightly elevate extractable ALS activity in the etiolated sorghum shoots. However, two observations in the experiments with the etiolated tissue made the increased ALS activity of doubtful significance. First, the protectants were much more toxic themselves than normally observed in greenhouse experiments conducted in the light (Table VIII). Second, rather than safening from imazaquin toxicity, the combination of protectant and imazaquin treatment resulted in the same or more injury to the etiolated sorghum shoots (Table VIII). For these reasons, the protectant effect on ALS activity extracted from green shoots or the roots of green shoots was examined.

ALS activity extracted from shoots or roots of plants grown in the light and treated with protectants was generally not significantly different from the activity extracted from untreated plants (Table IX). There was significantly less ALS activity extracted from corn shoots following dichlormid treatment. Despite the lack of increased ALS activity in plants treated with the safeners, the preservation of ALS activity by the protectants after imazaquin treatment was evident.

Table VIII The Effect of Naphthalic Anhydride (NA) or CGA-92194 Treatment on Growth of Etiolated Sorghum Shoots Exposed to Imazaquin

Protectant	Seed treatment (%,w/w)	Imazaquin treatment (g/ha)	Shoot fresh weight[a] (% of control)
Untreated	—	0	100
Untreated	—	67	54
NA	1	0	45
NA	1	67	30
CGA-92194	0.1	0	48
CGA-92194	0.1	67	27

[a]Treated or untreated sorghum seed (15 g) was planted in sand and imazaquin was applied to the surface of the sand. Shoots were harvested 1 week after growth in the dark at 25°C.

This protection of ALS levels was observed for NA, CGA-92194, and flurazole in corn shoots and for NA in corn roots (Table IX). NA was also significantly better than CGA-92194 and flurazole in protecting ALS activity in the corn shoots. The higher efficacy of NA in protecting ALS relates to the activity of this safener in preventing imazaquin phytotoxicity in corn observed in greenhouse tests (Figs. 5 and 7).

The protective activity observed in the greenhouse tests (Figs. 5, 6, 7, and 8) was thus also seen at the biochemical site of action for imazaquin. The protection was not due to an increase in ALS activity caused by protectant treatment. Also, *in vitro* experiments with ALS extracted from plants not treated with a safener showed no effect of including the protectants in the ALS assay on the measured activity. The possibilities of reduced imazaquin uptake or translocation and/or increased detoxication were next investigated as a basis for the protectant effect.

Table IX Effect of Protectant Treatment on Extractable Acetolactate Synthase Activity from Imazaquin Treated and Untreated Corn Shoots and Roots[a]

	Shoots		Roots	
	Imazaquin (kg/ha)		Imazaquin (kg/ha)	
Protectant	0	0.01	0	0.01
	ALS activity (nmol acetoin/hr/mg)			
Untreated	590	249	250	91
NA	590	590	261	216
CGA-92194	522	454	182	159
Dichlormid	443	283	193	125
Flurazole	511	409	193	148
LSD(0.05)	102		125	

[a]Corn seeds treated with naphthalic anhydride (1% w/w), CGA-92194 (0.1% w/w), flurazole (0.03% w/w), or untreated were planted in quartz sand. Dichlormid treatment (0.56 kg/ha) was applied as a soil drench. Imazaquin was applied as a preemergence treatment and watered into the sand. One week after planting, the corn seedlings were washed from the sand and ALS activity of the shoots and roots was determined according to Ray (1984) and Shaner *et al.* (1984).

The protectants had little effect on the amount of radioactivity from [^{14}C]imazaquin applied to the soil found in the roots or shoots of corn 2 weeks after planting (Table X). This was particularly evident for the amount of radioactivity absorbed on a per plant basis. The ^{14}C concentration in the plants (dpm/g dry weight) was also not affected in a manner consistent with a protective mechanism (Tables X). While there were slight, but significant, reductions in the concentrations, most of these differences could be attributed to the greater growth of the protected plants. The amount of uptake was the same in both cases but the continued growth of the protected plants caused dilution of absorbed radioactivity.

Distribution of the absorbed radioactivity was not altered by the protectant treatments (Tables X). The majority (80–90%) of the absorbed radioactivity in corn was found in the root tissue. This contrasts to soybean, in which the bulk (65%) of the radioactivity was located in the shoot tissues. It is interesting to speculate whether this difference in translocation between the corn and soybeans contributes to the differential sensitivity of these plants to imazaquin.

The metabolism of imazaquin was followed using the methodology of Shaner and Robson (1985). A 24-hr exposure to [^{14}C]imazaquin (ring labeled) was followed by subsequent extraction and characterization of the imazaquin remaining in the plants. NA seed treatment significantly increased the metabolism of imazaquin in corn roots (Table XI). While all the protectants, except dichlormid, increased imazaquin metabolism in the corn shoots (Table XI), the NA was the most effective protectant for increasing imazaquin metabolism. These results are in agreement with the protectant effects on ALS levels in corn plants (Table IX). NA protected ALS activity in roots and all the safeners except dichlormid maintained ALS activity in the corn shoots with NA being the most effective.

Table X Effect of Protectant Treatment on Imazaquin Uptake and Radioactivity Distribution of ^{14}C in Shoots and Roots of Corn and Soybeans[a]

		Shoot levels			Root levels		
Plant species	Protectant treatment	dpm/plant	dpm/g	% of total dpm/plant	dpm/plant	dpm/g	% of total dpm/plant
Soybean	Untreated	4366 A	22 A	65 A	2398 B	52 A	35 A
Corn	Untreated	679 B	16 AB	14 B	3860 A	23 B	87 B
Corn	NA	781 B	6 C	13 B	4742 A	31 AB	87 B
Corn	CGA-92194	1086 B	6 C	18 B	4847 A	32 AB	82 B
Corn	Flurazole	955 B	8 BC	19 B	4317 A	27 B	81 B
Corn	Dichlormid	785 B	24 A	12 B	4294 A	26 B	88 B

[a]Corn seeds treated with naphthalic anhydride (1% w/w), CGA-92194 (0.1% w/w), flurazole (0.03% w/w), or untreated were planted in quartz sand. Dichlormid treatment (0.56 kg/ha) was incorporated into the soil prior to planting. Soybean seeds were also planted for comparative purposes. [^{14}C]Imazaquin (ring labeled) was applied to the soil surface and watered in. Two weeks after treatment, the plants were harvested and separated into roots and shoots. ^{14}C activity associated with the plant parts was determined by oxidation followed by liquid scintillation determination of released $^{14}CO_2$.

Table XI Effect of Protectant Treatments on Imazaquin Metabolism and Radioactivity Distribution in Corn Roots and Shoots[a]

	Root levels			Shoot levels		
Protectant	**Parent imazaquin (%)**	**Metabolites (%)**	**Insoluble residues (%)**	**Parent imazaquin (%)**	**Metabolites (%)**	**Insoluble residues (%)**
Untreated	65 A	12 D	23 A	64 A	28 D	9 C
NA	46 B	35 A	18 AB	17 D	66 A	17 A
CGA-92194	60 A	19 BC	21 A	35 C	49 B	16 A
Flurazole	63 A	18 C	19 AB	46 B	41 C	13 B
Dichlormid	63 A	23 B	15 B	65 A	25 D	10 C

[a]Roots of corn seedlings grown in sand for 7 days were placed into nutrient solution containing [^{14}C]imazaquin (ring labeled) for 24 hr. The corn plants were grown from seed treated with naphthalic anhydride (1% w/w), CGA-92194 (0.1% w/w), flurazole (0.03% w/w), or untreated. Dichlormid (0.56 kg/ha) treatments were applied as a sand drench. The plants were harvested at the end of 24 hr and separated into roots and shoots, and imazaquin metabolism was determined as in Shaner and Robson (1985).

V. CONCLUDING REMARKS

The most significant and consistently reported protection of grass crops from sulfonylurea toxicity occurred following treatment with naphthalic anhydride (NA) or naphthalic acid. The oxime protectants CGA-92194 and cyometrinil, although less studied than NA, also provided protection of grass species, particularly corn. Dichlormid was the least effective protectant tested. Protection could be achieved by seed treatment, soil treatment, or foliar treatment of the grass plants prior to application of the sulfonylurea.

In summarizing the basis for protection from chlorsulfuron toxicity, protectant alteration of ALS activity does not appear to play a major role. Neither *in vitro* ALS sensitivity to chlorsulfuron nor extractable ALS levels are altered sufficiently to account for the safening. The role of increased metabolism of chlorsulfuron, or other sulfonylureas, due to protectant treatment is more difficult to assess. The experiments of Sweetser (1985) certainly establish the potential for increased detoxication following protectant application as a basis for the protection. However, the results of Frear *et al.* (1987) cloud this issue. Further experimentation will be required to determine if two metabolic pathways for chlorsulfuron are active in different corn genotypes, whether both pathways could be active in one corn genotype, and if the two pathways respond differently to stimulation by protectant treatment. There is evidence for chlorsulfuron × protectant × genotype interactions (Parker *et al.*, 1980; Rubin and Casida, 1985).

The observation that auxins afford protection to corn from chlorsulfuron damage is interesting (Frear *et al.*, 1987). However, the basis for this effect and its relationship to activity of other herbicide protectants are matters for speculation.

Similar to the work with sulfonylureas, the most effective protection of grass species (corn and sorghum) from imidazolinone (imazaquin) injury occurred following seed treatment with NA. Flurazole and CGA-92194 also showed activity in preventing imazaquin toxicity to corn and sorghum, whereas dichlormid was the least effective protectant studied.

The evidence suggests that effective protectants for preventing imazaquin toxicity to corn and sorghum achieve the safening effect by increasing the rate of imazaquin metabolism in the protected plants. Hatzios (1983) suggested that altered herbicide uptake, translocation, metabolism, and/or herbicide interactions with the biochemical site of action due to protectant treatment could all serve as a basis for protectant activity. All these factors were studied for protection from imazaquin injury to corn and sorghum and only increased imazaquin metabolism was detected as a potential cause of the safening.

Additional work is required to establish the identity and potential phytotoxicity of the imazaquin metabolism products being produced in the corn and sorghum. Increased metabolism of imazaquin to phytotoxic materials would not serve as an efficient protection mechanism. The mechanism by which the protectants elevate metabolism is also of prime interest. The experiments of Sweetser (1985) suggest that protein synthesis is involved in the protection process, which could be at the transcriptional or translational level. Mozer *et al.* (1983) demonstrated that new forms of detoxifying enzymes were produced in response to flurazole treatment. The new form of glutathione-*S*-transferase found by Mozer *et al.* (1983) in corn following flurazole treatment was more efficient in converting alachlor to its glutathione conjugate. The same products of herbicide metabolism could be formed after protectant treatment but at a higher rate because of a higher level of detoxifying enzymes or more effective enzymes.

The commercial significance of increasing grass crop tolerance to sulfonylureas and imidazolinones through the use of protectants has yet to be established. The development of both these herbicide groups is at a relatively early stage and the evidence indicates that substantial increases in crop tolerances may yet be obtained through altered chemical structures. The development of crop varieties, including corn, possessing altered ALS forms resistant to imidazolinone and sulfonylurea inhibition is also continuing. Protectants may aid in expanding the use of these two herbicide families or protecting grass crops from soil residues of the herbicides, but it is too soon to evaluate this impact.

References

Anderson, P. C. and Georgson, M. (1986). Selection of an imidazolinone tolerant mutant of corn. *In* "VI International Congress of Plant Tissue and Cell Culture Abstracts" (D. A. Somers, B. G. Gengenbach, D. D. Biesboer, W. P. Hackett, and C. E. Green, eds.), pp. University of Minnesota, Minneapolis.

Anderson, P. C., and Hibberd, K. A. (1985). Evidence for the interaction of an imidazolinone herbicide with leucine, valine, and isoleucine metabolism. *Weed Sci.* **33**, 479–483.

Anonymous (1985a). "Arsenal Herbicide Label." Am. Cyanamid Co., Agric. Div., Wayne, New Jersey.

Anonymous (1985b). "Oust Herbicide Label." E. I. DuPont DeNemours & Co., Wilmington, Deleware.

Anonymous (1986a). "Classic Herbicide Label." E. I. DuPont DeNemours & Co., Wilmington, Deleware.

Anonymous (1986b). "Glean Herbicide Label." E. I. DuPont DeNemours & Co., Wilmington, Deleware.

Anonymous (1986c). "Scepter Herbicide Label." Am. Cyanamid Co., Agric. Div., Wayne, New Jersey.

Barrett, M., and Olson, G. L. (1986). Protection from imazaquin injury: Studies on the mode of action of herbicide protectants. *IUPAC Int. Congr. Pestic. Chem., 6th,* Abstr. No. 3D–14.

Brown, M. A., Chiu, T. Y., and Miller, P. (1987). Hydrolytic activation versus oxidative degradation of Assert herbicide, an imidazolinone aryl-carboxylate, in susceptible wild oat versus tolerant corn and wheat. *Pestic. Biochem. Physiol.* **27**, 24–29.

Chaleff, R. S. and Mauvais, C. J. (1984). Acetolactate synthase is the site of action of two sulfonylurea herbicides in higher plants. *Science* **224**, 1443–1445.

Chaleff, R. S., and Ray, T. B. (1984). Herbicide-resistant mutants from tobacco cell cultures. *Science* **223**, 1148–1151.

Claus, J. S. (1987). Chlorimuron-ethyl (Classic): A new broadleaf postemergence herbicide in soybean. *Weed Technol.* **1**, 114–115.

Falco, S. C., and Dumas, K. S. (1985). Genetic analysis of mutants of *Saccharomyces cerevisae* resistant to the herbicide sulfometuron methyl. *Genetics* **109**, 21–35.

Fedtke, C., and Trebst, A. (1987). Advances in understanding herbicide modes of action. *In* "Pesticide Science and Biotechnology" (R. Greenhalgh and T. R. Roberts, eds.) pp. 161–168. Blackwell, Oxford.

Frear, D. S., Swanson, H. R., and Mansager, D. (1987). 1,8-Naphthalic anhydride/auxin protection against chlorsulfuron inhibition of corn seedling growth. *In* "Pesticide Science and Biotechnology" (R. Greenhalgh and T. R. Roberts, eds.), pp. 499–503. Blackwell, Oxford.

Goodwin, T. W., and Mercer, E. I. (1983). "Introduction to Plant Biochemistry," 2nd ed. pp. 339–341. Pergamon, Oxford.

Groves, K. E. M., and Foster, R. K. (1985). A corn (*Zea mays* L.) bioassay technique for measuring chlorsulfuron levels in three Saskatchewan soils. *Weed Sci.* **33**, 825–828.

Hageman, L. H., and Behrens, R. (1984). Basis for response differences of two broadleaf weeds to chlorsulfuron. *Weed Sci.* **32**, 162–167.

Hatzios, K. K. (1983). Herbicide antidotes: Development, chemistry, mode of action. *Adv. Agron.* **36**, 265–316.

Hatzios, K. K. (1984). Interactions between selected herbicides and protectants on corn (*Zea mays*). *Weed Sci.* **32**, 51–58.

Hutchinson, J. M., Shapiro, R., and Sweetser, P. B. (1984). Metabolism of chlorsulfuron by tolerant broadleaves. *Pestic. Biochem. Physiol.* **22**, 243–247.

LaRossa, R. A., and Schloss, J. V. (1984). The sulfonylurea herbicide sulfometuron methyl is an extremely potent and selective inhibitor of acetolactate synthase in *Salmonella typhimurium*. J. Biol. Chem. **259**, 8753–8757.

Mersie, W., and Foy, C. L. (1984). Safening corn against chlorsulfuron and DPX-T6376 injury. *Proc., South. Weed Sci. Soc.* **36**, 328–334.

Mills, J. A., Witt, W. W., and Olson, G. L. (1987). Effects of tillage systems on imazaquin and AC 263, 499 persistence. *Proc., South. Weed Sci. Soc.* **40**, 378.

Mozer, T. J., Tiemeier, D. C., and Jaworski, E. G. (1983). Purification and characterization of corn glutathione *S*-transferase. *Biochemistry* **22**, 1068–1072.

Mueller, T. C., Witt, W. W., and Barrett, M. (1985). Comparison of bioassay techniques. *Proc.—North Cent. Weed Control Conf.* **40**, 51.

Muhitch, M. J., Shaner, D. L., and Stidham, M. A. (1987). Imidazolinones and acetohydroxyacid synthase from higher plants. Properties of the enzyme from maize suspension culture cells and evidence for the binding of imazapyr to acetohydroxyacid synthase *in vivo*. *Plant Physiol.* **83**, 451–456.

O'Leary, N. F., and Prendeville, G. N. (1985). Uptake and phytotoxicity of chlorsulfuron in *Zea mays* L. in the presence of 1,8-naphthalic anhydride. *Weed Res.* **25**, 331–339.

Parker, C., Richardson, W. G., and West, T. M. (1980). Potential for extending the selectivity of DPX 4189 by use of herbicide safeners. *Proc. Br. Crop Prot. Conf.—Weeds*, pp. 15–22.

Ray, T. B. (1982a). The mode of action of chlorsulfuron: A new herbicide for cereals. *Pestic. Biochem. Physiol.* **17**, 10–17.

Ray, T. B. (1982b). The mode of action of chlorsulfuron: The lack of direct inhibition of plant DNA synthesis. *Pestic. Biochem. Physiol.* **18**, 262–266.

Ray, T. B. (1984). Site of action of chlorsulfuron: Inhibition of valine and isoleucine biosynthesis in plants. *Plant Physiol.* **75**, 827–831.

Ray, T. B. (1986). Sulfonylurea herbicides as inhibitors of amino acid biosynthesis in plants. *Trends Biochem. Sci.* **11**, 180–183.

Riden, J. R., and Asbell, W. J. (1975). The fate of Protect (1,8-naphthalic anhydride) in corn plants. *Environ. Qual. Saf., Suppl.* **3**, pp. 175–179.

Rost, T. L. (1984). The comparative cell cycle and metabolic effects of chemical treatments on root tip meristems. III. Chlorsulfuron. *J. Plant Growth Regul.* **3**, 51–63.

Rubin, B., and Casida, J. E. (1985). R-25788 effects on chlorsulfuron injury and acetohydroxyacid synthase activity. *Weed Sci.* **33**, 462–468.

Schloss, J. V. (1984). Interaction of the herbicide sulfometuron-methyl with acetolactate synthase: A slow binding inhibitor. *In* "Flavins and Flavoproteins" (R. C. Bray, P. C. Engel, and S. G. Mayhew, eds.), pp. 737–740. de Gruyter, Hawthorne, New York.

Shaner, D. L., and Reider, M. L. (1986). Physiological responses of corn (*Zea mays*) to AC 243, 997 in combination with valine, leucine, and isoleucine. *Pestic. Biochem. Physiol.* **25**, 248–257.

Shaner, D. L., and Robson, P. A. (1985). Absorption, translocation, and metabolism of AC 252 214 in soybean (*Glycine max*), common cocklebur (*Xanthium strumarium*), and velvetleaf (*Abutilon theophrasti*). *Weed Sci.* **33**, 469–471.

Shaner, D. L., Anderson, P. C., and Stidham, M. A. (1984). Imidazolinones Potent inhibitors of acetohydroxyacid synthase. *Plant Physiol.* **76**, 545–546.

Sweetser, P. B. (1985). Safening of sulfonylurea herbicides to cereal crops: Mode of herbicide antidote action. *Proc. Br. Crop Prot. Conf.—Weeds* **3**, 1147–1154.

Sweetser, P. B., Schow, G. S., and Hutchinson, J. M. (1982). Metabolism of chlorsulfuron by plants: Biological basis for selectivity of a new herbicide for cereals. *Pestic. Biochem. Physiol.* **17**, 18–23.

Witt, W. W., Wells, R. B., and Slack, C. H. (1985). Persistence of DPX-F6025, imazaquin, and dimethazone in a corn–soybean rotation. *Proc.—North Cent. Weed Control Conf.* **40**, 14.

Yadav, N., McDevitt, R., Bernard, S., and Falco, S. C. (1986). Single amino acid substitutions in the enzyme acetolactate synthase confer resistance to the herbicide sulfometuron methyl. *Proc. Natl. Acad. Sci. U.S.A.* **83**, 4418–4422.

Chapter 10

Terpenoid Biosynthesis as a Site of Action for Herbicide Safeners

Robert E. Wilkinson
Department of Agronomy, Georgia Station
Agricultural Experiment Stations
University of Georgia
Experiment, Georgia 30212

I. INTRODUCTION

The terpenoid carbon pathway is a major biosynthetic system for sterols, pigments, electron carriers, and gibberellins requisite to the growth of green plants (Fig. 1). Inhibition of any portion of the total biosynthetic process would result in decreased production of these requisite components for plant growth. For example, inhibition of synthesis or utilization of acetate would result in decreased total terpenoid synthesis along with decreased fatty acid synthesis. Geranylgeranylpyrophosphate (GGPP) is a major branch point in the synthesis of the various terpenoids and multiple steps are present in the formation of each product. For example, the conversion GGPP to β-carotene requires conjugation (two steps), desaturation (four steps), and cyclization (a minimum of two steps) (Fig. 2). Similarly, the synthesis of gibberellic acid (GA) requires cyclization to *ent*-kaurene (two steps), oxidation of kaurene to GA (at least four steps) (Fig. 3), and further modifications to the "active" GAs which are currently poorly understood. Inhibition of any one of these biosynthetic reactions results in the accumulation of precursors. Thus, depletion of one product and accumulation of

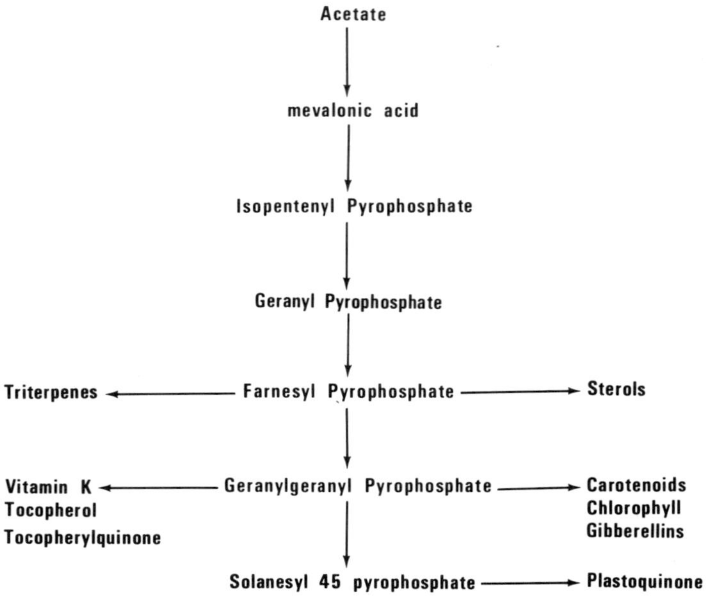

Fig. 1. Isoprenoid synthesis and relationships of terpenoid products.

the immediately preceding precursors is presumptive evidence for an inhibition at one reactive step. Axiomatically, inhibition of kaurene biosynthesis (Fig. 2) would result in a surplus of GGPP and increased chloroplast pigment accumulation would result (Figs. 1 and 2).

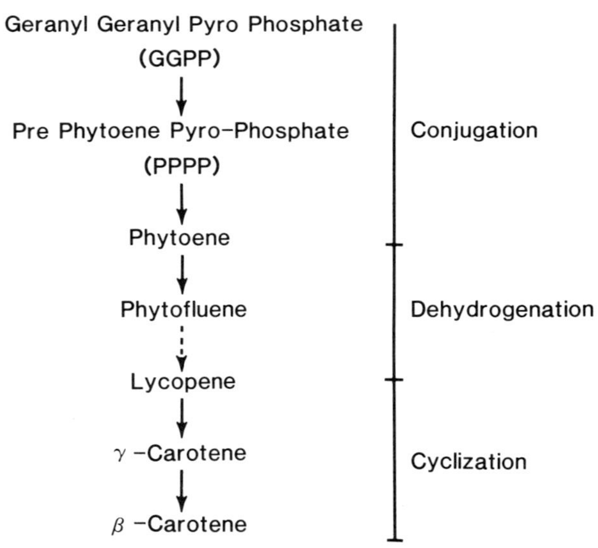

Fig. 2. Conversion of geranylgeranyl pyrophosphate → β-carotene.

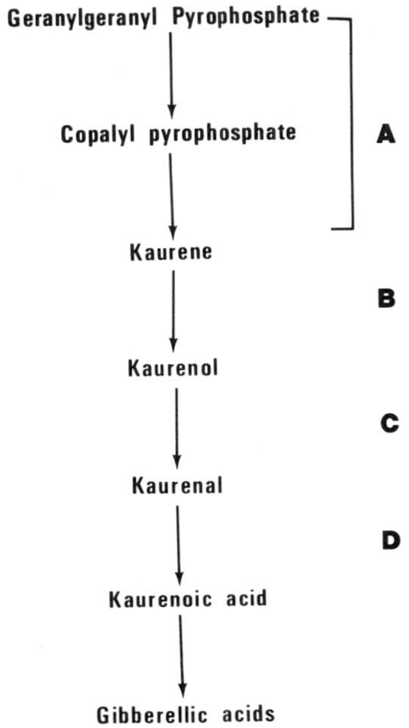

Fig. 3. Kaurene synthesis (A) and kaurene oxidation (B, C, D).

Thus, growth responses must be explained by biochemical responses, and many of the observed morphological results of carbamothioate and chloroacetanilide applications are closely aligned with modification of terpenoid biosynthesis. Consequently, evaluation of the influence of these herbicides on the production of terpenoids may help explain carbamothioate and chloroacetanilide modes of action.

II. PLANT GROWTH RESPONSES TO CARBAMOTHIOATES AND CHLOROACETANILIDE HERBICIDES

In dicots, growth retardation by herbicides whose activity is reversed by safeners results in shortened internodes and small leaves (Oster and West, 1968; Coolbaugh and Hamilton, 1976; Frost and West, 1977; Coolbaugh *et al.*, 1978). These responses are influenced by age or physiological stage of subject tissue, application method, plant species, toxicant concentration, toxicant volatility, toxicant degradation or metabolism, and environmental conditions. Similar response limitations exist in monocots, in which the seedling responses are a progression from rolled leaves to decreased leaf length until the primary leaf does not penetrate through the coleoptile. Stasis at approximately 1 cm coleoptile length is followed by either dead (brown) seedlings or the development of dark green seedlings which ultimately resume growth by the development of secondary buds which burst through the side of the coleoptile. Often, in the latter case, the tip of the developing leaf is trapped in the coleoptile and characteristic "loop leaves" are produced.

In wheat, EPTC (Fig. 4) and vernolate induce stasis at 125 ppbw (a.i.) incorporated into sand (Wilkinson, 1985). [^{14}C]Vernolate (125 ppbw, approximately 3.3 μmol/liter in the water in the sand) was present at 1.8 μmol/L in the

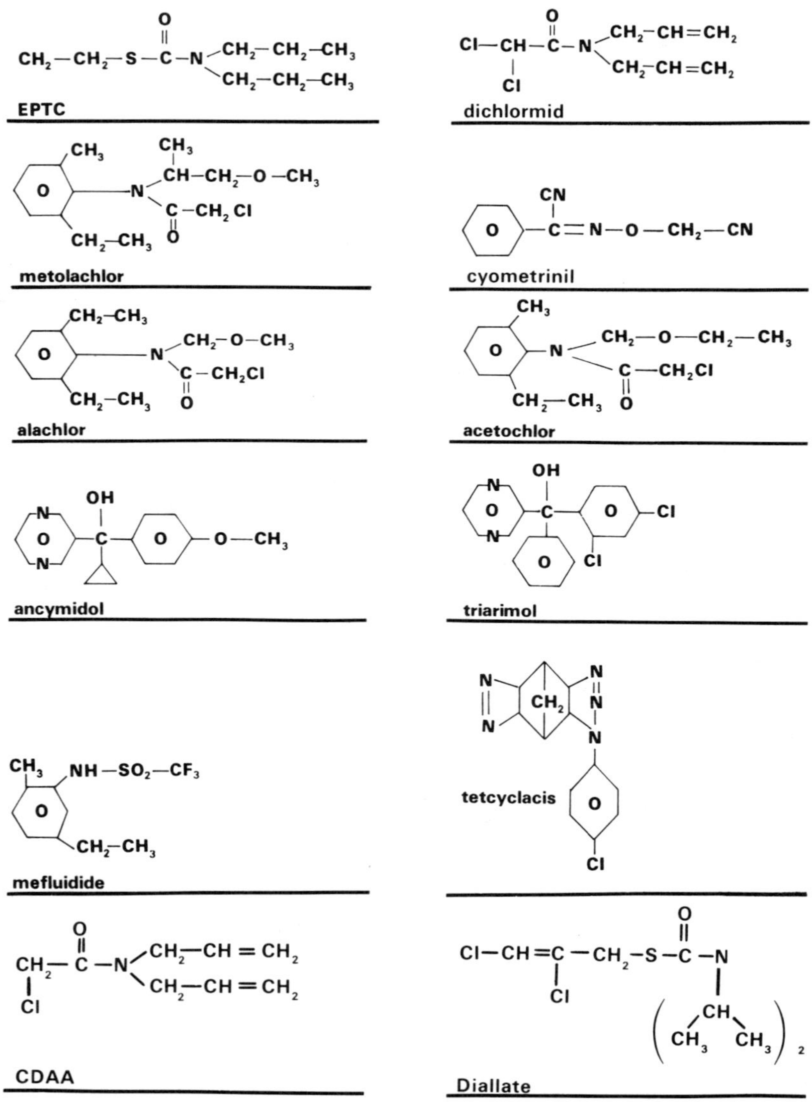

Fig. 4. Structures of various herbicides and their respective safeners.

coleoptile at stasis (96 hr). If degradation or other metabolic processes which modify herbicide activity can be ignored, 1-2 μmol/liter was sufficient to inhibit growth. In wheat, EPTC (62.5 ppbw) induced increased chl *a*, chl *b*, and carotenoid contents as well as β-carotene, lutein epoxide, lutein, violaxanthin, antheraxanthin, and zeaxanthin contents (Wilkinson, 1977) when measured on a micrograms per gram fresh weight basis. This corroborates previous reports of EPTC-induced increases in protochlorophyllide in cucumber (*Cucumis sativus* L.) cotyledons (Weinberg and Castelfranco, 1975). Thus, some influence on terpenoid synthesis resulted in decreased total terpenoid synthesis and increased synthesis of chloroplast pigments at the same time that growth stasis developed.

III. TOTAL TERPENOID SYNTHESIS

EPTC induced decreases in total chl *a*, chl *b*, carotene, and violaxanthin cycle component contents of wheat (*Triticum aestivum* L.) on a micrograms per pot basis (Table I). In the same tissue, the concentration (μg/g fresh weight) of these constituents increased (Table II). If increase in size were the only parameter influenced by this herbicide, then the total content of geranylgeranyl (GG) equivalents should have been the same at all EPTC concentrations. Obviously, there was a massive decrease in total GG equivalents produced (Table I) while growth was inhibited even more than GG equivalent biosynthesis and an apparent increase in chloroplast pigment concentration occurred (Table II). Therefore, there has to be one inhibition site in the early stages of terpenoid biosynthesis plus a second inhibition site in a later portion of the terpenoid biosynthesis scheme which results in greater decrease in size (fresh weight) than accounted for by the first inhibition site alone.

One of the earliest documentations of a biochemical response of plants to carbamothioate (previously known as thiocarbamate) herbicides was a decreased

Table I Influence of EPTC on Total Terpenoid Chloroplast Pigment Contents in Wheat[a]

EPTC (ppbw)	Total terpenoids (μmol geranylgeranyl equivalents)				
	Chl *a*	Chl *b*	Carotene	Quinones[b]	Total
0	5.66	2.01	3.55	2.33	13.55 (100.0)[c]
15.6	4.77	1.35	3.60	1.33	11.05 (81.5)
31.2	2.07	1.25	1.13	0.69	5.14 (37.9)
62.5	0.51	0.43	0.29	0.20	1.43 (10.6)

[a]From Wilkinson (1977). Reprinted by permission of the University of Chicago Press.
[b]Violaxanthin + antheraxanthin + zeaxanthin.
[c]Values in parentheses are percentages of the untreated control/pot.

Table II Terpenoid Contents of Wheat Grown During Exposure to EPTC[a]

EPTC (ppbw)	μmol geranylgeranyl equivalents/g fresh weight				
	Chl *a*	Chl *b*	Carotene	Quinones[b]	Total
0	0.62	0.33	0.58	0.39	1.92 (100.0)[c]
15.6	1.53	0.44	1.16	0.43	3.56 (185.4)
31.2	1.58	0.95	0.86	0.53	3.92 (204.2)
62.5	1.46	1.24	0.83	0.56	4.09 (213.0)

[a]From Wilkinson (1978). Reprinted by permission of Academic Press, Inc.
[b]Violaxanthin + antheraxanthin + zeaxanthin.
[c]Values in parentheses are percentages of the untreated control.

epicuticular wax content in EPTC-treated cabbage (*Brassica oleracea* L.) (Gentner, 1966). This response has been extended to other carbamothioates in other plants (Wilkinson and Hardcastle, 1969, 1970; Wilkinson, 1974; Wilkinson and Smith, 1975a,b, 1976a,b) as well as to metolachlor in sorghum [*Sorghum bicolor* (L.) Moench] (Ebert and Ramsteiner, 1984). Since fatty acids are the basic units synthesized by chloroplasts, which are then converted to fatty alcohols and alkanes found in the epicuticular waxes, some common precursor must be present that is utilized by fatty acid and terpenoid biosynthesis.

Acetyl-CoA is utilized for both fatty acid and terpenoid biosyntheses (Loomis and Croteau, 1980; Stumpf, 1980). Synthesis of acetyl-CoA can occur in three ways: (a) β-oxidation of fatty acids from seed triglycerides, (b) mitochondrial pyruvate dehydrogenase (EC 1.2.2.2) complex (PDC), and (c) chloroplastic PDC (Stumpf, 1980). Wheat mitochondrial PDC was inhibited by EPTC (Oswald and Wilkinson, 1985). Since CoA does not penetrate through either the mitochondria or chloroplast membranes (Stumpf, 1980), utilization of acetyl-CoA acetate produced in mitochondria requires the hydrolysis of acetyl-CoA, the migration of acetate to sites of utilization, and reformation of acetyl-CoA within the organelle where acetyl-CoA is utilized. The chloroplast envelope is a major site of acetyl-CoA synthetase (EC 6.2.1.1) (Stumpf, 1980). Thus, an influence of EPTC on chloroplastic acetyl-CoA synthesis would fit both fatty acid and terpenoid biosynthesis inhibitions. Acetate utilization for chloroplastic fatty acid synthesis was inhibited by EPTC (Fig. 5) and that inhibition was competitively reversed by acetate. A second source of chloroplastic acetyl-CoA is from chloroplast PDC. EPTC inhibited chloroplast fatty acid biosynthesis from [2-^{14}C]pyruvate, which is converted to acetyl-CoA, and the inhibition was competitively reversed by pyruvate (Fig. 6) (Wilkinson and Oswald, 1987). Thus,an inhibition of the biosynthesis of very early precursors (i.e. acetyl-CoA) utilized in both fatty acid and terpenoid synthesis occurs. Additionally, meto-

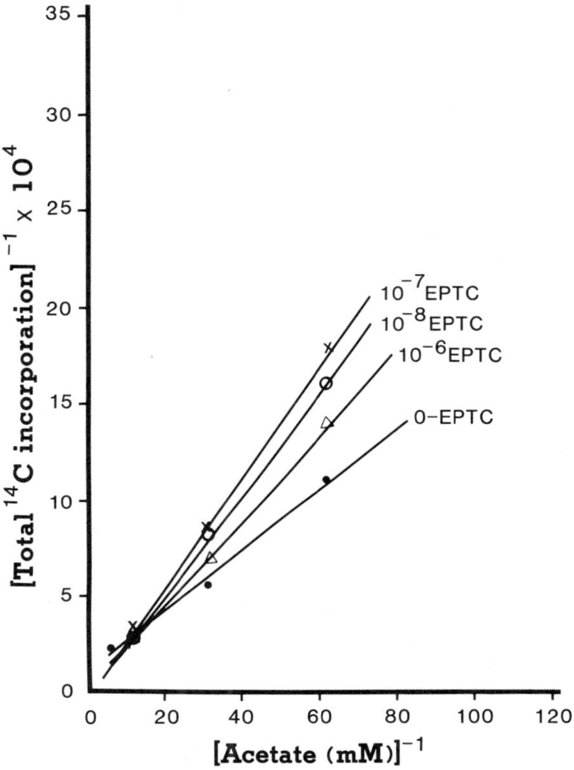

Fig. 5. Lineweaver–Burk plots of the incorporation of [2-^{14}C]acetate into fatty acids by intact wheat chloroplasts. (●) Influence of various concentrations of [^{14}C]acetate alone; (△) + 1.0 μM EPTC; (×) + 0.1 μM EPTC; (○) + 0.01 μM EPTC. (Reprinted by permission of Academic Press, Inc.)

lachlor inhibits epicuticular wax biosynthesis and that inhibition was reversed by the application of cyometrinil (Ebert and Ramsteiner, 1984). Thus, an influence of metolachlor on acetyl-CoA synthesis via PDC is postulated. What then is the site of the second type of inhibition in terpenoid biosynthesis which results in increased chloroplast pigment synthesis and even greater decreased growth in fresh weight?

IV. INHIBITION OF GIBBERELLIC ACID PRECURSOR BIOSYNTHESIS BY HERBICIDES

A. *ent*-Kaurene Synthesis Inhibition

As shown in Fig. 3, GA biosynthesis is a progression from GGPP to kaurene to GA. The original descriptions of kaurene synthesis and kaurene oxidation were

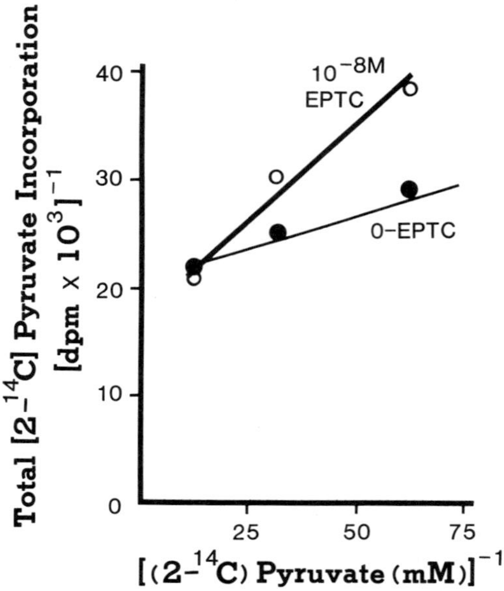

Fig. 6. Lineweaver–Burk plot of the incorporation of [2-^{14}C]pyruvate into fatty acids by intact chloroplasts. (●) Influence of various concentrations of [2-^{14}C]pyruvate, (○) + 0.01 μM EPTC. (Reprinted by permission of Academic Press, Inc.)

compilations of extensive and excellent work in microsome enzyme systems from seed of "half-mature" endosperm of wild cucumber (*Marah macrocarpus* Greene) (Graebe *et al.*, 1965; Dennis and West, 1967; Upper and West, 1967; Oster and West, 1968; Schechter and West, 1969; Hasson and West, 1973; Murphy and West, 1975; Hasson and West, 1976a,b; Frost and West, 1977; Knotz *et al.*, 1977) which have since been extended to pea (*Pisum sativum* L.) (Coolbaugh and Hamilton, 1976), tomato [*Lycopersicon esculentum* (L.) Mill.] (Yafin and Schechter, 1975), tobacco (*Nicotiana tabacum* L.) (Yafin and Schechter, 1975), corn (*Zea mays* L.) (Hedden and Phinney, 1975; Mellon and West, 1975), and barley (*Hordeum distichum* L.) (Murphy and Briggs, 1973, 1975). Conversion of GGPP to copalyl pyrophosphate (Fig. 3) commits the diterpene molecule to GA biosynthesis, and various chemicals utilized as plant growth retardants (e.g., AMO-1618, phosphon D, and phosphon S) have been shown to inhibit kaurene synthesis (Frost and West, 1977), whereas ancymidol inhibited kaurene oxidation (Coolbaugh and Hamilton, 1976; Frost and West, 1977). Application of these kaurene synthesis and kaurene oxidation inhibitors results in dwarfed plants.

Three lines of evidence are utilized to classify exogenous chemicals as plant growth retardants: (a) treated plants have lowered endogenous GA contents, (b) the morphological effects of treatment are relieved by simultaneous application of GA, and (c) GA biosynthesis in cell-free systems from higher plants is inhibited (Graebe, 1987). EPTC induced decreased GA, kaurene, and kaurenol contents in wheat (Wilkinson and Ashley, 1979); this report was corroborated by the incorporation of [^{14}C]mevalonic acid (MVA) into kaurene in a cell-free enzyme system from etiolated, unruptured sorghum coleoptiles in which EPTC inhibited kaurene synthesis (Table III) (Wilkinson, 1983). The inhibition was reversed by dichlormid. With EPTC alone, GGPP concentrations increased. With EPTC + dichlormid, GGPP concentration was equivalent to the untreated system. Thus, a second inhibition in terpenoid synthesis occurs which results in decreased GA biosynthesis, and GA deficiency results in rolled leaves and stunted plants. Proof of the concept that EPTC is a plant growth retardant is shown in Fig. 7, where concomitant application of GA_3 completely reversed the growth inhibitions of EPTC. Thus, all three criteria for classification of EPTC as a plant growth retardant have been published.

Metolachlor (0.1 μM) inhibited kaurene synthesis 90% (Table III) (Wilkinson, 1981) and cyometrinil reversed the inhibition. GGPP concentrations were not increased in the metolachlor treated system or metolachlor + cyometrinil treatments. Metolachlor tended to decrease total [^{14}C]MVA incorporation whereas metolachlor + cyometrinil treatments tended to induce an increase in [^{14}C]MVA incorporation. Thus, metolachlor possibly induced reactivities in isoprenoid synthesis other than simple kaurene synthesis inhibition. Yet, inhibitions of height by metolachlor were competitively reversed by concomitant applications of GA_3 in a sand potting medium (Fig. 8).

In sorghum, mefluidide appeared to have two activities in GA synthesis. GGPP accumulated over threefold at 0.01 μ*M* concentrations whereas kaurene was increased by 40% and kaurenol decreased by 80% (Table III) (Wilkinson, 1982a). The inhibition of kaurene synthesis was accompanied by major inhibitions of kaurene oxidation so that kaurene concentrations were not decreased even though there was an apparent small inhibition of kaurene synthesis. Thus, inhibition of GA precursor biosyntheses are major modes of action for EPTC and metolachlor, which explains the morphological responses of plants to these herbicides.

Within the limitations of [^{14}C]MVA incorporation into kaurenoids, the accumulation of GGPP could be accounted for by inhibitions of kaurene synthesis by CDAA and diallate (Wilkinson, 1985 a,f) (Table III).

B. *ent*-Kaurene Oxidation Inhibition

Kaurene oxidation is a mixed function oxidase (MFO, EC 1.14.14.1) enzyme system which contains SH groups and requires reduced nucleotides and O_2 (Hasson and West, 1973, 1976a,b). Incorporation of O_2 occurs twice. Conversion of kaurenol to kaurenal is inhibited by PCMBS

Table III Influence of Plant Growth Regulators on [^{14}C]Mevalonic Acid Incorporation into Kaurenoids in a Cell-Free Enzyme System from Unruptured, Etiolated Sorghum Coleoptiles

GG →(A) K →(B) K-ol →(C) K-al →(D) K-oic

Chemical	^{14}C Product	Concentration (μ*M*) 0.01	0.1	1.0	10.0	Reaction inhibited
		% Incorporation				
EPTC	GG	140	128	129	153	A[a]
	K	98	63	48	16	
	K-ol	60	41	33	34	C
	K-al	84	44	63	83	
	K-oic	76	66	68	69	
EPTC + dichlormid	GG	100	123	200	—	
	K	100	98	115	—	
	K-ol	42	46	140	—	
	K-al	139	112	86	—	
	K-oic	89	101	80	—	
CDAA	GG	91	316	181	178	A?[b,c]
	K	94	112	100	125	
	K-ol	159	135	142	110	
	K-al	151	152	110	83	
	K-oic	85	75	50	60	D
Metolachlor	GG	—	80	100	100	
	K	—	15	10	20	A[d]
Metolachlor + cyometrinil	GG	—	106	130	120	
	K	—	300	300	310	
Alachlor	GG	—	146	157	—	
	K	—	96	138	—	
	K-ol	—	41	21	—	B[e]
Alachlor + cyometrinil	GG	—	141	554	—	
	K	—	38	20	—	
	K-ol	—	284	808	—	B
Ancymidol	GG	—	110	170	140	
	K	—	160	375	375	
	K-ol	—	70	30	20	B[f]
Mefluidide	GG	340	280	220	215	
	K	115	185	170	160	A?[f]
	K-ol	20	15	15	18	B
Diallate	GG	200	300	100	85	A?
	K	100	100	100	70	
	K-ol	100	100	120	300	
	K-al	100	100	100	100	
	K-oic	100	90	80	50	D[g]

[a] From Wilkinson (1983). Reprinted by permission of Academic Press, Inc.
[b] From Wilkinson (1985a). Reprinted by permission of Academic Press, Inc.
[c] Inhibition is small and may be reversed by excess kaurene.
[d] From Wilkinson (1981). Reprinted by permission of Academic Press, Inc.
[e] From Wilkinson (1982b). Reprinted by permission of Academic Press, Inc.
[f] From Wilkinson (1982a). Reprinted by permission of Springer-Verlag, Inc., New York.
[g] From Wilkinson (1985b). Reprinted by permission of Academic Press, Inc.

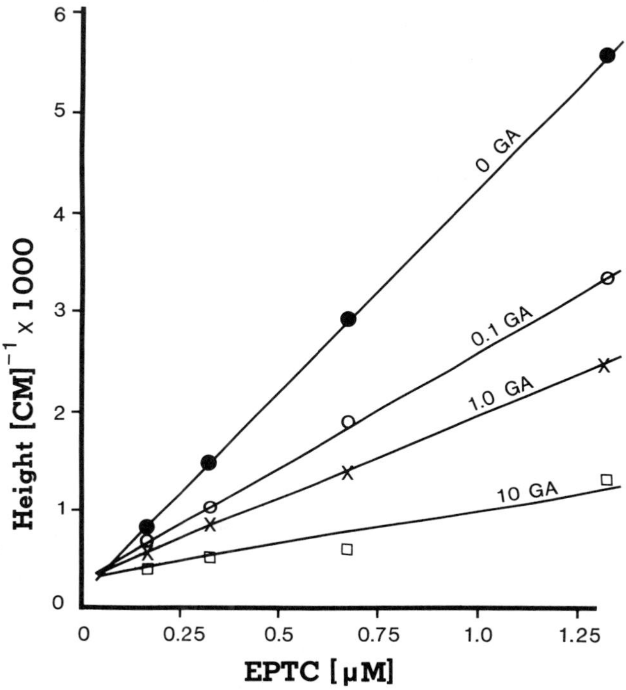

Fig. 7. Influence of EPTC (μ*M* concentrations) incorporated into sand on the growth of sorghum (14 days) and the reversal of the EPTC-induced growth retardation by 0.1, 1.0, and 10 ppmw GA_3 concomitantly incorporated into the sand. (●) EPTC + 0 GA; (○) EPTC + 0.1 ppmw GA; (×) EPTC + 1 ppmw GA; (□) EPTC + 10 ppmw GA. Slopes were significantly different at the <1% level. (Reprinted by permission of Academic Press, Inc.)

(*p*-chloromercuribenzenesulfonic acid), which reacts with SH groups; but the other reactions in kaurene oxidation are not influenced by PCMBS (West and Upper, 1969).

Data on [^{14}C]MVA incorporation show an EPTC influence on conversion of kaurenol to kaurenal (reaction C in Table III). Since this is an SH enzyme response, it is, possibly, similar to the inhibition of fatty acid synthesis, which is also an SH enzyme (Hitchcock and Nichols, 1971). Since EPTC did not inhibit the accumulation of [UL-^{14}C]sucrose into phloem sieve tube members, which is dependent on enzymes containing extracellular SH groups (Wilkinson, 1985b), the reactive portion of this system may not be the SH group.

Alachlor, ancymidol, and mefluidide inhibited reaction B (oxidation of kaurene to kaurenol) (Table III) (Wilkinson, 1982a,b). The alachlor inhibition was reversed by cyometrinil (Table III) (Wilkinson, 1982b). Thus, the reversal by cyometrinil of an alachlor-induced growth inhibition of sorghum (Nyffeler *et al.*, 1980) is explicable as an inhibition of cell enlargement. This would fit previous reports on the influence of GA precursor biosynthesis inhibitors on cell enlargement (Coolbaugh *et al.*, 1978; Deal and Hess, 1980; Coolbaugh and

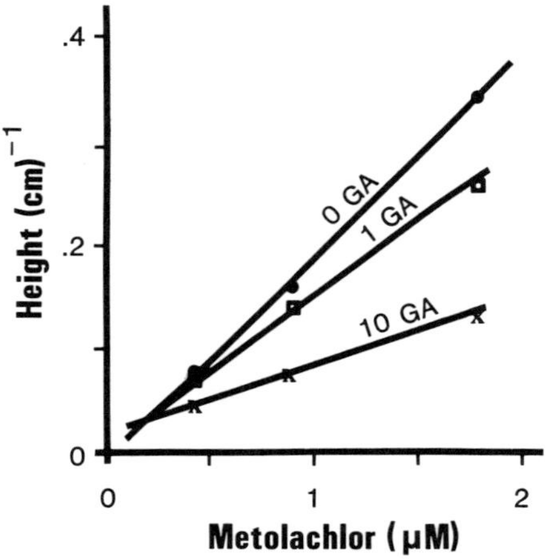

Fig. 8. Metolachlor (μM concentrations) incorporated into sand influence on the height of 14-day-old sorghum and the reversal of the metolachlor growth retardation by 1 and 10 ppmw GA_3 concomitantly incorporated in the sand. (●) Metolachlor + 0 GA; (□) metolachlor + 1 ppmw GA; (×) metolachlor + 10 ppmw GA. Slopes were significantly different at the $< 1\%$ level. (Reprinted by permission of Academic Press, Inc.)

Hamilton, 1986) (Figs. 7 and 8). Diallate and CDAA inhibited reaction D (Table III) (Wilkinson, 1985a,f). Since the growth inhibition by diallate was competitively reversed by exogenous GA (Fig. 9), the influence of diallate and CDAA on reaction D explains the morphological growth limitations by these compounds. Reactions B and D (Table III) are mixed-function oxygenase systems requiring NADPH and O_2. Additionally, they are inhibited by CO and cytochrome P-450 has been implicated (West, 1980).

Tetcyclasis, which inhibits kaurene oxidation (Rieber and Jung, 1983), is a plant growth regulator (PGR) utilized to inhibit growth of transplanted rice (*Oryza sativa* L.) (Rieber and Jung, 1983). The degradation time appears to be quite rapid because growth retardation of this compound lasts for a few days but is sufficient for utilization on mechanically transplanted rice and extends the period of use for rice seedlings by 4–5 days. Pretilachlor is utilized for weed control in rice. Elongation rates of individual rice leaves were inhibited by pretilachlor and this leaf elongation inhibition was reversed by CGA 123407 (4,6-dichloro-2-phenyl-pyrimidine) (Christ, 1985). Because of (a) close similarity of chemical structure between pretilachlor and alachlor and (b) the inhibition of GA precursor biosynthesis by alachlor (Table III) (Wilkinson, 1982b), this inhibition may be explicable as an inhibition of GA precursor biosynthesis. Data to substantiate this hypothesis have not been published. Yet analogies drawn from close similarities in chemical structure are perilous. Ancymidol is utilized as a dwarfing agent in horticultural greenhouse plants. Triarimol, a very close analog, is used as a fungicide but does not induce dwarfing of pea plants following postemergence application (Coolbaugh *et al.*, 1978). Studies of the

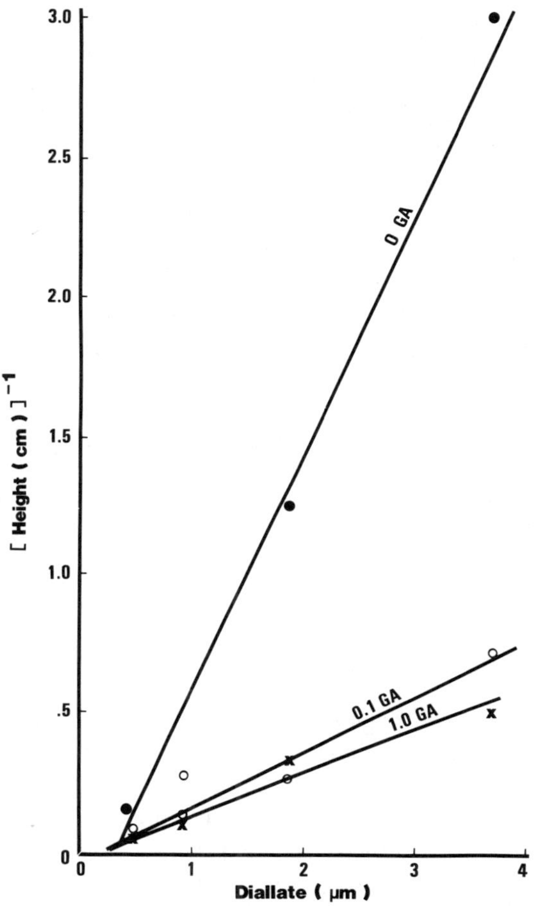

Fig. 9. Influence of sand-incorporated diallate (μM concentrations) on the height of 14-day-old sorghum and the reversal of the growth retardation by 0.1 and 1 ppmw GA_3 concomitantly incorporated into the sand. (●) Diallate + 0 GA; (○) diallate + 0.1 ppmw GA; (×) diallate + 1 ppmw GA. Slopes were significantly different at the $<5\%$ level. (Reprinted by permission of Academic Press, Inc.)

differences in response to ancymidol, triarimol, and several other analogs led to the suggestion that mediation of penetration of the compound into the plant as well as species differences in enzyme specificity were determining factors in the establishment of GA biosynthesis inhibition (Coolbaugh *et al.*, 1978).

V. STRUCTURE–ACTIVITY RELATIONSHIPS

Comparison of activities and structural configuration does not reveal any clear conclusions. EPTC inhibits kaurene synthesis and conversion of kaurenol to kaurenal (Table III) (Fig. 4) (Wilkinson and Ashley, 1979; Wilkinson, 1983).

Dichlormid reverses the EPTC responses; yet CDAA, a compound almost identical to dichlormid except for one Cl, has a questionable kaurene synthesis influence and inhibits kaurene oxidation (Table III) (Wilkinson, 1985a). Diallate inhibits both kaurene synthesis and oxidation but is structurally different from EPTC, dichlormid, and CDAA (Table III) (Wilkinson, 1985f). Metolachlor inhibits kaurene synthesis and cyometrinil reverses this activity (Table III) (Wilkinson, 1981). Yet alachlor is very similar to metolachlor structurally and alachlor inhibits kaurene oxidation to kaurenol whereas metolachlor does not (Table III) (Wilkinson, 1981, 1982b). Ancymidol and triarimol are very close structurally and have major differences in reactivity (Coolbaugh and Hamilton, 1976). These few comparisons indicate the variability of responses found for these compounds. Additionally, dichlormid is an active safener in corn for EPTC, alachlor, barban, molinate, pebulate, perfluidone, CDEC, and vernolate while only partially active in barley (*Hordeum vulgare* L.), field bean (*Phaseolus vulgaris* L.), and wheat (Stephenson and Chang, 1978). Naphthalic anhydride (NA) showed similar variations in safening capacity but reversed different herbicides in different crops (Stephenson and Chang, 1978). Thus, structure–activity relationships are not readily apparent.

VI. ENZYME REQUIREMENTS

Biochemical requirements for the reactions in terpenoid biosynthesis are not well defined. Wild cucumber (*Marah macrocarpus* L.) kaurene synthesis is regulated by adenylate energy charge above 0.80. Energy utilizing biosynthetic sequences show a sharp increase in rate as the energy charge is increased from 0.80 to 0.95 (Knotz *et al.*, 1977), and the rate-limiting step in terpenoid synthesis seems to be in the formation of isopentenylpyrophosphate (Fig. 1), where maximum reactivities were attained at 2 m*M* ATP. Additionally, divalent cations influenced kaurene synthesis; Mn^{2+} had a maximum influence at 3 m*M* and reactivity decreased above and below this concentration while the maximum reactivity of Mg^{2+} was at 5 m*M* concentration and combinations of $Mg^{2+} + Mn^{2+}$ (2:1) were maximal at 1–6 m*M*. Mixed function oxidation of kaurene to kaurenoic acid requires O_2, NADPH, NADH, ATP, FAD, and FMN (Hasson and West, 1976a); kaurene synthesis had pH optima of 6.9–7.3 (Frost and West, 1977).

GA precursor biosynthesis inhibition explains the morphological responses of these herbicides (EPTC, CDAA, metolachlor, alachlor, ancymidol, mefluidide, and diallate) and safeners (dichlormid and cyometrinil) in shoots. But with the exception of metolachlor and alachlor, root response to these compounds is negligible at concentraions which greatly inhibit shoot growth (Wilkinson and Karunen, 1977). Thus, the transport of energy to the roots was not decreased even though shoot length was decreased by 75–85%. An increased synthesis of chloroplast pigments (μg/g fresh weight) after application of GA precursor biosynthesis inhibitors indicates a commonality between pigment and GA synthesis. Thus, demonstration of kaurene synthesis (Moore and Coolbaugh, 1976) and pyruvate dehydrogenase complex (Camp and Randall, 1985) in chloroplasts eliminates troublesome problems of syntheses and utilization of acetate which was previously believed to be synthesized exclusively in mitochondria and utilized in microsomes for GA synthesis and in chloroplasts for terpenoid synthesis. Since some of these GA precursor biosyntheses require NADPH, modification of NADPH metabolism within the chloroplast is not unexpected; inhibitions of zeaxanthin epoxidation (Wilkinson, 1977), quinone oxidation (Wilkinson, 1978), conversion of GG-chlorophyll to phytol-chlorophyll (Wilkinson, 1985c), and inhibition of fatty acid desaturation (Karunen and Wilkinson, 1975; Wilkinson and Smith, 1975a,b) may be examples of these or other types of reactions. Reversal of these responses by safeners has been shown for kaurene oxidation (Wilkinson, 1983) and fatty acid desaturation (Wilkinson and Smith, 1976a,b).

VII. CAROTENOGENIC INHIBITION

Carotenogenesis includes conjugation (e.g., 2 GGPP to phytoene), dehydrogenation (phytoene to lycopene), and cyclization (lycopene to β-carotene) reactions (Fig. 2). Inhibition of cyclization and/or dehydrogenation by amitrole, pyriclor, and dichlormate (Burns *et al.*, 1971) results in the accumulation of phytoene, phytofluene, and/or α-carotene. Pyridazinone herbicides and analogs inhibit GGPP conjugation and phytoene dehydrogenation (Table IV). Thus, norflurazon (<16 ppbw) inhibited 2 GGPP conjugation to phytoene and GGPP and β-carotene contents were increased (Wilkinson, 1985d) (Table IV). At higher norflurazon concentrations (>32 ppbw) phytoene desaturation was inhibited and

Table IV Influence of Norflurazon on Geranylgeranyl Pyrophosphate (GGPP), Phytoene, and β-Carotene Contents in 14-Day-Old Wheat Grown in a Growth Chamber[a]

Norflurazon (ppbw)	μmol/g dry weight		
	GGPP	Phytoene	β-Carotene
0	75 c[b]	0.15 c	0.11 b
1	220 b	0.24 c	0.22 a
4	290 b	0.22 c	0.24 a
16	280 b	0.18 c	0.17 a
62	270 b	0.65 b	0.13 ab
250	530 a	0.94 a	0.04 c
1000	550 a	1.03 a	0.03 c

[a] From Wilkinson (1985d). Reprinted by permission of Academic Press, Inc.

[b] Values in a column followed by the same letter are not significantly different at the 5% level.

carotene concentrations decreased (Table IV). Thus, the possibility existed of a reversal of low concentrations (i.e., residual) of norflurazon by some compound which would result in a further accumulation of GGPP. Unpublished results in the field and laboratory have shown CDAA to partially reverse residual norflurazon activity (Fig. 10). Consequently, higher plant terpenoid biosynthesis around the branch point compound GGPP has been shown to be amenable to manipulation.

Inhibition of carotenogenesis is, currently, a highly active area of research for herbicidal activity (Sandmann *et al.*, 1984) (Table V). Consequently, many compounds have been evaluated for activity and only a few have been reported herein. New compounds will be reported in the future, and this will assist in complete description of carotenogenesis enzymes. Currently, oxadiazon inhibits chlorophyll synthesis; fluridone and norflurazon inhibit carotene synthesis, and acifluorfen-methyl inhibits carotene + chlorophyll synthesis (Sandmann *et al.*, 1984).

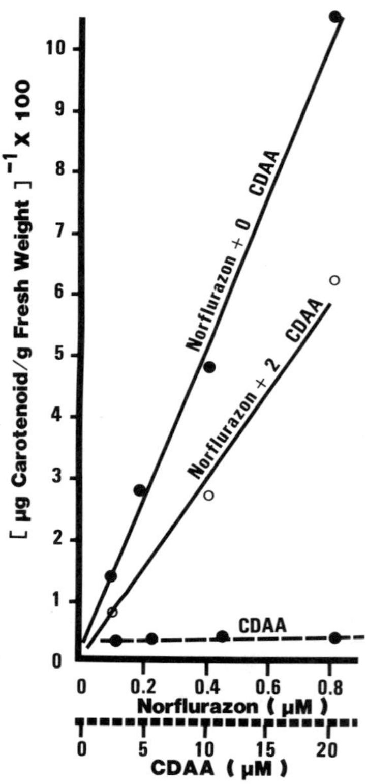

Fig. 10. Dixon plots of carotenoid content of 14-day-old wheat seedlings grown in nutrient solution containing norflurazon (μM concentrations) and CDAA. (●) Norflurazon + 0 CDAA; (○) norflurazon + 2 ppmw CDAA. Dash lines = CDAA at μM concentrations. Slopes were significantly different at the <1% level. (Reprinted by permission of Academic Press, Inc.)

Table V Carotenogenesis Inhibition by Selected Compounds

Compound	Concentration (μM)	Carotenogenesis inhibition (%)
Fluridone	0.1	83
R-40244	1	52
Norflurazon	1	95

VIII. CONCLUDING REMARKS

Terpenoid synthesis is influenced by several herbicides which induce development of morphological symptoms consistent with the hypothesis of an inhibition of total terpenoid and GA biosyntheses. Reversal of these morphological symptoms by utilization of safeners further substantiates the concept of GA synthesis as an explanation for these responses. Exogenous GA also alleviates these growth retardations. Yet problems remain. GA deficiency induces stunting (morphological response) but is not lethal if the plants protrude above ground level. For these herbicides, lethality appears to be associated with chlorophyll photooxidation. Many of these herbicides inhibit some reaction(s) in NADPH metabolism. But there is a time lag in response of the plants to the herbicides. Untreated chloroplasts do not show response to these plant growth regulators whereas chloroplasts extracted from plants grown during exposure to these compounds show very great modifications in e^- transport (R. E. Wilkinson, unpublished data). Data on the influence of these herbicides and safeners on H^+ transfer are not extant and the mode of lethality may be entirely different from the enzymatic reactions which modify plant morphology.

Carotenogenesis is another terpenoid biosynthetic process which is attacked by some herbicides, resulting in bleached plants. Inhibitors of conjugation, dehydrogenation, and cyclization are known. Extended research in this area will further elucidate the biochemistry and herbicide activity. For example, prevention of carotenogenesis by norflurazon results in chlorophyll photodegradation and bleached plants with accumulation of GGPP. Low norflurazon concentrations are reversed by some compounds which induce additional GGPP accumulations.

Thus, terpenoid biosynthesis appears to be an area for successful research on the interaction of exogenous compounds on plant growth. Much remains to be learned.

References

Burns, E. R., Buchanan, G. A., and Carter, M. C. (1971). Inhibition of carotenoid synthesis as a mechanism of action for amitrole, dichlormate, and pyrichlor. *Plant Physiol.* **47**, 144–148.

Camp, P. J., and Randall, D. O. (1985). Purification and characterization of the pea chloroplast pyruvate dehydrogenase complex. A source of acetyl-CoA and NADH for fatty acid biosynthesis. *Plant Physiol.* **77**, 571–577.

Christ, R. A. (1985). Effect of CGA 123407 as a safener for pretilachlor in rice (*Oryza sativa* L.). Recordings of elongation rates of single rice leaves. *Weed Res.* **25**, 193–200.

Coolbaugh, R. C., and Hamilton R. (1976). Inhibition of *ent*-kaurene oxidation and growth by α-cyclopropyl-α-(*p*-methoxyphenyl)-5-pyrimidine methyl alcohol. *Plant Physiol.* **57**, 245–248.

Coolbaugh, R. C., Hirano, S. S., and West, C. A. (1978). Studies on the specificity and site of action of α-cyclopropyl-α-(*p*-methoxyphenyl)-5-pyrimidine methyl alcohol (ancymidol), a plant growth regulator. *Plant Physiol.* **62**, 571-576.

Deal, L. M., and Hess, F. D. (1980). An analysis of the growth inhibitory characteristics of alachlor and metolachlor. *Weed Sci.* **28**, 168–175.

Dennis, D. T., and West, C. A. (1967). Biosynthesis of gibberellins. III. The conversion of (−)-kaurene to (−)-kauren-19-oic acid in endosperm of *Echinocystis macrocarpa* Greene. *J. Biol. Chem.* **242**, 3203–3300.

Ebert, E., and Ramsteiner, K. (1984). Influence of metolachlor and the metolachlor protectant CGA 43089 on the biosynthesis of epicuticular waxes on the primary leaves of *Sorghum bicolor* Moench. *Weed Res.* **24**, 383–389.

Frost, R. G., and West, C. A. (1977). Properties of kaurene synthetase from *Marah macrocarpus*. *Plant Physiol.* **59**, 22–29.

Gentner, W. A. (1966). The influence of EPTC on external foliage wax development. *Weeds* **14**, 27–31.

Graebe, J. E. (1987). Gibberellin biosynthesis and control. *Annu. Rev. Plant Physiol.* **38**, 419–465.

Graebe, J. E., Dennis, D. T., Upper, C. D., and West, C. A. (1965). Biosynthesis of gibberellins. I. The biosynthesis of (−)-kaurene, (−)-kauren-19-ol, and *trans*-geranylgeraniol in endosperm nucellus of *Echinocystis macrocarpa* Greene. *J. Biol. Chem.* **240**, 1847–1854.

Hasson, E. P., and West, C. A. (1973). A microsomal ATP-activated pyridine nucleotide transhydrogenase. *Arch. Biochem. Biophys.* **155**, 258–269.

Hasson, E. P., and West, C. A. (1976a). Properties of the systems for the mixed function oxidation of kaurene and kaurene derivatives in microsomes of the immature seed of *Marah macrocarpus*. Cofactor requirements. *Plant Physiol.* **58**, 473–478.

Hasson, E. P., and West, C. A. (1976b). Properties of the systems for the mixed function oxidation of kaurene and kaurene derivatives in microsomes of the immature seed of *Marah macrocarpus*. Election transfer components. *Plant Physiol.* **58**, 479–484.

Hedden, P., and Phinney, B. O. (1975). Comparison of *ent*-kaurene and *ent*-isokaurene synthesis in cell-free systems from etiolated shoots of normal and dwarf-5 maize seedlings. *Phytochemistry* **18**, 1475–1479.

Hitchcock C., and Nichols, B. W. (1971). "Plant Lipid Biochemistry." Academic Press, New York.

Karunen, P., and Wilkinson, R. E. (1975). Influence of *S*-ethyl dipropylthiocarbamate (EPTC) on wheat root phospholipid fatty acid metabolism. *Physiol. Plant.* **35**, 228–231.

Knotz, J., Coolbaugh, R. C., and West, C. A. (1977). Regulation of the biosynthesis of *ent*-kaurene from mevalonate in the endosperm of immature *Marah macrocarpus* seeds by adenylate energy charge. *Plant Physiol.* **60**, 81–85.

Loomis, W. D., and Croteau, R. (1980). Biochemistry of terpenoids. *In* "The Biochemistry of Plants" (P. K. Stumpf, ed.), Vol. 4, pp 368–418. Academic Press, New York.

Mellon, J. E., and West, C. A. (1975). Diterpene biosynthesis in maize seedlings in response to fungal infection. *Plant Physiol.* **64**, 406–410.

Moore, T. C., and Coolbaugh, R. C. (1976). Conversion of geranylgeranyl pyrophosphate to *ent*-kaurene in enzyme extracts of sonicated chloroplasts. *Phytochemistry* **15**, 1241–1247.

Murphy, G. J. P., and Briggs, D. E. (1973). *Ent*-kaurene, occurrence and metabolism in *Hordeum distichum*. *Phytochemistry* **12**, 2597–2605.

Murphy, G. J. P., and Briggs, D. E. (1975). Metabolism of *ent*-kaurenol-[17-^{14}C], *ent*-kaurenal-[17-^{14}C] and *ent* kaurenoic acid-[17-^{14}C] by germinating *Hordeum distichum* grains. *Phytochemistry* **14**, 429–433.

Murphy, G. J. P., and West, C. A. (1975). The role of mixed function oxidases in kaurene metabolism in *Echinocystis macrocarpa* Greene endosperm. *Arch. Biochem. Biophys.* **133**, 395–407.

Nyffeler, A., Gerber, H. R., and Hensley, J. R. (1980). Laboratory studies on the behavior of the herbicide safener CGA-43089. *Weed Sci.* **28**, 6–10.

Oster, M. O., and West, C. A. (1968). Biosynthesis of *trans*-geranylgeranyl-pyrophosphate in endosperm of *Echinocystis macrocarpa* Greene. *Arch. Biochem. Biophys.* **127**, 112–123.

Oswald, T. H., and Wilkinson, R. E. (1985). *S*-Ethyl dipropylthiocarbamate inhibition of acetyl-CoA derivatives in isolated wheat mitochondria. *Plant Physiol., Suppl.* **77**, 303.

Rieber, N., and Jung, J. (1983). Derivatives of 3,4-diaza-tricyclo-[4,2,1,$0^{2,5}$]-nona-3,7-diene and nona-7-enes, a new class of bioregulators. *Pestic. Chem.: Hum. Welfare Environ., Proc. Int. Congr. Pestic. Chem., 5th, 1982,* Abstr., 5 (II c), 13.

Sandmann, G., Clarke, I. E., Bramely, P. M., and Böger, P. (1984). Inhibition of phytoene desaturase—the mode of action of certain bleaching herbicides. *Z. Naturforsch. C: Biosci.* **39C**, 443–449.

Schechter, I., and West, C. A. (1969). Biosynthesis of gibberellins. IV. Biosynthesis of cyclic diterpenes from *trans*-geranylgeranyl pyrophosphate. *J. Biol. Chem.* **244**, 3200–3209.

Stephenson, G. R., and Chang, F. Y. (1978). Comparative activity and selectivity of herbicide antidotes *In* "Chemistry and Action of Herbicide Antidotes" (F. M. Pallos and J. E. Casida, eds.), pp. 35-61. Academic Press, New York.

Stumpf, P. K. (1980). Biosynthesis of saturated and unsaturated fatty acids, *In* "The Biochemistry of Plants" (P. K. Stumpf, ed.), Vol. 4, pp. 177–204. Academic Press, New York.

Upper C. D., and West, C. A. (1967). Biosynthesis of gibberellins. II. Enzymic cyclization of geranylgeranylpyrophosphate to kaurene. *J. Biol. Chem.* **242**, 3285–3292.

Weinberg, M. B., and Castelfranco, P. A. (1975). Effect of EPTC on plastid membrane constituents in germinating cucumber cotyledons. *Weed Sci.* **23**, 185–187.

West, C. A. (1980). Hydroxylases, monooxygenases, and cytochrome P-450. *In* "The Biochemistry of Plants" (D. D. Davies, ed.), Vol. 2, pp. 317–364. Academic Press, New York.

West, C. A., and Upper, C. D. (1969). Enzymatic synthesis of (−)-kaurene and related diterpenes. *In* "Methods in Enzymology" (R. B. Clayton, ed.), Vol 15, pp. 481–490. Academic Press, New York.

Wilkinson, R. E. (1974). Sicklepod surface wax response to photoperiod and *S*-(2,3-dichloroallyl) diisopropylthiocarbamate (Diallate). *Plant Physiol.* **53**, 269–273.

Wilkinson, R. E. (1977). Zeaxanthin epoxidation inhibition by EPTC (*S*-ethyl dipropylthiocarbamate). *Bot. Gaz. (Chicago)* **138**, 270–275.

Wilkinson, R. E. (1978). Quinone accumulation in *S*-ethyl dipropylthiocarbamate treated wheat. *Pestic. Biochem. Physiol.* **8**, 208–214.

Wilkinson, R. E. (1981). Metolachlor [2-chloro-N-(2-ethyl-6-methylphenyl)-*N*-(2-methoxy-1-methylphenyl) acetamide] inhibition of gibberellin precursor biosynthesis. *Pestic. Biochem. Physiol.* **16**, 199–205.

Wilkinson, R. E. (1982a). Mefluidide inhibition of sorghum growth and gibberellin precursor biosynthesis. *J. Plant Growth Regul.* **1**, 85–94.

Wilkinson, R. E. (1982b). Alachlor influence on sorghum growth and gibberellin precursor synthesis. *Pestic. Biochem. Physiol.* **17**, 177–184.

Wilkinson, R. E. (1983). Gibberellin precursor biosynthesis inhibition by EPTC and reversal by R-25788. *Pestic. Biochem. Physiol.* *19*, 321–329.

Wilkinson, R. E. (1985a). CDAA inhibition of kaurene oxidation in etiolated sorghum coleoptiles. *Pestic. Biochem. Physiol.* **23**, 19–23.

Wilkinson, R. E. (1985b). Ethyl dipropylthiocarbamate; *N,N*-diallyl-2-chloroacetamide; and 2-chloroallyl diethylcarbamate influence on sulfhydryl-dependent sucrose accumulation. *Pestic. Biochem. Physiol.* **23**, 95–101.

Wilkinson, R. E. (1985c). Inhibition of conversion of geranylgeranyl-chlorophyll to phytolchlorophyll by *S*-ethyl dipropylthiocarbamate (EPTC). *Pestic. Biochem. Physiol.* **23**, 289–293.

Wilkinson, R. E. (1985d). Carotenogenic inhibition by norflurazon in wheat. *Pestic. Biochem. Physiol.* **23**, 370–377.

Wilkinson, R. E. (1985e). Gibberellin influence on membrane disruption by thiocarbamate herbicides. *Pestic. Biochem. Physiol.* **24**, 103–111.

Wilkinson, R. E. (1985f). Diallate inhibition of gibberellin biosynthesis in sorghum coleoptiles. *Pestic. Biochem. Physiol.* **25**, 93–97.

Wilkinson, R. E., and Ashley, D. (1979). EPTC induced modification of gibberellin biosynthesis. *Weed Sci.* **27**, 270–274.

Wilkinson, R. E., and Hardcastle, W. S. (1969). EPTC effects on sicklepod petiolar fatty acids. *Weed Sci.* **17**, 335–338.

Wilkinson, R. E., and Hardcastle, W. S. (1970). EPTC effects on total leaflet fatty acids and hydrocarbons. *Weed Sci.* **18**, 125–128.

Wilkinson, R. E., and Karunen, P. (1977). Water utilization by *S*-ethyl dipropylthio-carbamate treated wheat. *Weed Res.* **17**, 335–338.

Wilkinson, R. E., and Oswald, T. H. (1987). *S*-Ethyl dipropylthiocarbamate (EPTC) and 2,2-dichloro-*N,N*-di-2-propylene-acetamide (dichlormid) inhibitions of synthesis of acyl-coenzyme A derivatives. *Pestic. Biochem. Physiol.* **28**, 38–43.

Wilkinson, R. E., and Smith, A. E. (1975a). Reversal of EPTC induced fatty acid synthesis inhibition. *Weed Sci.* **23**, 90–92.

Wilkinson, R. E., and Smith, A. E. (1975b). Thiocarbamate inhibition of fatty acid biosynthesis in isolated spinach chloroplasts. *Weed Sci.* **23**, 100–104.

Wilkinson, R. E., and Smith, A. E. (1976a). Butylate, pebulate, and vernolate inhibition of plant fatty acid biosynthesis. *Phytochemistry* **15**, 841–842.

Wilkinson, R. E., and Smith, A. E. (1976b). EPTC altered beet disc betacyanin efflux and fatty acid synthesis. *Weed Sci.* **24**, 235–238.

Yafin, Y., and Schechter, I. (1975). Comparison between biosynthesis of *ent*-kaurene in germinating tomato seeds and cell suspension cultures of tomato and tobacco. *Plant Physiol.* **56**, 671–675.

Part Three

Alternative Approaches for Crop Safening against Herbicide Injury

Chapter 11

The Use of Activated Carbon and Other Adsorbents as Herbicide Safeners

Robert E. Hoagland
USDA, ARS, Southern Weed Science Laboratory
Stoneville, Mississippi 38776

I. INTRODUCTION

Chemicals such as salts and industrial by-products such as smelter wastes were applied to roadsides and pathways for centuries to rid them of vegetation. However, chemical weedkilling may be considered to have been born in 1896 when a French grape grower, Bonnet, observed that the copper sulfate mixture he applied to his vines for protection against downy mildew turned the leaves of yellow charlock (*Sinapis arvensis* L.) black. The first organic chemical for weed control was 2-methyl-4,6-dinitrophenol (DNOC), followed by the development of the synthetic hormone weedkillers 2,4-D and MCPA during the early years of America's involvement in World War II. Although these compounds, and the many that followed, proved to be a boon to agriculture, they also presented problems regarding removal of chemical residues from equipment and spray tanks and confining their effects to the target species. The long-known purifying and adsorptive properties of activated carbon (AC; also referred to as activated charcoal) suggested the possibility of its use for these purposes. AC was first used as a herbicide safener when Hartley and Sutcliff, Speakman and Co. (1945)

applied it as a seed coating to protect seedlings from preemergent herbicide injury. Lucas and Hamner (1947) first reported the use of AC to eliminate 2,4-D residues in spray equipment. Arle *et al.* (1948) and Weaver (1948a,b) used AC directly on crop plants to protect them from 2,4-D injury, hence making AC the first herbicide antidote or safener. Later, chemical antidotes were developed as herbicide safeners and various aspects of those compounds have been presented and discussed in this volume.

The purpose of this chapter is to review adsorbents used as herbicide safeners. Since AC is generally the most efficient and widely used of these adsorbents, particular attention has been given to the physical, chemical and adsorptive properties and nature of AC and its use as a herbicide safener. Some relevant historical perspectives on the use of carbon by man, including its applications in agriculture, are also presented.

II. HISTORY

The use of carbon for purification and various other uses dates back many centuries. Historical points concerning the development of carbon, including its use in some agricultural situations, are outlined in Table I. Charcoal was used for the purification of potable water for centuries before the birth of Christ. The written record of the use of charcoal for purposes other than fuel and metallurgy dates back to the Egyptians in 1550 B.C. (Deitz, 1944) when various carbons were administered for medicinal uses.

The adsorptive properties of charcoal were clearly recognized (Scheele, 1773, in Deitz, 1944) in experiments with gases. Later, these properties were utilized

Table I Development and Uses of Carbon through History[a]

Year	Description of event
1550 B.C.	Early recorded use of charcoal
1811 A.D.	Bone char used for sugar processing
1852	First granular charcoal filter
1906	First commercial production of AC (Eponite) in Europe
1910	First application of AC in drinking water treatment (England)
1913	First commercial production of AC in the U.S.
1916	AC shown to promote plant growth (Skinner and Beattie, 1916)
1945	"Seed wafer" process patented (England)
1947	AC used to deactivate herbicide residue in spray equipment (Lucas and Hamner, 1947)
1948	AC applied directly to sweet potatoes to protect them from 2,4-D injury (Arle *et al.*, 1948)
	AC shown to reduce toxicity of 2,4-D in soils (Weaver, 1948a)

[a] Modified from Hassler (1974), by permission of Chemical Publishing Co., Inc.

to decolorize many solutions (Lowitz, 1785, in Deitz, 1944). Soon after, wood charcoal was used for purification in the cane and beet sugar industries in France.

The need for carbon with higher adsorptive capacities led to the development of AC in 1900, but commercial quantities of this material did not become available in the United States until 1913. The fundamentals of many of the laboratory-scale processes used in the early days of charcoal manufacture are similar to those in current industrial use. However, their large-scale or commercial use was not immediate because of the lack of need for a carbon more adsorptive than bone char and engineering difficulties caused by the lack of available structural materials resistant to the corrosive action of the activation conditions.

In the early twentieth century, inventions by Ostrejko (1900 and 1901, in Hassler, 1974) paved the way for the development of modern commercial AC production. These processes basically consisted of (1) incorporating metallic chlorides with vegetable substances followed by carbonization at a suitable temperature or (2) heating vegetable charcoal to red heat with the simultaneous addition of carbon dioxide.

Eponite, the first powdered AC, was produced in Europe in 1909 from wood using Ostrejko's methods. A second well-known carbon, Norite, was first produced in Holland in about 1911. The advent of World War I in 1915 and the use of chlorine gas by the Germans on the battlefront prompted the Allies to develop protection in the form of gas masks containing AC as the adsorbent. The soft powdered AC's then available, however, were unsuitable for the masks and new materials were examined and new processing equipment developed to find a suitable adsorbent. Granular coconut char was found suitable because of its great adsorptive capacity and its resistance to abrasion. This event and newly developed technology and equipment stimulated research to extend the use of AC into new areas. Since then, its use has grown tremendously in various areas such as air purification, solvent recovery, flavor and odor removal of foods, water purification, and other important uses.

The adsorptive properties of AC as an agricultural aid were recognized as early as 1916 when Skinner and Beattie showed the beneficial effects of carbon on cowpea [*Vigna unguiculata* (L.) Walp. subsp. *unguiculata*] growth, which they attributed to the removal, via adsorption on the carbon, of harmful soil components.

Hartley and Sutcliff, Speakman and Co. (1945) were the first to use AC as a herbicide safener when they used it as a seed coating to protect seedlings against injury from preemergent herbicides. The utility of AC as a herbicide safener in removing or inactivating 2,4-D in spray equipment prior to use of the equipment to spray plants sensitive to residual herbicide levels was shown by Lucas and Hamner (1947). This was followed by two simultaneous reports in March of 1948 on the use of AC directly on plants to provide herbicide protection. Arle *et al.* (1948) applied AC directly to sweet potato plants [*Ipomoea batatas* (L.) Lam.]

to protect this sensitive species from 2,4-D injury. At the same time, Weaver (1948a) reported that addition of AC to soil containing 2,4-D reduced or eliminated toxicity to plants and that AC, dusted or sprayed in aqueous suspension on several species, protected the plants from direct sprays of various 2,4-D salts and esters. In December 1948, another report (Weaver, 1948b) showed the utility of AC as a protectant when (1) applied to soils, (2) applied as a concentrated band above germinating seeds, (3) coated directly on seeds, or (4) sprayed directly on plants.

III. MANUFACTURE OF ACTIVATED CARBON

ACs are a group of materials in which individual members cannot be characterized by a definite structural formula or separately identified by chemical analysis. Individual ACs are characterized only by their adsorptive and catalytic properties. Carbons prepared from different sources and/or different activation processes will have different absorptive properties. Through the years, various materials have been used as sources for carbon (Hassler, 1974). Among the agriculturally related sources are beet sugar sludge, cereals, corncobs, cottonseed hulls, and rice hulls. Comparison of specific adsorptive properties of different ACs in aqueous solution demonstrates a wide range of properties among nine various carbons (Table II). It is beyond the scope of this chapter to give a detailed appraisal of the manufacture of AC. What is given here is a brief

Table II Specific Adsorptive Power of Different Activated Carbons in Aqueous Solutions[a]

Carbon code	Methylene blue[b]	Phenol[c]	Iodine[d]	Molasses[e]	Caramel[e]
A	0.84	1.35	0.65	0.8	1.3
B	0.74	1.24	0.59	0.6	0.9
C	0.70	1.30	0.57	1.1	1.6
D	0.37	0.60	0.36	0.7	1.2
E	0.30	0.60	0.32	0.6	1.3
F	0.44	1.20	0.55	0.2	0.4
G	0.83	0.83	0.39	1.0	1.0
H	0.73	0.57	0.37	1.1	1.0
J	0.50	1.60	0.59	nil	nil

[a]Data from *Ind. Eng. Chem.* **37,** 645 (1945); reprinted by permission of the American Chemical Society.

[b]Millimoles adsorbed per gram carbon at an equilibrium concentration of 0.1 mmol per liter.

[c]Millimoles adsorbed per gram carbon at an equilibrium concentration of 1.0 mmol per liter.

[d]Grams adsorbed per gram carbon at an equilibrium concentration of 0.127 g per liter.

[e]Color units adsorbed per gram carbon at 90% decolorization.

overview of some of these procedures. The reader is referred to Smisek and Cerny (1970), Mattson and Mark (1971), and Hassler (1974) for detailed accounts of these processes.

A. The Process of Carbonization

The process of carbonization, or pyrolysis, is usually conducted in the absence of air. Temperatures below 600°C are preferable for producing carbons suitable for steam activation, but these conditions are not always used (Hassler, 1974). AC from sugar, carbonized at 900°C, has been produced, but in an atmosphere of steam at 850°C, anthracite carbonization yielded a more porous product which was more readily activated (Hassler, 1958). Carbonization effectiveness is enhanced when the carbonaceous source is impregnated with a solution of a metallic chloride (i.e., calcium, magnesium, zinc). This mixture is dried and carbonized at 400–900°C. The char, after washing with acid and water, has an adsorptive capacity suitable for many applications.

B. Activation with Oxidizing Agents

Many processes are available for reacting carbon with oxidizing gases such as steam, air, and carbon dioxide at elevated temperatures. The adsorptive properties obtained during activation are dependent on (1) the chemical nature and concentration of the oxidizing gas, (2) the reaction temperature, (3) time and extent of activation, and (4) the type and concentration of mineral ingredients in the carbon source. Steam is preferable to carbon dioxide and is much better than air for the activation of many types of carbon. The relative adsorptivities of one carbon (pine char) activated under different conditions are outlined in Table III. The oxidizing action of the activating gases does not involve an indiscriminate removal of successive layers of atoms on the surface. Rather, the surface is

Table III Adsorption by Activated Pine Wood Char[a]

Activation conditions		Substances adsorbed (g/g carbon)			
Gas	Temperature (°C)	Chrysoidine R	Ponceau R	Aniline	Iodine
Air	600	0.34	0.10	0.05	0.36
Air	740	0.16	0.08	0.05	0.40
Air	790	0.15	0.08	0.06	0.42
Air	860	0.14	0.08	0.06	0.42
Air	910	0.13	0.10	0.06	0.40
Steam	770	0.37	0.19	0.16	0.60
Steam	825	0.37	0.17	0.17	0.60
Steam	880	0.36	0.16	0.21	0.62
CO_2	880	0.32	0.12		

[a]From Hassler (1974). Reprinted by permission of Chemical Publishing Co., Inc.

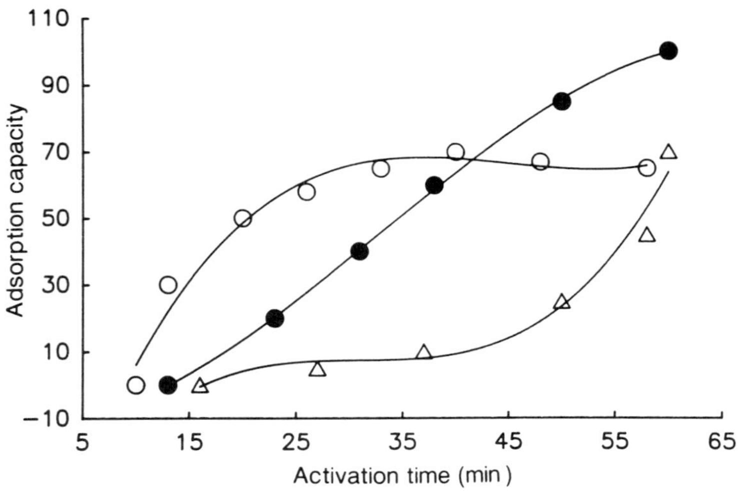

Fig. 1. Rate of development of specific adsorptive capacities during steam activation of chemically treated peat. Open circles = phenol; solid circles = methylene blue; open triangles = molasses color. (Redrawn from Hassler, 1974, by permission of Chemical Publishing Co., Inc.)

selectively eroded, which increases the surface area, develops greater porosity, and preserves the remainder of atoms in configurations having specific affinities for other molecules. Furthermore, all adsorptive properties do not develop simultaneously or at a uniform rate. This can be demonstrated by comparison of the development of adsorptive capacities during activation of a carbon (Fig. 1).

IV. CHARACTERISTICS OF CARBON AS AN ADSORBENT

A pure carbon surface is considered to have a nonpolar character, but carbon–oxygen complexes are often present which can render the surface polar. This nonpolar character facilitates adsorption of organics and hinders adsorption in inorganic electrolytes. ACs have been produced which possess virtually every type of chemical functional group known in organic chemistry. The types most frequently found on the AC surface (Fig. 2) are carbonyl, phenolic hydroxyl, quinone-carbonyl, normal lactone, fluorescein-type lactone, carboxylic acid anhydride, and cyclic peroxide groups. ACs generally have much greater adsorption affinity for organic molecules than for inorganic molecules. One commercial agricultural carbon producer (Anonymous, 1981) states that its product is ineffective for inorganic pesticides such as compounds of arsenic and lead, sodium chlorate, sulfur, borax, etc. and water-soluble organic herbicides such as amitrole, paraquat, diquat, MSMA, and DSMA. An important property of this particular product is its lack of adsorption of fertilizers. There are,

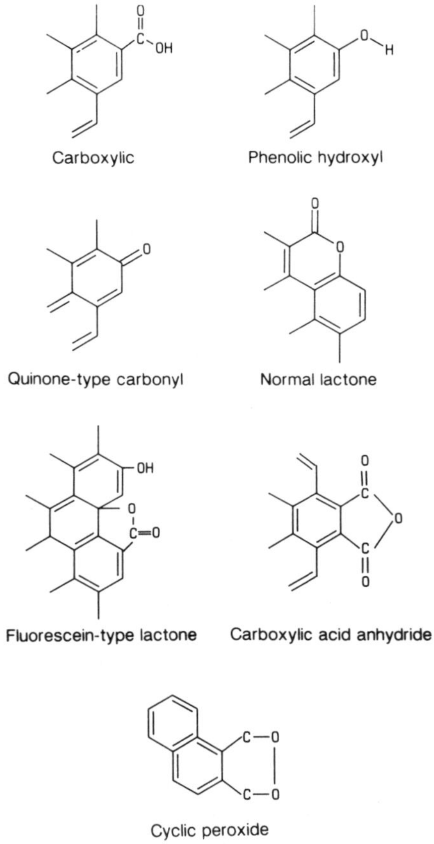

Fig. 2. Functional groups most frequently found on the surface of AC. (Redrawn from Cookson, 1978.)

however, many carbons available with a variety of properties and some of these carbons can remove a wide range of inorganic chemicals from solution (Table IV). Chemical interactions between inorganics and AC have been extensively reviewed by Huang (1978).

The adsorption process results in the net accumulation of a substance at the interface of two phases. This can involve liquid–solid, gas–liquid, gas–solid, liquid–liquid, or solid–solid interfaces. Carbon applied to soils or to plants for adsorption of herbicides or other compounds primarily involves liquid–solid interfaces; i.e., adsorption of chemicals onto carbon from the soil solution or herbicide application solution. Two adsorption processes, chemisorption and physisorption, may be operative when molecules in the liquid phase become attached to the carbon surface. These forces result when the attractive forces of the adsorbent (carbon) are greater than the kinetic energy of the contaminant molecules (adsorbate) in solution.

Table IV Some Inorganic Chemicals Which May Be Removed by AC[a]

Alkali metals	Hydrogen sulfide
Ammonia	Hypochlorous acid
Arsenates	Iodine
Calcium hypochlorite	Lead
Chlorine	Mercuric chloride
Chromium	Molybdates
Cobalt	Permanganates
Copper	Selenium
Cyanide	Silver salts
Dichromates	Sodium hypochlorite
Ferric salts	Zinc
Gold chloride	

[a]From Lyman (1978).

Physisorption results when adsorbate molecules become physically attached to the adsorbent as a result of energy differences or electrical attractive forces, i.e., van der Waals forces. Adsorption of this type results in multilayers, with each molecular layer forming atop the previous layer. The number of layers is proportional to the adsorbate concentration in the solution. This process is reversible and reversibility is dependent on the strength of attractive forces between adsorbate and adsorbent. Chemisorption occurs when a chemical reaction (i.e, production of a chemical compound) takes place between the adsorbate and adsorbent. This process differs from physisorption in that only a monomolecular layer is formed and the reaction is irreversible since energy is required to form the new chemical at the surface and is also necessary to reverse the process. Several factors are known to affect adsorption processes, including (1) chemical and physical characteristics of the adsorbent (chemical composition, pore size, surface area, etc.) and adsorbate (chemical composition, molecular size, polarity, etc), (2) adsorbate concentration in the liquid phase, (3) pH, ionic strength, and temperature of the liquid phase, and (4) the adsorption time. Cookson (1978) has recently presented a detailed treatise on adsorption mechanisms pertaining to the chemistry of organic adsorption on AC.

The effectiveness of AC in the adsorption of herbicides and other chemicals is accentuated by its extremely large surface area and the number and size of pores. Typically, the surface area of AC can range from 500 to 1400 m^2/g, but some are as high as 2500 m^2/g (Table V). While surface area is quite critical to the adsorption process, the chemical nature or polarity of the carbon surface is much less important.

Pores exist throughout the structure of an AC particle as illustrated in Fig. 3. The large surface:size ratio of AC is influenced by pore structure. While macropores (>1000 Å) provide passageways to the interior of the particle and its micropores (10–1000 Å), they do not appreciably alter total surface area. Micropores (formed primarily during the activation process), on the other hand,

Table V Surface Area of Some Commercially Available ACs[a]

Commercial name	Origin	Surface area (m^2/g)
PCC SGL	Bituminous coal	1000–1200
PCC BPL	Bituminous coal	1000–1200
PCC RB	Bituminous coal	1200–1400
PCC GW	Bituminous coal	800–1000
Columbia CXA/SXA	Coconut shell	1100–1300
Columbia AC	Coconut shell	1200–1400
Columbia G	Coconut shell	1100–1500
Darco S 51	Lignite	500–550
Darco G 60	Lignite	750–800
Darco KB	Wood	950–1000
Hydro Darco	Lignite	550–600
Nuchar Aqua	Pulp mill residue	550–650
Nuchar C	Pulp mill residue	1050–1100
Nuchar (various)	Pulp mill residue	500–1400
Norit (various)	Wood	700–1400

[a]From Cheremisinoff and Morresi (1978).

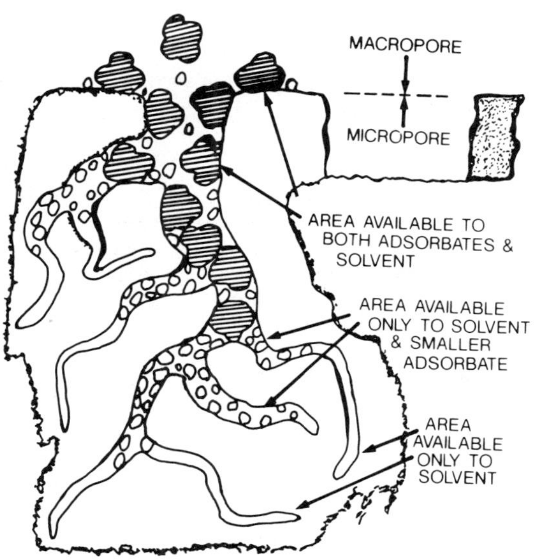

Fig. 3. Molecular discrimination effects of carbon pores. (Redrawn from Cheremisinoff and Morresi, 1978.)

result in large surface area formation. These pore structures are a major factor in adsorption and pore distribution regulates the size range of adsorbate molecules that can enter the carbon particle. Certain large molecules can block micropores, causing reduced surface area availability, but smaller molecules generally penetrate and are adsorbed in the micropores. To obtain optimal adsorption, the proper choice of carbon for the removal of a specific chemical is a prerequisite, a point that has not been adequately considered in herbicide safening research. Carbon properties vary widely (Tables VI and VII) and AC can be separated into two main types: granular (>300 μm, 50 mesh or greater) and powdered (<300 μm). Carbon adsorption capacity is dependent on total surface area while the rate of adsorption is related to particle size. The use of adsorption isotherms (relation of adsorptivity of an adsorbate and its concentration in solution) at constant conditions (temperature, pH, flow rate, etc.) helps in determining the amount of carbon necessary to reduce contamination to the desired level.

Various mathematical models have been used to describe adsorption. One such adsorption model is an equation developed to describe the adsorption of gases bysolids (Langmuir, 1916 and 1918, in Bailey and White, 1970). The reader is also referred to Brunauer *et al.* (1967) for a complete discussion of the equation. This model is based on the assumptions that (1) the energy of adsorption is constant and independent of surface coverage, (2) adsorption occurs at localized sites with no interaction between adsorbate molecules, and (3) adsorption is that of a monomolecular layer. These assumptions also mean that there is an upper limit to adsorption. Early adsorption studies indicated that there was often a lack of conformity to the Langmuir equation. As a result, another empirical adsorption model equation was developed (Freundlich, 1926, in Bailey and White, 1970). This equation is widely used in studies on adsorption of chemicals in soils and with carbon and other adsorbents. However, it has the disadvantage of less

Table VI Properties of Several Commerically Available ACs[a]

Physical properties	ICI Americas Hydrodarco 3000	Calgon Filtrasorb 300 (8 × 30)	Westvaco Nuchar W-L (8 × 30)	Witco 517 (12 × 30)
Surface area, m^2/g	600–650	950–1050	1000	1050
Apparent density, g/cc	0.43	0.48	0.48	0.48
Density, backwashed and drained, lb/ft^3	22	26	26	30
Real density, g/cc	2.0	2.1	2.1	2.1
Particle density, g/cc	1.4–1.5	1.3–1.4	1.4	0.92
Effective size, mm	0.8–0.9	0.8–0.9	0.85–1.05	0.89
Uniformity coefficient	1.7	≤1.9	≤1.8	1.44
Pore volume, cc/g	0.19	0.85	0.85	0.60
Mean particle diameter, mm	1.6	1.5–1.7	1.5–1.7	1.2

[a]From Cheremisinoff and Morresi (1978).

Table VII Typical Properties of Several DARCO Powdered Carbons[a]

Property	DARCO Powdered Carbons S-51	G-60	KB
Bulk density, lb/ft^3, 30 min tamp test	32	25	28
Storage space, ft^3/ton	80	100	75
Particle size			
through 100 mesh	98	95	99
through 325 mesh	70	70	70
Moisture, %	8	8	25
Decolorizing efficiency[b]	100	125	190
pH of water extract	5	5	5
Water-soluble ash, 4 boiling leaches, %	1	0.2	1.3
Rate of filtration[c]	25	45	20
Total surface area, m^2/g, dry basis	650	600	1450
Total pore volume, mg/g, dry basis	1	1	1.5
Pore volume distribution, <20 Å	0.25	0.1	0.3
Volume in pores of given radii, mg/g, dry basis			
20–50 Å	0.15	0.1	0.4
50–100 Å	0.1	0.3	0.2
100–500 Å	0.2	0.35	0.45
500	0.3	0.15	0.15
Mean pore radius, Å	30	25	23

[a]From Cheremisinoff and Morresi (1978).
[b]Based on decolorization of standard blackstrap molasses relative to a reference carbon with a decolorizing efficiency of 100.
[c]Gallons of water per hr/ft^2 filter area/inch of carbon filter cake (40 psig).

sensitivity and no assumptions for an upper limit on adsorption. For a more detailed theoretical discussion of these adsorption models and their applications, the reader is referred to Bailey and White (1970), Wauchope and Koskinen (1983) (soils), and Suffet and McGuire (1980) (carbon).

V. SOIL–HERBICIDE–ACTIVATED CARBON ADSORPTION INTERACTIONS

Herbicide residues in soils pose several problems: (1) injury to sensitive plants grown in rotations with treated crops, (2) accumulation of residues from application rates which exceed rates of dissipation, (3) unacceptably high residue levels in crops grown on contaminated soils, and (4) possible inhibition of beneficial soil microorganisms.

Several reviews have described the mechanisms of pesticide adsorption by soil colloids (Adams, 1973; Bailey and White, 1970; Helling *et al.*, 1971; Osgerby, 1970; White and Mortland, 1971). Various factors such as the physical and chemical nature of the soil colloid, temperature, pH, ionic strength, moisture content, the nature of the saturating cation on these colloids, and the nature of the pesticide formulation affect the adsorption–desorption of these chemicals in soils. The influence of temperature on the adsorption of herbicides on AC is demonstrated in Fig. 4. For example, AC adsorbed 2,4-D to a lesser degree than it adsorbed prometon, and significantly more 2,4-D was retained at 10°C than at 55°C. Prometon adsorption was not significantly influenced by temperature (Weber *et al.*, 1965).

One problem with dry AC is that it is difficult to get into suspension and maintain in suspension. Studies indicated that surfactants were useful for this purpose and that significant differences were found between rates and types of surfactants used. Low rates of surfactants (0.3%, v/v) did not alter AC protective properties, but at 0.9 and 2.7%, the surfactants significantly lowered protection (Burr, 1969). Surfactants and herbicide additives vary greatly. Although these additives obviously play a role in the efficacy of carbon as a herbicide safener, little research has been conducted in this area.

As indicated, the above factors are operative in AC adsorption mechanisms, and when field situations are considered, with heterogeneous soil types and

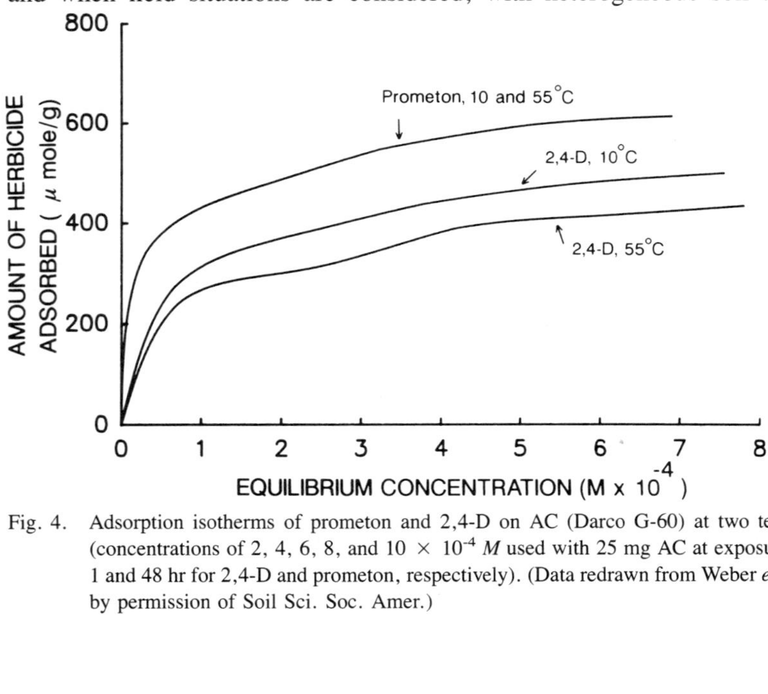

Fig. 4. Adsorption isotherms of prometon and 2,4-D on AC (Darco G-60) at two temperatures (concentrations of 2, 4, 6, 8, and 10 $\times$ 10^{-4} M used with 25 mg AC at exposure times of 1 and 48 hr for 2,4-D and prometon, respectively). (Data redrawn from Weber *et al.*, 1965, by permission of Soil Sci. Soc. Amer.)

composition, the possible interactions are numerous. Soil types have been tested to compare the rate of charcoal required to give adequate protection to ryegrass (*Lolium perenne* L.) from diuron. A higher rate of charcoal was required on a lighter, sandy loam soil than on a heavier, clay loam soil to give an equal degree of protection (Burr, 1969). Sandy loam soil required at least three times more AC than a clay loam soil to obtain equal protection against diuron (Burr *et al.*, 1972), probably because of the greater adsorption capacity of the clay loam. Soil moisture content at application time and subsequent irrigation did not influence AC protection.

Carbon affects temperature, texture, salt retention, and moisture retention of soils. These variables influence plant growth. Effects of carbon black on the properties of soils have also been investigated (Everson and Weaver, 1950a). When mixed with light colored soils in proportions of 1 to 250 parts or more, these soils were considerably darkened. It was concluded that, by darkening soils, carbon black increased the energy of adsorption, and differences in charges between colloidal clay and carbon led to flocculation and aggregation in some soils. Carbon applied at the surface and 2 inches deep in soil increased the maximum and minimum temperatures and the time available for plant growth (Everson and Weaver, 1950b). Coutts (1945) found that heating soil to 250–500°C caused about 20% reduction in buffering capacity, but a temperature lower than 250°C did not cause significant alteration of buffering capacity, indicating that veld-burning under field conditions (resulting in charred or carbonized plant material) had only a minor effect on buffering capacity.

In studies on the effects of charcoal on physical, chemical, and biological properties of forest soils, Tryon (1948) found that addition of charcoal reduced the evaporation rate and increased the time required to reach the wilting percentage. Total nitrogen was increased by charcoal from a hardwood source but was decreased by conifer charcoal. The pH and levels of phosphoric acid, potassium, calcium, and magnesium were all increased by charcoal treatment.

Numerous chemical reactions occur on surfaces. Charcoal was used as a model to study surface oxidation and behavior as an acidoid as related to soil properties and plant nutrition (Puri *et al.*, 1953). Charcoal, by virtue of formation of a surface layer of CO_2, was capable of undergoing surface ionization in water, yielding hydrogen ions which can neutralize alkaline compounds and decompose carbonates and sulfides.

Titration curves of hydrogen ion soils and of sodium ion soils with K^+, Na^+, NH_4^+, Ba^{2+}, and Ca^{2+} charcoals indicated that as neutralization takes place, a transfer of cations actually takes place from one surface to another (Puri *et al.*, 1958), which points to important mechanisms of adsorption and of soil acidity and alkalinity. Adsorption of herbicides on AC can be a function of pH. For example, the highest adsorption of ametryn occurred when the herbicide was in the molecular form, i.e., at pH 7.1 (Fig. 5) (Yamane and Green, 1972).

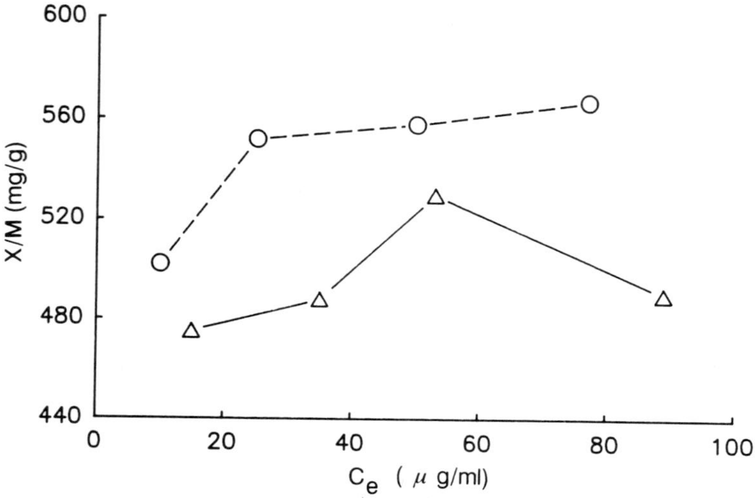

Fig. 5. Ametryn adsorption on AC (Darco G-60) as a function of pH. Open circles = pH 7.1; open triangles = pH 4.1. (Data redrawn from Yamane and Green, 1972, by permission of Soil Sci. Soc. Amer.)

Herbicide equilibrium in soils in relation to soil water content was examined using atrazine (Green and Obien, 1969). They concluded that only on low adsorption soils will water content variations significantly alter herbicide concentration in the soil solution. The principal effect of soil water content on herbicide phytotoxicity probably is associated with herbicide transport, which is more sensitive to changes in water content than is the concentration of herbicide in soil solutions. Since various herbicides have differing adsorptivities in soils with different water contents, and since carbon added to soil can influence water content, these interactions could be important in influencing herbicide transport and, hence, phytotoxicity. Little research has been done relating these areas.

Charcoal has been shown to reduce decomposition rates of atrazine, chlorthiamid, and linuron and to reduce available herbicide in the soil solution (Moyer *et al.*, 1972). Various herbicides, herbicide degradation or metabolic products, and some naturally occurring compounds (allelochemicals) are aromatic ring compounds. In standard chemical equilibrium adsorption studies, granular AC took up to 3 weeks to reach equilibrium with the aromatic compound phenol and up to 5 weeks to reach equilibrium with *o*-chlorophenol, whereas powdered AC isotherms indicated 3 to 5 days to attain equilibrium (Peel and Benedek, 1980). Eighty percent of the adsorptive equilibrium was reached during the first few hours, but the remaining capacity was utilized very slowly, suggesting a dual rate mechanism.More recently, the thermodynamic functions and the kinetics of adsorption of 2,4,5-T by AC from aqueous solutions have been reported (Gomez-Jimenez *et al.*, 1987).

VI. EFFECTS OF ACTIVATED CARBON ON PLANT GROWTH IN SOILS AND NUTRIENT CULTURE

As early as 1916, Skinner and Beattie found that carbon black mixed with soil did not increase productivity. However, carbon encased in porous tubes buried in test plots did increase the growth of cowpeas. This beneficial effect was attributed to the removal of harmful components of the soil solution by the adsorptive carbon. Charcoal was found to be an economical and highly efficient adsorbent for the removal of SO_2 from the soil solution (Anonymous, 1921). Miller (1925) outlined a role for charcoal in the nature and origin of soil acidity. Perotti and Ferretti (1930) found carbon to be effective in reducing harmful effects of the toxins produced by soil microorganisms and in activating the respiratory process. Holynski (1928, 1932) proposed that AC had an enhancing effect on plant growth when applied with fertilizers because it regulated the consumption of fertilizers and water by plants by altering the physical properties of soils and increasing activity of the nutritional constituents of the soil. Scheel (1934) reported that charcoal also improves the aeration and water capacity of soils.

Growth of tobacco (*Nicotiana tabacum* L.) crops was reported to be increased and leaves had a greater burning capacity when vegetable carbon was added to the soil (Bonuccelli and Cini, 1930). Swanback (1942) reported that AC used around the stems of tobacco plants caused increased absorption and retention of heat, building a reserve for the night and resulting in a slight increase in leaf yield. Plant growth in aqueous nutrient solutions (Zinzadze, 1932) and ammonification of organic matter (Verona and Ciriotti, 1935) were also improved by AC.

Scharrer and Schropp (1935) found that activated coal, mixed with various soils, caused only a slight increase in yield of corn (*Zea mays* L.) and spring barley (*Hordeum vulgare* L.). In contrast, Engels (1933) and Wyatt and Leahey (1936) reported that no beneficial effect was derived from the addition of charcoal to soils.

Addition of charcoal to rooting media is frequently practiced in horticulture, but there are few studies on the effects and mode of action of charcoal in this regard. Charcoal added to sand cultures of cereals increased yield in some cases up to three-fold and the mechanism was suggested to be maintenance of a favorable soil pH by AC's adsorbing properties (Prianeshnikov and Domontovitch, 1926). Zinzadze (1932) found that the addition of AC to aqueous cultures maintained at various pH levels resulted in improved plant growth of up to 3%. This beneficial effect was considered to be chiefly due to the adsorption of toxins, the beneficial oxidation catabolized by charcoal, and the addition of minor elements to the medium. Similar views were held by Papadakis (1941) and Holynski (1928). Gukova and Butkovich (1941) showed that addition of 6%

charcoal to sand cultures of inoculated soybean caused a two-fold increase in nodules and in plant dry weight. Charcoal was thought to improve aeration and thereby benefit plants. Tincker (1947) attributed improved aeration to the charcoal's beneficial effect on the growth of *Lolium* spp. Addition of low amounts of AC (0.5–2.0%) to the rooting medium of inoculated peas (*Pisum sativum* L.) in nitrogen-free sand culture resulted in marked increases in nitrogen fixation and plant dry weights, but lowered nodule number (Vantsis and Bond, 1950). AC also increased dry weight of noninoculated peas supplied with inorganic nitrogen. It was tentatively suggested that the favorable effects were due to adsorption of harmful excretions from roots or microorganisms or of excess nutrients, and from the maintenance of a more favorable pH in the rooting medium. Kjeldahl analyses of the rooting medium, however, revealed no evidence that the enhanced fixation was due to excretion of fixed nitrogen.

Experiments with AC on nitrogen fixation in clover (Vantsis and Bond, 1950) supported earlier conclusions of Nutman (1945) that nodulation was influenced by plant root secretions. Turner (1955) also showed that charcoal in rooting media of clover plants inoculated with effective and ineffective *Rhizobium* strains caused a stimulation of nodule production. Stimulation was suggested to be due to charcoal adsorption of inhibitory compounds secreted from clover roots. Some inhibitors were eluted from the charcoal and shown to affect nodule production.

The question of whether herbicides that are tightly bound to soil particles are accessible to metabolism by microorganisms has been posed (Anderson, 1968; Gast, 1962). For example, adsorbed simazine was found to be protected from microbial metabolism for long periods of time (Gast, 1962). Geissbuhler *et al.* (1963) found similar results with chloroxuron, i.e, the herbicide was more rapidly broken down in sandy soil than in a humic soil, where 8 and 1% of the herbicide was in the soil solution, respectively. Helweg-Andersen (1969) investigated the effect of charcoal adsorption of endothal and DNOC on microbial degradation of these compounds in soil. Endothal (weakly adsorbed by charcoal) was degraded by microorganisms irrespective of the presence of charcoal in the soil. Charcoal more firmly bound DNOC with a concomitant decrease in DNOC metabolism.

Nearly all the studies of AC in soils have shown beneficial effects on plant growth. Furthermore, there has been no indication that the rate of herbicide degradation is slowed in soils containing AC.

VII. USES OF ACTIVATED CARBON AS A HERBICIDE SAFENER

The effects of AC when applied as a herbicide safener are diverse. Many of these effects are presented in the following sections, grouped by herbicide class. Other publications have also summarized some situations where AC has been used as a herbicide safener (Anonymous, 1974, 1976; Blair *et al.*, 1976).

A. Aliphatic Acid Herbicides

When several rates of AC and dalapon were applied to oats (*Avena sativa* L.),carbon was able to provide protection against dalapon effects on oat stand and dry weight (Long and Scranton, 1969). There was significant protection by AC against tricalcium arsenate injury to turfgrass without affecting control of crabgrass (*Digitaria* spp.) (Jagschitz, 1974). AC in combination with cacodylic acid, dimethyl arsinic acid, sodium arsenite, or DSMA had no effect on turfgrass stand 9 weeks after seeding of Kentucky bluegrass (*Poa pratensis* L.) and red fescue (*Festuca rubra* L.). AC did not alter dalapon effects on ryegrass and red fescue stand. AC was found ineffective against TCA (trichloroacetic acid) toxicity to germinating sugar beet (*Beta vulgaris* L.) (Ripper, 1958).

B. Amide Herbicides

AC provided little or no safener effect against alachlor or cisanilide in forage crops, even at charcoal applications of 448 kg/ha (Bertges and Behrens, 1974). In greenhouse studies, propachlor was inactivated at carbon:herbicide ratios of about 200:1 in beans (*Phaseolus vulgaris* L., 'Black Valentine') and cucumber (*Cucumis sativus* L.) (Bovey and Miller, 1969).

Studies with diphenamid and AC (0.100 ppm) showed that 26.3 ppm herbicide was required to give 50% soybean root inhibition. The ratio of I_{50} with adsorbent to I_{50} without herbicide was 18:1, and 46% of the herbicide could be desorbed from AC via aqueous extraction (Coffey and Warren, 1969). When AC was added to vermiculite (1.56 lb ACHF/ft^3 vermiculite), the adsorption of diphenamide was increased 1000-fold over that of AC in silica sand. This resulted in a 10-fold safety margin if diphenamide was applied at 6.7 kg/ha (William and Romanowski, 1972).

Greenhouse experiments indicated that 224 kg/ha of AC banded over the row consistently protected seeded tomatoes (*Lycopersicon esculentum* Mill) from injury by diphenamid at 4 lb/acre. Toxicity to snap beans from alachlor was also prevented with surface band applications of 100 lb/acre charcoal (Putnam *et al.*, 1974). Field studies to evaluate preplant AC root dips in reducing herbicide injury to newly planted strawberries (*Fragaria ananassa* Duch.) showed that injury by diphenamid and napropamide was reduced by AC treatment. However, alachlor or metolachlor caused equal amounts of injury with or without charcoal applied to roots (Poling and Monaco, 1985).

AC at 122 kg/ha was sufficient to deactivate a fresh application of pronamide (1.12 kg ai/ha) when ryegrass was used as a bioassay species (Meyers *et al.*,1973). Annual bluegrass (*Poa annua* L.) could absorb a dose of pronamide in 2 weeks sufficient to result in death even after deactivation of residual herbicide with AC (Meyers and Sligh, 1973). AC at 122 kg/ha applied at 0- to 14-day intervals to established stands of annual bluegrass previously treated with lethal doses of pronamide showed that AC could prevent damage if applied 0–2 days after herbicide treatment. Annual bluegrass absorbed sufficient pronamide after 10 days to render AC ineffective if applied at that time (Meyers and McCarthy, 1974).

AC protected newly sprigged centipedegrass [*Eremochloa ophiuroides* (Munro) Hack.] from pronamide injury, resulting in significantly higher appearance and vigor ratings up to 7 weeks after planting. Rates used were two applications of 1.7 kg/ha at 0 and 3 weeks after sprigging plants with roots dipped in a slurry of 900 g charcoal in 3.8 liters of water. Little effect was noted after 7 weeks and weed control at the end of the growing season was not affected by AC (Johnson, 1976).

C. Benzoic Acid Herbicides

AC was found to improve turfgrass stands in seedbeds containing dicamba, and also to alleviate dicamba damage to established turfgrass (Jagschitz, 1974). Protection, although not complete, was obtained in cucumber receiving 3.0 and 6.0 chloramben methyl ester and AC at 224 kg/ha (Putnam *et al.*, 1974).

D. Benzonitrile Herbicides

Germination and early growth of cucumbers, sweet corn, beans, and tomatoes were greatly enhanced by applying charcoal (224 and 448 kg/ha) to soil treated with dichlobenil at 2.8, 5.6, and 12.0 kg/ha. At the two lower herbicide rates,carbon treatment resulted in yields of sweet corn and beans that were 80% and of cucumbers, about 50% compared to plants not receiving dichlobenil (Locascio, 1967). Increased dichlobenil rates required correspondingly increasing charcoal rates to provide protection to oats (Long and Scranton, 1969).

E. Bipyridillium Herbicides

Coffey and Warren (1969) reported that paraquat was not adsorbed by AC, and thus the biological activity of paraquat was not reduced by AC . In contrast, Faust and Zarins (1969) reported that both diquat and paraquat were adsorbed by carbon in studies on the interaction of these herbicides with clay and carbon in aqueous solutions. Although water-soluble herbicides such as paraquat are difficult to adsorb, this herbicide was effectively adsorbed by AC and other adsorbents (Fig. 6). The quantities of AC, H-bentonite, and calcium-bentonite necessary to reduce initial paraquat concentrations to 0.1 mg/liter indicated that

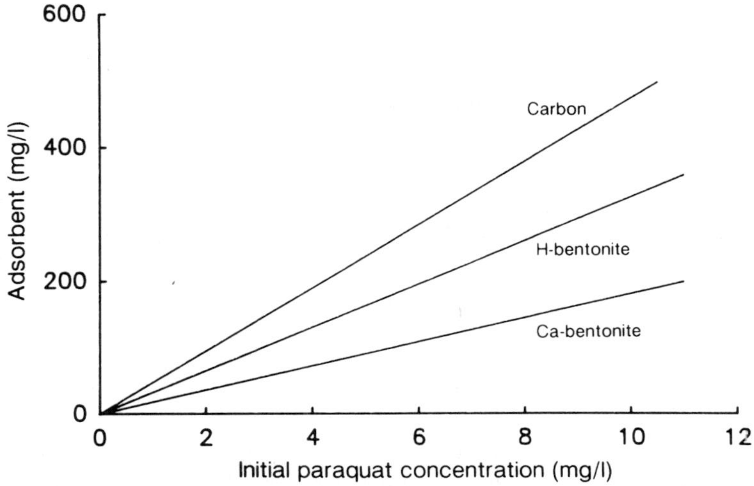

Fig. 6. Adsorbent dosages for removal of paraquat from aqueous solution. (Data redrawn from Faust and Zarins, 1969, by permission of Springer-Verlag.)

calcium-bentonite was most effective. Different conditions and carbon types may explain these results, as well as the fact that paraquat is a strongly cationic and water-soluble compound.

F. Dinitroaniline Herbicides

Trifluralin was not adequately absorbed by AC in field studies to be of value for establishing forages (Bertges and Behrens, 1974). However, AC was found to adsorb sufficient amounts of trifluralin to protect grain sorghum [*Sorghum bicolor* (L.) Moench]. Desorption studies indicated that this herbicide was 35% desorbable from AC (Coffey and Warren, 1969).

AC improved turfgrass stands treated with benefin and nitralin (Jagschitz, 1974). Benefin and butralin injury in centipedegrass has been reported to be reduced by AC (Johnson, 1976). Increasing AC rates generally decreased the activity of nitralin on grain sorghum. Furthermore, lower rates of charcoal were more effective after a period of time; i.e., 280 kg/ha AC inactivated nitralin at 1.12 kg/ha after 4 weeks and 140 kg/ha was effective after 8 weeks (Kennedy and Talbert, 1973). These results may be due to decreased amounts of herbicide residues via degradation and movement with time.

G. Diphenylether Herbicides

Surface band application of 224 kg/ha AC prevented toxicity of fluorodifen (4.48 kg/ha) to snap beans (Putnam *et al.*, 1974).

H. Phenoxyalkanoic Acids

One of the earliest reports on the potential of AC as a herbicide safener was reported in 1948 when Arle *et al.* used AC to protect sweet potato against 2,4-D. Weaver (1948a,b) also found that the toxic effects of 2,4-D on marigold (*Tagetes* spp.), mustard (*Brassica* spp.), kidney bean (*Phaseolus* spp.), and soybean could be eliminated or decreased by applying dry or aqueous suspensions of AC on the plants prior to herbicide application. Sweet potato sprouts were moistened and then dusted with AC (about 1 lb/1000 sprouts) and planted in soil previously treated with 2,4-D at 1000 to 4000 ppm. Yields were not different from those of untreated control, except at rates of 3000 to 4000 ppm herbicide where increased yields were observed. Weaver (1948a,b) found similar protection against 2,4-D injury when AC was added to the soil, coated on seeds, or sprayed directly on plant surfaces. 2,4-D was adsorbed by AC as evidenced by a root growth bioassay using cucumber. The herbicide was also readily desorbed from AC (Coffey and Warren, 1969). AC improved turf seedbeds and established stands which had been treated with 2,4-D or silvex (Jagschitz, 1974).

I. Phenylcarbamate Herbicides

In a series of adsorption/desorption tests of various herbicides, chlorpropham was found to be most readily adsorbed, as indicated by a root bioassay, and this herbicide showed little or no desorption (Coffey and Warren, 1969). This correlates well with the finding that chlorpropham was strongly adsorbed by organic matter (Roberts and Wilson, 1965; Warren, 1956). AC placed below the soil layer where most weed seeds germinate but above sugar beet seeds was effective in preventing phytotoxicity to germinating sugar beet seeds (Ripper, 1958). Coating roots of strawberry plants with AC prior to planting protected plants from chlorpropham (Schubert, 1967). AC improved turfgrass stands when terbutol was added to seedbeds (Jagschitz, 1974). When terbutol and AC were applied to centipedegrass, AC provided protection and increased appearance and vigor of newly sprigged grass (Johnson, 1976).

J. Thiocarbamate Herbicides

Ripper (1958) found that AC protected germinating sugar beets against EPTC phytotoxicity. EPTC and vernolate injury to AC-coated strawberry plant roots was also prevented (Schubert, 1967). AC was effective as a barrier to protect corn seedlings from injury by EPTC vapors (Gray and Joo, 1978). AC applied as a suspension in the transplant water protected sweet potatoes from vigor reductions caused by vernolate at rates of up to 6.7 kg/ha. However, dipping roots in a suspension of AC prior to planting failed to provide protection (Wilson *et al.*, 1974a).

K. Triazine Herbicides

Simazine injury was prevented in strawberry plants when the roots were coated with AC before transplanting, followed by herbicide treatment at 1.12 kg/ha several days after planting. Dipping strawberry roots in AC before planting was found more effective and practical than applying a similar quantity of charcoal around each plant at planting time (Robinson, 1965). Flagg and Bayer (1966) found that AC applied to the roots of strawberries did not prevent serious injury from simazine. However, injury from simazine and prometryn was prevented in strawberry plants where AC was coated on roots prior to transplanting and herbicide exposure (Schubert, 1967).

In another study, injury symptoms from simazine appeared quickly and the phytotoxic effect was quite apparent in oats, but AC produced a marked reduction in simazine injury (Anderson, 1968). Simazine adsorption required an AC:herbicide ratio of from 200:1 to 400:1, and AC tended to be less efficient in the inactivation of simazine at lower herbicide concentrations. A growth increase (above that with no simazine) was also noted with some treatments of simazine plus AC.

Simazine at 3.4 kg/ha was deactivated by charcoal at 244, 488, and 976 kg/ha, at 10 kg/ha by 488 and 976 kg/ha, and at 20 kg/ha only by a rate of 976 kg/ha of charcoal (Long and Scranton, 1969). Inactivation of simazine by AC was persistent over the course of 6 months, and the plant growth-inhibiting properties of the herbicide on white mustard (*Sinapis alba* L.) were highly reduced (Ostrowski, 1972). Charcoal improved stands in seedbeds and alleviated simazine damage to established turfgrass (Jagschitz, 1974). A 1-inch band of carbon protected seeded grasses and legumes from atrazine and simazine (Kay, 1972).

Charcoal treatments of 0.8 g/100 ml H_2O applied to tomato plants at transplanting increased plant vigor after treatments of atrazine at 1.12, 2.24, and 3.36 kg/ha (Hopen, 1967). Alfalfa (*Medicago sativa* L.) stand was reduced 95–100% on plots treated with atrazine (1.68 kg/ha) even when AC was applied up to 112 kg/ha. On the other hand, when bands of AC were sprayed on soil over rows of seeded alfalfa, the legume was protected from herbicide damage. Rates of 28 to 56 kg/ha gave adequate protection for 1.7 kg/ha of G-36393 (Linscott and Hagin, 1967). AC was shown to prevent the appearance of phytotoxic symptoms on snap beans, beets (*Beta vulgaris* L.), cabbage (*Brassica oleracea* L.), and tobacco caused by atrazine (Kring and Ahrens, 1968). AC (224 kg/ha) incorporated in soil completely reversed atrazine (1.8 and 5.4 kg/ha) injury to oats (Brenchley and Burrill, 1968). Oats are severely damaged by as little as 0.3 ppm atrazine and AC, at 10–30 ppm, was shown to alleviate toxic residues of the herbicide (Jordan and Smith, 1971).

Greenhouse experiments indicated that considerable protection against atrazine injury (as much as 0.50 ppm of the herbicide) could be provided for cucumbers by 100 ppm AC. Field studies showed that 168 kg/ha AC could protect against 0.17 kg/ha atrazine, but only when the AC was incorporated at the same depth as the herbicide. Snap beans were more tolerant to atrazine than cucumbers and 112 kg/ha AC offered effective protection against atrazine rates of up to 0.34 kg/ha (Wilson *et al.*, 1974b).

Persistence of atrazine residues under greenhouse conditions was affected by levels of AC added to a silica sand potting medium (Harvey, 1973). Atrazine degradation and uptake from the soil by corn crops were reduced by AC treatment at 0.4 g/kg soil. Alternate freezing and thawing of these soils decreased AC's ability to deactivate the herbicide. AC at 358 kg/ha provided only partial protection from atrazine (2.24 kg/ha) injury to watermelon [*Citrullus lanatus* (Thunb.) Matsum. & Nakai] (Glaze *et al.*, 1979).

AC rates of 640 and 1920 ppm inactivated propazine at initial rates of 9.6 and 28.8 ppm, respectively. The triazine herbicide G514260 at 9.6 and 28.8 ppm was inactivated by 640 ppm charcoal (Bovey and Miller, 1969).

Metribuzin was more adequately adsorbed by AC than was atrazine (Bertges and Behrens, 1974). AC was used to reduce phyotoxiciy of metribuzin to direct-seeded tomatoes in the greenhouse and the field (Henne and Guest, 1974). In contrast to Bertges and Behrens, Henne and Guest found that metribuzin applied preplant incorporated into the soil was not adequately adsorbed by charcoal treatments, thereby resulting in seedling injury. However, preemergence applications successfully detoxified the herbicide. AC was effective when applied as a dry mixture with vermiculite or when applied alone as a surface spray, but was less effective if applied directly with the seed.

In field tests, AC applied at 112 kg/ha in a 3.75 cm band over the row completely alleviated injury from metribuzin broadcast at 0.56 kg/ha on seeded tomatoes. A similar protection against metribuzin by AC was provided to seeded asparagus (*Asparagus officinalis* L.) (Putnam *et al.*, 1974).

Although with metribuzin there were significant differences between 56 and 112 kg/ha of AC for protecting asparagus, asparagus plants treated with this herbicide were stunted and chlorotic (Ogg, 1978).

Both preemergence and postemergence application of metribuzin to winter wheat, without the use of AC over the seeded row, caused considerable injury (Rydrych, 1985). AC at 84, 167, and 336 kg/ha in 5-cm bands over the rows provided protection.

Metribuzin, when applied pretransplant at 0.28 kg/ha, killed peppers (*Capsicum annuum* L.) that were transplanted bare-root or in plugs, but AC at 1 g per plant effectively reduced injury to peppers in soils treated with 0.56 kg/ha metribuzin (Majek, 1986). The level of protection for plug transplants was dependent on the amount of AC per plant rather than on percentage in the soil mixture by dry weight.

L. Uracil Herbicides

Brenchley and Burrill (1968) found that charcoal treatment (224 kg/ha) was sufficient to overcome bromacil residue from a 1.8 kg/ha rate. In tests to increase herbicide selectivity on rangeland with AC, seed species were afforded some protection from bromacil and terbacil injury when charcoal was banded over the seed rows (Kay, 1972).

AC was used to adsorb and inactivate terbacil in planted hills of direct-seeded watermelon. AC was incorporated at 1008 kg/ha when the soil was treated with terbacil at rates of 1.12, 2.24, or 4.48 kg/ha. Yields were significantly higher at all rates of terbacil when AC was incorporated at 1008 kg/ha than when AC was surface applied at 336 kg/ha or in the absence of AC (Elmstrom, 1972). Glaze *et al.* (1979) found that AC at 358 kg/ha applied as a spot treatment on soil over the seed of watermelon gave full protection from terbacil (1.12 kg/ha) injury.

Terbacil in soil (0.20 ppmw or higher) severely injured and reduced foliage dry weight of alfalfa, beans, corn, potatoes (*Solanum tuberosum* L.), spring wheat, and sugar beets. However, AC at 150 kg/ha prevented a significant weight loss in beans and corn in soil containing 0.10 ppm terbacil and fully protected potatoes grown in soils containing 0.19 ppm of this herbicide (Ogg, 1982). AC at 300 kg/ha fully protected sugar beets and prevented a significant weight loss in alfalfa grown in soils containing 0.19 ppm terbacil. Wheat grown in 0.19 ppm terbacil was not protected, even at 600 kg/ha AC.

Field studies to evaluate AC use as a root dip to reduce herbicide injury to newly planted strawberries showed that plant growth following application of 0.28 kg/ha terbacil was equal to that of plants receiving no herbicide (Poling and Monaco, 1985). Charcoal root dips partially protected crop plants at 0.28 kg/ha but did not protect from injury by terbacil at 0.56 or 1.11 kg/ha rates.

M. Urea Herbicides

The usefulness of AC for the inactivation of linuron in soil seeded with oats was investigated (Anderson, 1968). Linuron at 28.8 ppm was slightly injurious to beans, but AC rates of 640 ppm reduced this effect (Bovey and Miller, 1969). When cucumbers were the crop, an AC:herbicide ratio of 200:1 was adequate to inactivate linuron, whereas norea was deactivated at a ratio less than 200:1. AC alleviated linuron damage to established turfgrass (Jagschitz, 1974).

Diuron, linuron and neburon phytotoxicity to strawberry plants was prevented when their roots were coated with AC prior to planting (Schubert, 1967). Generally AC application to roots could protect plants from up to twice the basic herbicide rate. AC reduced injury to oat seedlings caused by diuron but was not sufficient to overcome high diuron residues in soil (Brenchley and Burrill, 1968). Several charcoals, when applied at 30 to 50 ppm with a diuron residue level of 1.0 ppm, produced significant protection to oats (Jordan and Smith, 1971). Protection against diuron at 0.84 kg/ha was obtained from AC banded over rows at 5.04 kg/ha (Kay, 1972). Broadcast AC at an AC:diuron ratio of $\geq$5:1 (w/w) prevented death of barley seedlings receiving a herbicide rate of 32 kg/ha, even though 0.8 kg/ha without AC killed plants (Toth and Milham, 1975). Diuron was adequately adsorbed by AC, indicating AC's protective value in the establishment of forage grasses (Bertges and Behrens, 1974).

The effectiveness of AC in protecting seedling cotton (*Gossypium hirsutum* L.) from diuron at 1.78 and 3.55 kg/ha was evaluated under field conditions (Chandler *et al.*, 1978). AC did not provide adequate protection when applied through the seed hopper box or when sprayed in the seed furrow over each hill. However, AC at 83 and 167 kg/ha applied as a spot over each hill of cotton provided adequate protection. Shallow incorporation of the AC significantly reduced the degree of protection. Narrow bands of AC sprayed at 100 kg/ha on the soil surface directly over rows seeded with rice (*Oryza sativa* L.), efficiently protected (90%) the rice crop from diuron (1.1 kg/ha) damage (de Andrade and Bauman, 1985).

Fluometuron (1.68 kg/ha) when simultaneously incorporated with AC (560 kg/ha) had no adverse effect on squash (*Cucurbita maxima* Duch.). When incorporated separately, this AC rate did not completely inactivate this urea herbicide, but AC rates as low as 69 kg/ha inactivated fluometuron residues after 4 weeks in soil. AC was effective in deactivating fluometuron effects on squash, and AC activity persisted for up to 12 weeks (Kennedy and Talbert, 1973).

AC at 56 and 112 kg/ha in a 3-cm band over the seeded row protected asparagus from chlorbromuron at 2.2 kg/ha (Ogg, 1978).

N. Miscellaneous Herbicides

AC has been used in studies with various miscellaneous herbicides to determine its usefulness as a herbicide safener. Long and Scranton (1969) found that charcoal could overcome the effects of DCPA (13.4, 26.9, and 40.3 kg/ha) in oats. In greenhouse experiments with Toa silty clay, a ratio of 3600:1 AC:picloram was required to inactivate the herbicide using a bean bioassay (Bovey and Miller, 1969). In field tests with this soil, AC protected oats from picloram (0.56 kg/ha) but did not protect cucumbers or beans at rates up to 672 kg/ha AC. Bensulide, picloram, oxadiazon, endothall, and MH injury to grasses was prevented by AC (Jagschitz, 1974). Forage grass seedlings were not adequately protected against perfluidone injury by application of AC (Bertges and Behrens, 1974). Dipping centipedegrass sprigs in AC improved early growth and appearance of the turfgrass when treated with bensulide and DCPA. The rate of turfgrass growth was also improved by AC for bensulide but not for DCPA (Johnson, 1976). Some herbicidal oils were tested for deactivation by AC (Jagschitz, 1974). Diesel fuel and hydraulic oil inhibited growth of grass seedlings but charcoal did not provide a reversal.

VIII. METHODS OF ACTIVATED CARBON APPLICATION

Extensive work has been performed to develop the most efficacious use of AC as a protective agent for plants against herbicides. Generally AC may be used in three instances: (1) to reduce chemical contamination, (2) for plant protection, and (3) for enhancing germination and plant growth. Various methods of application of AC are outlined below, and some of these application methods have been refered to in the discussion of the adsorption by AC of specific herbicides. Application methods have also been summarized and discussed by others (Clapp, 1975; Blair *et al.*, 1976; Holly, 1976; Anonymous, 1981).

A. AC Applied as Root Dip

The earliest reports of plant protection against a herbicide using AC were the application of carbon to protect against 2,4-D injury (Arle *et al.*, 1948; Weaver, 1948a,b). Later, various plants dipped into slurries of 0.12–0.24 kg/liter AC prior to transplant protected seedlings from various herbicides (Ahrens, 1964). AC has also been applied to roots via incorporation into the transplant water (Ahrens, 1964).

B. AC Applied as Broadcast Spray

AC can be applied as a dry powder or some carbons can be mixed with water to form a slurry which can be sprayed onto soil. Slurries require agitation to prevent settling of AC and to achieve equal distribution. For small areas, a sprinkling can or garden sprayer will suffice. AC is generally applied in a water slurry, using spray equipment designed for concentrated wettable powders, including centrifugal pumps, piston pumps with ceramic cylinders, and hardened stainless steel nozzles capable of handling abrasive powders. Various nozzles and pressures can be used to achieve the desired dosages. Concentrations of 1–2 lb AC/gallon water or less are commonly used. Continuous agitation during spraying is important to maintain a constant application rate (Anonymous, 1981).

C. AC Applied as Incorporation

AC must be incorporated after broadcast application so as to expose it to the herbicide or other compound(s) to be adsorbed. In small areas, this may be accomplished by raking, rotary tilling, disking, plowing, etc. In some instances, such as established turfgrass, AC can be incorporated by thorough watering to avoid mechanical incorporation which would damage the turf. In some cases and under certain conditions, deactivation may require several days (3–4 days up to a week).

D. AC Applied as Banded Spray

This technique is efficient in providing herbicide protection to newly emerging seedlings and allows herbicide weed control in areas without carbon. Lee (1973) was one of the earliest to experiment with the banded spray technique by applying AC in a narrow band directly over the seed row of ryegrass; AC protected the emerging plants.

E. AC Applied to Seeds

Various researchers have examined the use of AC as a seed coating and seeds coated with AC have been made commercially available. As previously indicated, the coating or pelleting of seeds with charcoal to protect them from herbicides was patented in 1945 by Hartley and Sutcliff, Speakman and Co. Disadvantages of this method are that crop seedlings rapidly grow out of the protected zone and weeds in the zone are also protected (Croxford *et al.*, 1975). The method has been successful in some cases, for example, rice (International Rice Research Institute, 1969; Nangju *et al.*, 1976), maize and cowpea (Gupta and Niranwal, 1976), soybeans (Iswaren *et al.*, 1971), and *Rosa* spp. (Stryckers and van Himme, 1967). The advantages and uses of seed pelleting have been reviewed (Plucknett, 1971). A seed wafer technique has also been developed which has had some success (DeTar, 1973; Kratky and Warren, 1971). The seed wafer is composed of compressed vermiculite, seeds, AC, and fertilizer. The wafer provides anticrustant properties and the AC provides a herbicide safening effect. Some commercial potting soils also have incorporated AC (Carpenter, 1973). AC is marketed for use in ball and burlap transplants of nursery stock (Anonymous, 1981).

F. AC Applied as Dry Powder

AC may also be applied from a spreader designed for application of finely ground powders. Spreader calibration is required. This practice has been successful in some turf application to adsorb herbicide or oil contamination (Clapp, 1975). A machine was developed which in a single operation could sow crop seeds in a newly opened furrow, partly fill in and consolidate the furrow on top of the seed, spray a narrow band of AC as adsorbent into the furrow, complete the filling, and finally spray preemergence herbicide on the soil surface (Ripper, 1956; Ripper and Scott, 1957) (Fig. 7).

IX. OTHER AGRICULTURALLY RELATED APPLICATIONS OF ACTIVATED CARBON AS A PESTICIDE ADSORBENT

The adsorptive properties of AC have been studied in various situations for adsorption of unwanted amounts of pesticides. Some examples of these applications are briefly presented.

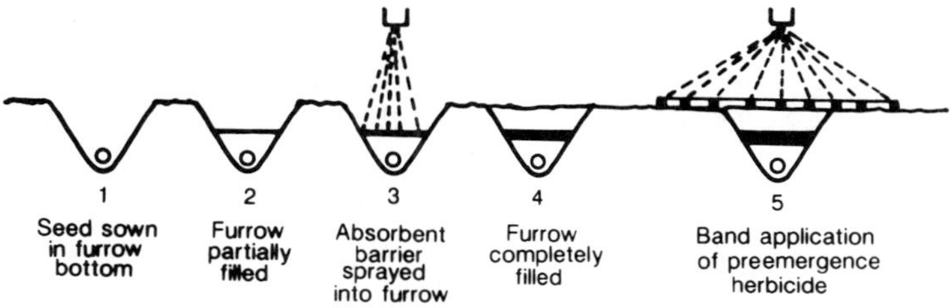

Fig. 7. Stages in the application of a barrier of AC to improve the selectivity of a residual preemergence herbicide.(Redrawn from Holly, 1976; modification of Ripper, 1956, by permission of Academic Press.)

AC has been applied in experiments to reduce insecticide levels and their associated phytotoxicities in crop plants. In 1950, Glass showed that AC could reduce parathion injury to apples. Protection from DDT by AC used on various crop plants was also demonstrated (MacCallum, 1956; Rodriguez *et al.*, 1960).

AC lowered the phytotoxicity of various insecticides to wheat, white mustard, and sugar beet seedlings when AC and insecticides were applied as seed dressings prior to planting. AC reduced the systemic action of Thimet against mustard beetle, *Phaedon cochleoriae* F., and the bird-cherry aphid, *Rhopalosiphum padi* (L.). Formulations permitting the slow release of insecticides have a practical value in that they can combine decreased phytotoxicity with increased systemic insecticidal activity (Bardner, 1960).

Laboratory and field studies showed that AC could substantially reduce insecticidal residues of aldrin, dieldrin, heptachlor, heptachlor epoxide, and chlordane in various vegetable crops. AC in soil prevented the total extraction of the insecticides. Residue binding to the carbon–soil complex increased with time, and differences in insecticide uptake caused by carbon were attributed to these different binding rates and/or to different penetration rates of the insecticides into the crop tissues (Lichtenstein *et al.*, 1968, 1971).

The ability of AC to adsorb herbicides has also been applied to animal systems. Bovine ingestion of atrazine-contaminated hay is unlikely (Johnson *et al.*, 1972), but cattle have been known to ingest lethal amounts of formulated atrazine. Animals receiving 454 g AC per animal at 4, 24, 48, and 72 hr after lethal atrazine doses survived, indicating the utility of AC in increasing bovine survival chances (Kobel *et al.*, 1985).

Increased microbial growth by AC has been attributed to its ability to adsorb toxic fatty acids (Foster *et al.*, 1950). Inhibitory plant phenolic compounds found in plants and soils are also known to be adsorbed by AC (Fridborg *et al.*, 1978; Johansson, 1983) and therefore not available to inhibit microbial and plant growth. Addition of charcoal, especially in the presence of glucose or pectin, greatly stimulated ericoid mycorrhiza formation by *Pezizella ericae* Read on *Vaccinium angustifolium in vitro* (Duclos and Fortin, 1983). AC was suggested to act by adsorption of toxic compounds from the rhizosphere. However, inhibitory effects of AC in culture media have also been reported. Even the inhibitory effect of AC has been demonstrated to be due to adsorption of some components of culture media, e.g., chelated iron (Heberle-Bors, 1980), hormones (Martineau *et al.*, 1981) and other constituents (Babbar and Gupta, 1986).

Fridborg and Eriksson (1975) also ascribed the effects of charcoal in plant tissue culture systems to its ability to adsorb growth-active compounds. Later, Fridborg *et al.* (1978) suggested the identity of these inhibitors as phenylacetic acid and benzoic acid derivatives. Various organic acids and phenolic compounds such as these are exuded from plant roots or are metabolic products of soil microbial activity, and these chemicals influence the interactions of the rhizosphere and plant roots (Hoagland and Williams, 1985). AC could exert a strong influence on the levels of these compounds in soils. Studies by Duclos and Fortin (1983) with mycorrhizae indicated that AC can stimulate microbial growth via adsorption of toxic compounds from the rhizosphere. However, as yet definitive studies have not been reported in this area.

X. HERBICIDE ADSORBENTS OTHER THAN ACTIVATED CARBON

Calcium polysulfide (a fungicide for fruit trees and ornamentals) added to soil accelerated the breakdown of simazine as indicated by an oat bioassay (Castelfranco and Deutsch, 1962). This chemical catalyst of herbicide degradation yields a metabolite which is chloroform-insoluble and is readily adsorbed on charcoal. Calcium polysulfide has been used as a soil-corrective agent with no problems regarding animal toxicity or agricultural practices. Calcium polysulfide had no effect on residual atrazine phytotoxicity (Zemanek, 1964) or on simazine residues (Sheets and Harris, 1965).

A mixture of AC and vermiculite was used to increase direct-seeded cucumber and tomato tolerance to simazine. Also, the tolerance of cucumbers to DCPA and nitralin was increased. The mixture was placed in a hole (1.9 cm deep × 2.54 cm diameter) over the crop seed to the level of the soil surface, which detoxified the preemergence herbicides without injury to the crop plants. A wafer consisting of AC–vermiculite–fertilizer–seed provided protection to direct seeded crops (Kratky and Warren, 1971).

Exchangeable magnesium cations normally present in vermiculite could be replaced by various other cations with concomitant alteration on the organization of the structural water layers. Of the ions which entered this structure, only potassium and ammonium ions were not associated with layers of water molecules in the vermiculite lattice. The quantity of structured water present was sensitive to humidity changes, but less sensitive than in montmorillonites (Walker and Milne, 1950). Powdered AC increased the adsorptive capacity of a vermiculite anticrustant mixture and protected tomato and weed seedlings (William and Romanowski, 1972).

Several studies have been performed using ionic exchange resins and various clays as herbicide adsorbents. Some of these results are presented below to demonstrate their comparative adsorption efficacies with various herbicides.

The effectiveness of different adsorbents was determined by comparing the concentrations of herbicides required to give 50% root inhibition of a test plant. Of eight herbicides tested, six were more strongly adsorbed by AC than by muck soil, bentonite clay, a cation exchange resin (Chelex 100), or an anion exchange resin (Dowex 1-X8). Paraquat biological activity was not reduced by AC, but was reduced by bentonite clay and the cation exchange resin. DNBP was more strongly adsorbed by the anion exchange resin than by AC (Coffey and Warren, 1969). Similar studies showed that bentonite was more effective than AC in paraquat adsorption (Faust and Zarins, 1969) (Fig. 6).

The adsorption–desorption of trifluralin, triallate, and diallate was examined using several adsorptive materials. Generally, the herbicides were strongly adsorbed on the materials as follows: AC > peat moss > wheat straw = cellulose triacetate > cation exchange resin > anion exchange resin > silica gel = cellulose powder > kaolinite = montmorillonite. Trifluralin and triallate could be readily desorbed from montmorillonite by water, with minimal desorption from peat moss, cellulose triacetate and wheat straw (Grover, 1974).

Adsorption and desorption of picloram, trifluralin, and paraquat by ionic and nonionic exchange resins revealed that an anionic resin (Cl-form) desorbed 375, 0.08, and 0.67 mg of picloram, paraquat, and trifluralin, respectively (per gram dry resin). A nonionic resin adsorbed 4.0, 0.34, and 10.0 mg of picloram, paraquat, and trifluralin, respectively (per gram dry resin) (McCall *et al.*, 1972), whereas the cationic resin (H-form) adsorbed 2.3, 226, and 0.17 mg picloram, paraquat, and trifluralin, respectively. Desorption studies indicated that picloram was adsorbed mainly in the anionic form by coulombic or electrostatic interactions and to a lesser degree by weak physical bonding, i.e., van der Waals interaction. Paraquat adsorption was as a cation via coulombic forces, whereas with trifluralin there were no coulombic forces and adsorption was via physical binding on resin sites.

Model adsorbents and soil organic preparations were used to study the adsorption of linuron, atrazine, and EPTC (Hance, 1969). The model adsorbents were prepared by treating cellulose phosphate powder with a series of alkyltrimethylammonium compounds in which the size of the alkyl group was varied from C_8 to C_{18}. On these materials, linuron, atrazine, and EPTC adsorption increased logarithmically with increasing chain length. The amount of herbicide adsorption on these model adsorbents was large compared to that on humic acid and on a preparation of peat soil which had inorganic constituents removed. These materials provide useful insights into herbicide adsorption mechanisms, since soil organic matter is thought to contain alkyl groups. This information could be useful in the development of new adsorbents for use as herbicide safeners.

Clays such as illite, kaolinite, and montmorillonite suspended in dilute pesticide solutions absorbed little 2,4-D or chlorpropham, whereas adsorption of CIPC (2,4-D not tested) with powdered AC was extensive (Schwartz, 1967). Physical or weak chemical bonding between chlorpropham and the AC surface was suggested and equilibrium conditions for the chlorpropham–AC system can be represented by a Freundlich isotherm.

Chlorsulfuron was more strongly adsorbed on a strong anion exchange resin than on a weak anion resin or aluminum oxide anionotropic adsorbent. Hydrogen bonding was suggested for the adsorption on the strong anion exchange resin. Chlorsulfuron was not adsorbed on aluminum oxide cationotropic absorbent, montmorillonite, illite, or kaolinite. The herbicide was stongly adsorbed on AC, and had much less affinity for organic matter and only little affinity for cellulose. No significant adsorption was noted on mineral soils low in organic matter. Hydrogen bonding and charge transfer bonds were postulated as the major mechanisms responsible for chlorsulfuron adsorption in soil (Shea, 1986).

The effects of temperature and exposure time on the adsorption of four herbicides (diquat, paraquat, 2,4-D, and prometon) by montmorillonite, kaolinite, charcoal, and an anion exchange resin in phosphate buffer (pH 6.0) were investigated (Weber *et al.*, 1965). The organic cations diquat and paraquat were preferentially adsorbed by the clays and the adsorption mechanism appeared to be ion exchange. The organic anion 2,4-D was negatively adsorbed by the clays. Prometon in molecular form was adsorbed by montmorillonite but not kaolinite. The mechanism here was suggested to be physical and was temperature-dependent. Prometon was adsorbed (physically bound) by the anion exchange resin to a much lesser degree than was 2,4-D (ion exchange). All these herbicides were adsorbed by charcoal but to varying degrees. The organic cations were adsorbed least, but at high temperatures (55°C) continued to be adsorbed with time such that equilibrium was never attained. The triazine, prometon, was the most strongly adsorbed and was unaffected by temperature. Nearly as much 2,4-D was adsorbed, but significantly more was retained at 10°C than at 55°C (Fig. 4).

Ametryn was adsorbed to a greater extent than atrazine on soil and montmorillonite, but Freundlich isotherm slopes were generally greater for atrazine than for ametryn, suggesting that adsorption intensity of atrazine may exceed that of ametryn (Yamane and Green, 1972). With montmorillonite and soil materials, a decrease in pH protonized the triazines and increased adsorption (Fig. 5).

Various adsorbents used in column cleanup of pesticide residues may have potential for use as herbicide safeners. Materials frequently used are polar, such as magnesium silicate, alumina, silica gel, cellulose, and magnesium oxide, or nonpolar, like charcoal. Basically, these materials adsorb unwanted substances and/or take advantage of selective desorption and elution by a suitable solvent (Morley, 1966; Romer, 1986).

The controlled release of herbicides has been extensively studied. This methodology offers another herbicide safening technology. Polysaccharides have been used as control release agents with metribuzin (McCormick *et al.*, 1981), chloramben (Raboy and Hopen, 1982), and 2,4-D (Mehltretter *et al.*, 1974). Alachlor has been encapsulated in plaster of Paris for use in ornamentals (Smith and Verma, 1977). Metribuzin formulated in polyvinyl alcohol polymers was found to have reduced mobility and release of herbicide was primarily dependent on soil microbial metabolism of covalent linkages of the herbicide and polymers.

The use of kraft lignins in controlled-release herbicides has been shown to be effective with 2,4-D (DelliColli, 1980) and atrazine (Dunigan and MacIntosh, 1971). Generally, little is known about the physical properties of kraft lignins, but they possess relatively high numbers of hydroxyl, carbonyl, and keto groups, (DelliColli, 1980) which may participate in hydrogen bonding with herbicides. Furthermore, kraft lignins consist of cross-linked polymers (Marton, 1971) with potential to entrap herbicides. Release of metribuzin and alachlor from pine kraft lignin was examined using radiolabeled herbicides and lignin with soil TLC techniques (Riggle and Penner, 1987). A series of kraft lignins was used and one material was shown to have potential to provide controlled release of metribuzin and alachlor.

AC was the most suitable of seven adsorbents (including kieselguhr, vermiculites, manure, grass, unactivated carbon, and soot) for use as protectant against simazine damage to newly planted strawberries (Robinson, 1965).

It was established that chlorinated hydrocarbons in soil could affect the nitrogen, phosphorous and potassium composition of beans, soybeans, and cotton and the mite populations on these host plants (Rodriguez *et al.*, 1957). Several carbon powders, peat moss, calcium hydroxide, and AC pellets were examined for effects of soil insecticides (DDT) on plant sugars and mite nutrition (Rodriguez *et al.*, 1960). The ACs were more efficient adsorbents (except for AC pellets) and alleviated the effects of DDT on plants.

Performance characteristics of eight synthetic adsorbents and two ACs were compared for adsorption efficiency with respect to the removal of a specific organic compound, phenol (van Vliet *et al.*, 1980). Generally the ACs performed best, followed by carbonaceous and resinous adsorbents. This modeling procedure showed good agreement between predicted and experimental curves up to about 80% of saturation.

XI. CONCLUDING REMARKS

As seen from this presentation, AC is an excellent adsorbent for herbicides and some other agriculturally related chemicals. In nearly all applications of AC to soils or plants, protection of crops from herbicide injury was achieved, if adequate carbon was applied. Although the research literature contained many reports on the use of AC for 30 years following the discovery of its use as a herbicide safener (1945), the number of these reports has declined in recent years. Most of the reports pertained to AC application or application techniques. In many instances, inadequate descriptions of the AC types were given. As has been shown, carbon type can play an important role in the adsorption of a specific compound, e.g., a specific herbicide. Many types of carbon are available and many have diverse adsorptive properties. Presently, AC is not used to a large extent as a herbicide safener, and perhaps its greatest use is for protection of turfgrass and nursery stock from various herbicide treatments. The decline in the use of AC as a general herbicide safener can be attributed to the fact that it does require application techniques and equipment that differ from those required for conventional herbicide application, it is an added expense, and in certain circumstances crops are injured and/or weed control is not adequate. This latter situation occurs when crop roots grow through the carbon protective barrier and come into contact with the herbicide, or when weed roots penetrate, or weed seeds germinate in carbon bands and are afforded protection.

Protecting fluid drilled lettuce from herbicides by incorporating AC into gels such as hydroxyethyl-cellulose or polyacrylamide is one of the more recent innovations regarding AC as a herbicide safener (Taylor and Warholic, 1987). In greenhouse studies, this technique has thus far shown that 5.2 kg/ha of AC would be satisfactory for lettuce when incorporated into 20 g/liter of gel, but field studies are as yet marginal.

AC is presently in wide use for water purification, and its use under specified conditions was proposed by the U.S. Environmental Protection Agency as the primary choice for the control of "synthetic organic chemicals" including pesticides in drinking water (National Academy of Sciences, 1980). A more recent advance is the manufacture of carbonaceous solids derived from synthetic polymeric precursors termed "polymeric carbons." Generally the chemical composition, pore structure, physical form, and surface chemistry of these carbonaceous adsorbents differ from those of AC (Jenkins and Kawamura, 1976). These materials have the advantage for adsorption of many chemicals, including pesticide residues in ground and surface water (Neely and Isacoff, 1982). Thus far, their utility as herbicide safeners has not been reported.

Clear advantages of AC for herbicide safening exist, especially as more selective and better quality carbons become available. Few or no greenhouse, field, or laboratory studies have compared the more diverse carbon types for adsorption and safening efficiency. Studies such as these are necessary for a fuller evaluation that could lead to increased utility of AC as a herbicide safener.

References

Adams, R. S., Jr. (1973). Factors influencing soil adsorption and bioactivity of pesticides. *Residue Rev.* **47,** 1–54.

Ahrens, J. F. (1964). Antidote for herbicides. *Front. Plant Sci.* **17,** 5–7.

Anderson, A. H. (1968). The inactivation of simazine and linuron in soil by charcoal. *Weed Res.* **8,** 58–60.

Anonymous (1921). Prevention of injury to arable land by adsorption of sulfur dioxide by means of charcoal. *Chem.-Ztg.* **45,** 96–97.

Anonymous (1974). "Gro–Safe Activated Charcoal References." ICI Americas, Inc., Wilmington, Deleware.

Anonymous (1976). "Selected References to the Use of Activated Charcoal and Other Adsorbents in Conjunction with Herbicides, 1954–1975," ARC Weed Res. Organ. Annot. Bibliogr. No. 91. Begbroke Hill, Yartron, Oxford.

Anonymous (1981). "Gro-Safe Activated Charcoal: General Recommendations." ICI Americas, Inc., Wilmington, Deleware.

Arle, H. R., Leonard, O. A., and Harris, V. C. (1948). Inactivation of 2,4-D on sweet-potato slips with activated carbon. *Science* **107,** 247–248.

Babbar, S. B., and Gupta, S. C. (1986). Promotory and inhibitory effects of activated charcoal on microspore embryogenesis in *Datura metel. Physiol. Plant.* **66,** 602–604.

Bailey, G. W., and White, J. L. (1970). Factors influencing the adsorption, desorption, and movement of pesticides in soils. *Residue Rev.* **32**, 29–92.

Bardner, R. (1960). Effect of formulation on toxicity to plants and insects of some systemic insecticidal seed dressings. *J. Sci. Food Agric.* **11**, 736–744.

Bertges, W. J., Jr., and Behrens, R. (1974). Establishing forages using activated charcoal with herbicides. *Proc.—North Cent. Weed Control Conf.* **29**, 33.

Blair, A. M., Parker, C., and Kasasian, L. (1976). Herbicide protectants and antidotes—A review. *Proc. Natl. Acad. Sci. U. S.A.* **22**, 65–74.

Bonuccelli, G., and Cini, V. (1930). Effect of carbon on tobacco culture. *Boll. Ist. Super. Agrar. Pisa* **6**, 339–354.

Bovey, R. W., and Miller, F. R. (1969). Effect of activated carbon on the phytotoxicity of herbicides in a tropical soil. *Weeds* **17**, 189–192.

Brenchley, R. G., and Burrill, L. C. (1968). The use of charcoal to overcome persistent herbicides. *West. Soc. Weed Sci. Proc. Prog. Rep.*, pp. 104–105.

Brunauer, S., Copeland, L. E., and Kantro, D. L. (1967). The Langmuir and BET theories in the solid–gas interface. *In* "The Solid–Gas Interface" (E. A. Flood, ed.). Dekker, New York.

Burr, R. J. (1969). Some factors influencing the use of activated charcoal as a crop protectant. *West. Soc. Weed Sci. Res. Prog. Rep.*, pp. 81–82.

Burr, R. J., Lee, W. O., and Appleby, A. P. (1972). Factors affecting use of activated carbon to improve herbicide selectivity. *Weed Sci.* **20,** 180–183.

Carpenter, P. L. (1973). Chemical weed control in container-grown nursery stock. *HortScience* **8,** 385–386.

Castelfranco, P., and Deutsch, D. B. (1962). Action of polysulfide ion on simazine in soil. *Weeds* **10,** 244–245.

Chandler, J. M., Wooten, O. B., and Fulgham, F. E. (1978). Influence of placement of charcoal on protection of cotton (*Gossypium hirsutum*) from diuron. *Weed Sci.* **26**, 239–244.

Cheremisnoff, P. N., and Morresi, A. C. (1978). Carbon adsorption applications. In "Carbon Adsorption Handbook." (P.N. Cheremisnoff and F. Ellerbusch, eds.), pp 1–53. Ann Arbor Sci. Publ., Ann Arbor, Michigan.

Clapp, K. E. (1975). Methods of applying activated charcoal. *Proc. West. Soc. Weed Sci.* **28**, 31–34.

Coffey, D. L., and Warren, G. F. (1969). Inactivation of herbicides by activated carbon and other adsorbents. *Weed Sci.* **17**, 16–19.

Cookson, J. T. (1978). Adsorption mechanisms: The chemistry of organic adsorption on activated carbon. *In* "Carbon Adsorption Handbook" (P. N. Cheremisinoff and F. Ellerbusch eds.) , pp. 241–279. Ann Arbor Sci. Publ., Ann Arbor, Michigan.

Coutts, J. R. H. (1945). Effect of veld burning on the base-exchange capacity of a soil. *S. Afr. J. Sci.* **41**, 218–224.

Croxford, D. E., Elkins, D. M., and Kapusta, G. (1975). Crop protectants and herbicides for orchardgrass–alfalfa establishment. *Weed Sci.* **23**. 414–418.

deAndrade, V. A., and Bauman, T. T. (1985). Activated carbon and herbicides in the control of red rice in rice. *Lavoura Arrozeira* **38**, 27–29.

Deitz, V. R. (1944). History of commercial adsorbents in relation to the sugar refining industry. *In* "Bibliography of Solid Adsorbents, 1900–1942" (V. R. Deitz, ed.), pp. ix–lxxxi. U.S. Cane Sugar Refiners and Bone Char Manufacturers, and the National Bureau of Standards, Washington, D. C.

DelliColli, H. T. (1980). Pine kraft lignin as a pesticide delivery system. *In* "Controlled Release Technologies: Methods, Theory, and Application" (A. F. Kydonieus, ed.), Vol. 2, pp. 225–234. CRC Press, Boca Raton, Florida.

DeTar, W. R. (1973). Seed wafer research, 1971–1973. *Proc. Am. Soc. Agric. Eng.*, pp. 1–31.

Duclos, J. L., and Fortin, J. A. (1983). Effect of glucose and active charcoal on *in vitro* synthetase of bacteroid mycorrhiza with *Vaccinium* spp. *New Phytol.* **94**, 95–102.

Dunigan, E. P., and MacIntosh, T. (1971). Atrazine–soil organic matter interaction. *Weed Sci.* **19**, 279–282.

Elmstrom, G. W. (1972). Protection of watermelon from terbacil injury activated charcoal. *Proc., South. Weed Sci. Soc.* **25**, 204–208.

Engels, O. (1933). The use of charcoal as a fertilizer medium. *Dsch. Landwirtsch. Presse* **60,** 658.

Everson, J. N., and Weaver, J. B. (1950a). Effects of carbon black on the properties of soils. II. Effects on humid soils. *Soil Sci.* **69**, 369–376.

Everson, J. N., and Weaver, J. B. (1950b.) Effect of carbon black on properties of soils. Effect on soil temperatures. *Ind. Eng. Chem.* **41**, 1798–1801.

Faust, S. D., and Zarins, A. (1969). Interaction of diquat and paraquat with clay minerals and carbon in aqueous solutions. *Residue Rev.* **29**, 151–170.

Flagg, C. V., and Bayer, G. H. (1966). Weed control in strawberries. *Proc. Northeast. Weed Control Conf.* **20**, 157–160.

Foster, J. W., Hardwick, W. A., and Guirard, B. (1950). Antisporulation factors in complex organic media. I. Growth and sporulation studies on *Bacillus larvae*. *J. Bacteriol.* **59**, 463–470.

Fredrickson, D. R., and Shea, P. J. (1986). Effect of soil pH on degradation, movement, and plant uptake of chlorsulfuron. *Weed Sci.* **34**, 328–332.

Fridborg, G., and Eriksson, T. (1975). Effects of activated carbon on growth and morphogenesis in cell cultures. *Physiol. Plant.* **34**, 306–308.

Fridborg, G., Pederson, M., Landstrom, L. E., and Ericksson, T. (1978). The effect of activated carbon on tissue cultures: adsorption of metabolites inhibiting morphogenesis. *Physiol. Plant.* **43**, 104–106.

Gast, H. W. (1962). Contributions to the knowledge of behavior of triazines in soil. *Meded. Landbouwhoogesch. Opzoekings stn. Staat Gent* **27**, 1252–1274.

Geissbuhler, H., Haselbach, C., Aebi, H., and Abner, L. (1963). The fate of *N'*-(chlorophenoxy)-phenyl-*N'*, *N*-dimethyl urea (C–1938) in soils. III. Breakdown in soils and plants. *Weed Res.* **3**, 277–297.

Glass, E. H. (1950). Parathion injury to apple foliage and fruit. *J. Econ. Entomol.* **43**, 146–151.

Glaze, N. C., Phatak, S. C., and Threadgill, E. P. (1979). Spot application of activated charcoal to increase herbicide selectivity on watermelon. *HortScience* **14**, 632–633.

Gomez-Jimenez, L., Garcia-Rodriguez, A., de Dios Lopez-Gonzalez, J., and Navarreti-Guijosa, A. (1987). Study of the kinetics of the adsorption by activated carbons of 2,4,5-trichlorophenoxyacetic acid from aqueous solution. *J. Chem. Technol. Biotechnol.* **37**, 271–280.

Graves, W. L., Kay, B. L., and Williams, W. A. (1975). Seed treatment of mojave desert shrubs. *Agron. J.* **67**, 773–777.

Gray, R. A., and Joo, G. K. (1978). Site of uptake and action of thiocarbamate herbicides and herbicide antidotes in corn seedlings. *In* "Chemistry and Action of Herbicide Antidotes" (F. M. Pallos and J. E. Casida, eds.), pp. 67–84. Academic Press, New York.

Green, R. E., and Obien, S. R. (1969). Herbicide equilibrium in soils in relation to soil water content. *Weed Sci.* **17**, 514–519.

Grover, R. (1974). Adsorption and desorption of trifluralin, triallate, and diallate by various adsorbents. *Weed Sci.* **22**, 405–408.

Gukova, M. M., and Butkovich, W. S. (1941). Influence of aeration and soil temperature on the development of inoculated and non-inoculated soya plants. *C. R. Acad. Sci. URSS* **31**, 937–940.

Gupta, O. P., and Niranwal, N. K. (1976). Increasing herbicide selectivity in maize and cowpeas by seed treatment with activated carbon and NA. *Proc. Natl. Acad. Sci. U. S.A.* **22**, 86–89.

Hance, R. J. (1969). The adsorption of linuron, atrazine and EPTC by model aliphatic adsorbents and soil organic preparations. *Weed Res.* **9**, 108–113.

Hartley, G. S., and Sutcliff, J. F. Speakman and Co. (1945). Protecting seeds. British Patent 599, 445.

Harvey, R. G. (1973). Influence of cropping and activated carbon on persistence of atrazine in sand. *Weed Sci.* **21**, 204–206.

Hassler, J. W. (1958). Active carbon in America—1914 to 1957. *For. Prod. J.* **8**, 25A–27.

Hassler, J. W. (1974). "Purification with Activated Carbon: Industrial, Commercial, Environmental." Chem. Publ. Co., New York.

Heberle-Bors, E. (1980). Interaction of activated charcoal and iron chelates in anther cultures of *Nicotiana* and *Atropa belladonna*. *Z. Pflanzenphysiol.* **99**, 339–347.

Helling, C. S., Kearney, P. C., and Alexander, M. (1971). Behavior of pesticides in soils. *Adv. Agron.* **23**, 147–240.

Helweg-Andersen, A. (1969). Influence of charcoal-adsorbed herbicides on microorganisms. *Weed Res.* **9**, 254–257.

Henne, R. C., and Guest, R. T. (1974). Activated carbon as a method of reducing metribuzin phytotoxicity to seeded tomatoes. *Proc. Annu. Meet. Northeast. Weed Sci. Soc.* **28**, 242–248.

Hoagland, R. E., and Williams, R. D. (1985). The influence of secondary plant compounds on the associations of soil microorganisms and plant roots. *ACS Symp. Ser.* **268**, 301–325.

Holly, K. (1976). Selectivity in relation to formulation and application methods. *In* "Herbicides: Physiology, Biochemistry, Ecology" (L. J. Audus, ed.), 2nd ed., Vol. 2, pp. 249–277. Academic Press, London.

Holynski, S. (1928). Application of activated charcoal as fertilizer for grains. *Przemysl. Chem.* **12**, 190–196.

Holynski, S. (1932). Activated charcoal as an adjunct in fertilizer for staple plants. II. *Przem. Chem.* **16**, 4–8.

Hopen, H. J. (1967). Tomato protection from triazine damage with activated carbon. *Res. Rep.—North Cent. Weed Control Conf.*, pp. 39–40.

Huang, C. P. (1978). Chemical interactions between inorganics and activated carbon. *In* "Activated Carbon Handbook" (P. N. Cheremisinoff and F. Ellerbusch, eds.,), pp. 281–329. Ann Arbor Sci. Publ., Ann Arbor, Michigan.

International Rice Research Institute (1969). "Annual Report for 1968." IRRI, Los Banos, Philippines.

Iswaran, V., Vimal, O. P., and Apte, R. (1971). Pelleting of soybean (*Glycine max*) seeds with activated charcoal and humus. *Zentralbl. Bakteriol., Parasitenkd., Infektionskr. Hyg., Abt. 2, Naturwiss.: Allg., Landwirtseh. Tech. Mikroviol.* **126**, 45–57.

Jagschitz, J. A. (1974). Use of activated charcoal to nullify the harmful effects of chemicals in turfgrass. *Proc. Int. Turfgrass Res. Conf., 2nd, 1973,* pp. 399–409.

Jenkins, G. M., and Kawamura, K. (1976). "Polymeric Carbons—Carbon Fibre, Glass and Char." Cambridge Univ. Press, London and New York.

Johansson, L. (1983). Effects of activated carbon on anther cultures. *Physiol. Plant.* **54**, 397–403.

Johnson, B. J. (1976). Effect of activated charcoal on herbicide injury during establishment of centipedegrass. *Agron. J.* **68**, 802–805.

Johnson, E. A., VanKampen, K. R., and Binns, W. (1972). Effects on cattle and sheep of eating hay treated with the triazine herbicides atrazine and prometon. *Am. J. Vet. Res.* **33**, 1433–1438.

Jordan, P. D., and Smith, L. W. (1971). Adsorption and deactivation of atrazine and diuron by charcoals. *Weed Sci.* **19**, 541–544.

Kay, B. L. (1972). Increasing herbicide selectivity on rangeland with activated charcoal. *Proc., Annu. Calif. Weed Conf.* **24**, 122–125.

Kennedy, J. M., and Talbert, R. E. (1973). Activated charcoal as a herbicide antidote. *Arkansas Farm Res.* March–April, p. 12.

Kobel, W., Sumner, D. D., Campbell, J. B., Hudson, P. B., and Johnson. J. L. (1985). Protective effect of activated charcoal in cattle poisoned with atrazine. *Vet. Hum. Toxocol.* **27**, 185–188.

Kratky, B. A., and Warren, G. F. (1971). Activated carbon–vermiculite mixture for increasing herbicide selectivity. *Weed Sci.* **19**, 79–81.

Kring, J. B., and Ahrens, J. F. (1968). Activated carbon, pesticide residues, and potatoes. *Am. Potato J.* **45**, 405–408.

Lee, W. O. (1973). Clean grass seed crops established with activated carbon bands and herbicides. *Weed Sci.* **21**, 537–541.

Lichtenstein, E. P., Fuhremann, T. W., and Schulz, K. R. (1968). Use of carbon to reduce the uptake of insectidical soil residues by crop plants. Effects of carbon on insecticide adsorption and toxicity in soils. *J. Agric. Food Chem.* **16**, 348–355.

Lichtenstein, E. P., Schulz, K. R., and Fuhremann, T. W. (1971). Long-term effects of carbon in reducing uptake of insecticidal soil residues by crops. *J. Econ. Entomol.* **64**, 585–588.

Linscott, D. L., and Hagin, R. D. (1967). Protecting alfalfa seedlings from a triazine with activated charcoal. *Weeds* **15**, 304–306.

Locascio, S. J. (1967). Effect of activated charcoal on the toxicity of dichlobenil to vegetables. *Proc. South. Weed Conf.* **20**, 157–163.

Long, C. E., and Scranton, R. F. (1969). The action of charcoal on the herbicidal activity of several herbicides. *Proc.—North Cent. Weed Control Conf.* **24**, 55–56.

Lucas, E. H., and Hamner, C. L. (1947). Inactivation of 2,4–D by adsorption on charcoal. *Science* **105**, 340.

Lyman, W. J. (1978). Applicability of carbon adsorption to the treatment of hazardous wastes. *In* "Carbon Adsorption Handbook" (P. N. Cheremisinoff and F. Ellerbusch, eds.), pp. 131–165. Ann Arbor Sci. Pub., Ann Arbor, Michigan.

MacCallum, G. B. (1956). Effects and correction of DDT phytotoxicity to cucumbers. *Mem. N. Y., Agric. Exp. Stn. (Ithaca).* **339**, 1–48.

McCall, H. G., Bovey, R. W., McCully, M. G., and Merkle, M. G. (1972). Adsorption and desorption of picloram, trifluralin, and paraquat by ionic and nonionic exchange resins. *Weed Sci.* **20**, 250–255.

McCormick, C. L., Anderson, K. W., Pelezo, J. A., and Lichatowich, D. K. (1981). Controlled release of metribuzin, 2,4–D, and model aromatic amines from polysaccharides and polyvinyl alcohol. *In* "Controled Release of Pesticides and Pharmaceuticals" (D.H. Lewis, ed.), pp. 147–158. Plenum, New York.

Majek, B. A. (1986). Activated carbon for safening peppers (*Capsicum annuum* L.) in soils treated with metribuzin. *Weed Sci.* **34**, 467–470.

Martineau, B., Hanson, M. R., and Ausubel, F. M. (1981). Effect of charcoal and hormones on anther culture of *Petunia* and *Nicotiana*. *Z. Pflanzenphysiol.* **102**, 109–116.

Marton, J. (1971). Reactions in alkaline pulping. *In* "Lignins: Occurrence, Formation, Structure, and Reactions" (K. V. Sarkanen and C. H. Ludwig, eds.), pp. 636–689 Wiley (Interscience), New York.

Mattson, J. S., and Mark, H. B., Jr. (1971). "Activated Carbon: Surface Chemistry and Adsorption from Solution." Dekker, New York.

Mehltretter, C. L., Roth, W. B., Weakley, F. B., McGuire, T. A., and Russell, C. R. (1974). Potential controlled-release herbicides from 2,4-D esters of starches. *Weed Sci.* **22**, 415–418.

Meyers, H. G., and McCarthy, W. J. (l974). Effects of activated charcoal and time of application on established kerb treated *Poa annua*. *Proc. Fla. State Hortic. Soc.* **87**, 513–515.

Meyers, H. G., and Sligh, C. M. (1973). Deactivation effects of activated charcoal on kerb treated *Poa annua*. *Proc. Fla. State Hortic. Soc.* **86**, 445–447.

Meyers, H. G., Currey, W. L., and Barnes, D. E. (1973). Deactivation of kerb with sewage sludge, topdressing and activated charcoal. *Proc. Fla. State Hortic. Soc.* **86**, 442–444.

Miller, E. J. (1925). Adsorption of activated sugar charcoal with particular reference to soil acidity. *Mich. Agric. Exp. Stn. Bull.* **73.**

Morley, H. V. (1966). Adsorbents and their application to column cleanup of pesticide residues. *Residue Rev.* **16**, 1–29.

Moyer, J. R., Hance, R. J., and McKone, C. E. (1972). The effect of adsorbents on the rate of degradation of herbicides incubated with soil. *Soil Biol. Biochem.* **4**, 307–311.

Nangju, D., Plucknett, D. L., and Obien, S. R. (1976). Performance of carbon-coated rice seed on flooded soil. *Weed Sci.* **24**, 73–80.

National Academy of Sciences, Safe Drinking Water Committee (1980). "Drinking Water and Health," Vol. 2. National Academy Press, Washington, D. C.

Neeley, J. W., and Isacoff, E. G. (1982). "Carbonaceous Adsorbents for the Treatment of Ground and Surface Waters." Dekker, New York.

Nutman, P. S. (1945). A factor in clover nodule formation associated with the volume of the medium occupied by the roots. *Nature (London),* **156**, 20.

Ogg, A. G., Jr. (1978). Herbicides and activated carbon for weed control in direct-seeded asparagus (*Asparagus officinalis*). *Weed Sci.* **26**, 284–286.

Ogg, A. G., Jr. (1982). Effect of activated carbon on phytotoxicity of terbacil to several crops. *Weed Sci.* **30**, 683–687.

Osgerby, J. M. (1970). Sorption of un-ionized pesticides in soils. *SCI Monogr.* **37**, 630.

Ostrowski, J. (1972). Further studies on the elimination of herbicides' carryover effect by their adsorption by activated carbon. *Rocz. Glebozn.* **23**, 77–82.

Papadakis, J. S. (1941). An important effect of soil colloids on plant growth. *Soil Sci.* **52**, 283–291.

Peel, R. G., and Benedek, A. (1980). Attainment of equilibrium in activated carbon isotherm studies. *Environ. Sci. Technol.* **14**, 66–71.

Perotti, R., and Ferretti, C. (1930). Effect of carbon on vegetation. III. *Boll. Ist. Super. Agrar. Pisa* **6**, 147–169.

Plucknett, D. L. (1971). "Use of Pelleted Seed in Crop and Pasture Establishment," Coop. Ext. Serv. Circ. **446**, University of Hawaii, Honolulu.

Poling, E. B., and Monaco, T. (1985). Activated charcoal root dips enhance herbicide selectivity in strawberries. *HortScience* **20**, 251–252.

Prianeshnikov, D. N., and Domontovitch, M. K. (1926). The problem of a proper nutrient medium. *Soil Sci.* **21**, 327–348.

Puri, B. R., Lakhanpal, M. L., and Verma, B. (1953). Acidoid behavior of charcoal in relation to soil properties. *Soil Sci.* **75**, 209–218.

Puri, B. R., Nath, T., and Sharma, L. R. (1958). Interaction of soils with charcoal. *Soil Sci.* **86**, 18–23.

Putnam, A. R., Love, A. P., and Lockerman, R. H. (1974). Efficacy of activated charcoal as an herbicide antidote for seeded vegetable crops. *Proc.—North Cent. Weed Control Conf.* **29**, 69.

Raboy, V., and Hopen, H. J. (1982). Effectiveness of starch xanthide formulations of chloramben for weed control in pumpkin (*Cucurbita moschata*). *Weed Sci.* **30**, 169–174.

Riggle, B. D., and Penner, D. (1987). Evaluation of pine kraft lignins for controlled release of alachlor and metribuzin. *Weed Sci.* **35**, 243–246.

Ripper, W. E. (1956). A new method of selective weed control for related plants, in particular broadleaf weeds in beet. *Proc. Br. Weed Control Conf.* **3**, 225–232.

Ripper, W. E. (1958). Chemical control of broad-leaved weeds in beet. *Br. Sugar Beet Rev.* **26**, 171–176.

Ripper, W. E., and Scott, J. K. (1957). *Z. Pflanzenkr. (Pflanzenpathol.) Pflanzenschutz* **64**, 469–477.

Roberts, H. A., and Wilson, B. I. (1965). Adsorption of chlorpropham by different soils. *Weed Res.* **5**, 348–350.

Robinson, D. W. (1965). The use of adsorbents and simazine on newly planted strawberries. *Weed Res.* **5**, 43–51.

Rodriguez, J. G., Chen, H. H., and Smith, W. T., Jr. (1957). Effects of soil insecticides on beans, soybeans, and cotton and resulting effect on mite nutrition. *J. Econ. Entomol.* **50**, 587–598.

Rodriguez, J. G., Maynard, D. E., and Smith, W. T., Jr. (1960). Effects of soil insecticides and absorbents on plant sugars and resulting effect on mite nutrition. *J. Econ. Entomol.* **53**, 491–495.

Romer, T. R. (1986). Use of small charcoal/alumina cleanup columns in determination of trichothecine mycotoxins in foods and feeds. *J. Assoc. Off. Anal. Chem.* **69**, 699–703.

Rydrych, D. J. (1985). Inactivation of metribuzin in winter wheat by activated carbon. *Weed Sci.* **33**, 229–232.

Scharrer, K., and Schropp, W. (1935). The action of different adsorbents on plant growth. *Landwirtsch. Vers.-Sta.* **122**, 321–338.

Scheel, W. C. (1934). Use of charcoal as a manure. *Biedermanns Zentralbl, Abt. A* [n.s.] **4,** 558.

Schubert, O. E. (1967). Can activated charcoal protect crops from herbicide injury? *Crops Soil Mag. Aug.–Sept.*, pp. 10–11.

Schwartz, H. G., Jr. (1967). Adsorption of selected pesticides on activated carbon and mineral surfaces. *Environ. Sci. Technol.* **1,** 332–337.

Shea, P. J. (1986). Chlorsulfuron dissociation and adsorption on selected adsorbents and soils. *Weed Sci.* **34**, 474–478.

Sheets, T. J., and Harris, C. I. (1965). Herbicide residues in soils and their phytotoxicities to crops grown in rotations. *Residue Rev.* **11**, 119–140.

Skinner, J. J., and Beattie, J. H. (1916). A study of the action of carbon black and similar absorbing materials in soils. *Soil Sci.* **2**, 93–101.

Smisek, M., and Cerny, S. (1970). "Activated Carbon—Manufacture, Properties, and Applications." Elsevier. Amsterdam.

Smith, A. E., and Verma, B. P. (1977). Weed control in nursery stock by controlled release of alachlor. *Weed Sci.* **25**, 175–178.

Stryckers, J., and van Himme, J. (1967). Protection against soil–acting herbicides in seedbeds of *Rosa* wildings. *Meded. Rijksfac. Landbouwwet., Gent* **32**, 964–976.

Suffet, I. H., and McGuire, M. J., eds. (1980). "Activated Carbon Adsorption of Organics from the Aqueous Phase," Vol 1. Ann Arbor Sci. Publ., Ann Arbor, Michigan.

Swanback, T. R. (1942). Use of carbon black to activate growth early in the season. *Bull.—Conn. Agric. Exp. Stn., New Haven.* **457**, 244–246.

Taylor, A. G., and Warholic, D. T. (1987). Protecting fluid drilled lettuce from herbicides by incorporating activated carbon into gels. *J. Hortic. Sci.* **62**, 31–37.

Tincker, M. A. H. (1947). Experiments with lilies at Wisley. VI. *In* "The Lily Year Book," pp. 82–86. R. Hortic. Sci., London.

Toth, J., and Milham, P. J. (1975). Activated-carbon and ash-carbon effects on the adsorption and phytotoxicity of diuron. *Weed Res.* **15**, 171–176.

Tryon, E. H. (1948). Effect of charcoal on certain physical, chemical, and biological properties of forest soils. *Ecol. Monogr.* **18**, 81–115.

Turner, E. R. (1955). The effect of certain adsorbents on the nodulation of clover plants. *Ann. Bot. (London)* [N. S.] **19**, 149–160.

Vantsis, J. T., and Bond, G. (1950). The effect of charcoal on the growth of leguminous plants in sand culture. *Ann. Appl. Bot.* **37**, 159–168.

van Vliet, B. M., Weber, W. J., and Hozumi, H. (1980). Modeling and prediction of specific compound adsorption by activated carbon and synthetic adsorbents. *Water Res.* **14**, 1719–1728.

Verona, O., and Ciriotti, P. (1935). Action of charcoal on vegetation. *Boll. Ist. Super. Agar. Pisa* **11**, 401–420.

Walker, G. F., and Milne, A. (1950). Hydration of vermiculite saturated with various cations. *Trans. Int. Congr. Soil Sci., 4th, 1950*, Vol. 4, pp. 58–59.

Warren, G. F. (1956). The relative adsorption of several herbicides by widely differing soils. *Proc.—North Cent. Weed Control Conf.* **13**, 5.

Wauchope, R. D., and Koskinen, W. C. (1983). Adsorption–desorption equilibria of herbicides in soil: A thermodynamic perspective. *Weed Sci.* **31**, 504–512.

Weaver, R. J. (1948a). Contratoxification of plant growth regulators in soils and on plants. *Bot. Gaz. (Chicago)* **109**, 72–84.

Weaver, R. J. (1948b). Some uses of activated carbon in contratoxification of plant growth regulators. *Bot. Gaz. (Chicago)* **110**, 300–312.

Weber, J. B., Perry, P. W., and Upchurch, R. P. (1965). The influence of temperature and time on the adsorption of paraquat, diquat, 2,4-D and prometone by clays, charcoal,and an anion-exchange resin. *Soil Sci. Soc. Am. Proc.* **29**, 678–688.

White, J. L., and Mortland, M. M. (1971). Pesticide retention by soil minerals. *Pestic. Soil: Ecol. Degradation, Mov., Int. Symp. 1970*, p. 95.

William, R. D., and Romanowski, R. R. (1972). Vermiculite and activated carbon adsorbents protect direct-seeded tomatoes from partially selective herbicides. *J. Am. Soc. Hortic. Sci.* **97**, 245–249.

Wilson, H. P., Graves, B., and Savage, C. P., Jr. (1974a). Effects of incorporation depth and activated carbon on sweet potato response to vernolate. *J. Am. Soc. Hortic. Sci.* **99**, 245–246.

Wilson, H. P., Hines, T. E., and Belote, J. N. III. (1974b). Relationship between activated carbon and atrazine on cucumbers and snap beans. *Proc. Annu. Meet. Northeast. Weed Sci. Conf.* **28**, 235–241.

Wyatt, F. A., and Leahey, A. (1936). Activated carbon as a fertilizer. *Sci. Agric.* **17**, 1–9.

Yamane, V. K., and Green, R. E. (1972). Adsorption of ametryne and atrazine on an oxisol, montmorillonite, and charcoal in relation to pH and solubility effects. *Soil Sci. Soc. Am. Proc.* **36**, 58–64.

Zemanek, J. (1964). A study of various means of inactivating atrazine in the soil. *Rostl. Vyroba* **37**, 959–972.

Zinzadze, C. (1932). Nutrition artificielle des plantes cultivées. I. Mélanges nutritifs à pH stable. *Ann. Agron. [N. S.]* **2**, 809–853.

Chapter 12

Controlled Release as a Factor for Protection of Crop Species from Herbicide Injury

Bruce D. Riggle
ICI Americas, Inc.
Eastern Research Center
Goldsboro, North Carolina 27533

Donald Penner
Department of Crop and Soil Sciences
Michigan State University
East Lansing, Michigan 48824

I. INTRODUCTION

Preemergence herbicides that are mobile in the soil may lose efficacy rapidly following heavy rainfall. Even though water is essential for the mobilization of many of the surface-applied herbicides into the soil for effective weed control, the addition of large amounts of water over time or large initial influxes of water can establish diffusion gradients of the applied chemicals. As this occurs, the herbicide is gradually removed from the top portion of the soil with potential reduction in weed control efficacy. If the crop plants are reasonably deep-rooted, herbicide movement of the herbicide into the region of the developing root system of the crop plant may result in crop phytotoxicity.

A strategy amenable to reducing herbicide-related crop phytotoxicity would be the modification of herbicide release. Herbicide availability would be modulated in a manner which would reduce leaching as well as potential crop phytotoxicity while maintaining weed control. This could be achieved if an effective herbicide concentration was maintained in the weed-seed zone near the soil surface. In essence, the strategy is an attempt to restrict herbicide leaching by a manipulation of its formulation to obtain crop protection.

Since herbicide efficacy is usually dependent on a dose response, a reduction in the dose or amount of freely available herbicide present in the soil by complexing a portion of the applied herbicide with a controlled release agent could reduce weed control. The amount of herbicide that will be degraded microbiologically or chemically is generally dependent on the first-order rate law (Collins and Doglia, 1973), and thus herbicide persistence may be increased by a reduction of the first order rate constant (Collins and Doglia, 1973). This could be achieved by reducing the amount of free herbicide in the soil as compared to the amount of herbicide that would be incorporated in the controlled release agent. The rate of herbicide release from the controlled release agent would also have a first-order rate constant (McCormick *et al.*, 1981) which would determine, depending on the magnitude, how quickly or slowly the herbicide would be released. Ideally for herbicides with short half-lives, a larger rate constant for release is desirable. This approach may require more herbicide than the conventional approach (Collins *et al.*, 1973); however, the additional cost incurred could be offset by the potential for reducing crop injury and improving weed control.

The theoretical amount of herbicide that would be needed in conjunction with a controlled release agent would not be less than the conventional amount necessary to achieve the required weed control (Collins *et al.*, 1973). Furthermore, the amount needed could be more than that used for multiple applications but less than that needed for a single application (Collins *et al.*, 1973). The predicted savings would be the possible elimination of multiple or split application of herbicide as well as potentially enhanced yields from the prevention of crop injury and the greater weed control.

II. CONTROLLED RELEASE TECHNOLOGY

Various approaches have been tried to find effective controlled release agents, generally polymers, for water-mobilized preemergence herbicides. These approaches can be classified as involving either covalent bonding of the herbicide and a controlled release agent or physical entrapment of the herbicide by the controlled release agent. The following is a summary of controlled release research which involved the testing of herbicides in both the field and laboratory.

Covalent bonding of herbicides to a controlled release agent has been done with various herbicides and both man-made and natural occurring polymers. Combinations of cross-linked corn starch and cyanoethylate with 2,4-dichlorophenoxy-acetyl-chloride in pyridine, which formed 2,4-D esters of high acyl content, have resulted in the controlled release of varying amounts of 2,4-D and water-soluble 2,4-D esters (Mehltretter *et al.*, 1974). Ester combina tions of 2,4-D on an α-cellulose backbone showed controlled release potential at pH 7 (Allan *et al.*, 1977). When 2,4-DB was attached to bark as a polymeric backbone by acylation of the hydroxylated polymer with the 2,4-DB acid chloride form (Allan *et al.*, 1971), there was no injury at the 10 g/pot (140 cm^2 surface area) rate to Douglas fir [*Pseudotsuga menziesii* (Mirb.) Franco], but there was significant injury to Western red alder (*Alnus rubra* Bong.) (Allan *et al.*, 1972). DCBA (2,6-dichlorobenzaldehyde) has been attached to polyvinyl alcohol by an acid-catalyzed condensation (Schacht *et al.*, 1978). Amitrole has been attached to polyacrylic acids, which resulted in a slow hydrolytic release rate (Hartmann *et al.*, 1985a,b). Amylose linked metribuzin has been demonstrated to release metribuzin in a zero order manner as measured by HPLC analysis (McCormick *et al.*, 1981); however, this combination was not tested with either soils or plants. Linear and cross-linked copolymers of polyvinyl alcohols have been linked with pendant metribuzin, which has resulted in varying release rates of the herbicide (Harris, 1980; McCormick and Fooladi, 1980).

Various entrapment strategies have also been used to control the release of water-mobile herbicides. Starch xanthide entrapment of chloramben methyl ester salts has been tested, but this combination appeared to have little effect on herbicide efficacy (Raboy and Hopen 1982). A starch xanthide formulation appeared to release alachlor at an unacceptable level but did improve the residual level of metolachlor. Results of this work suggest that acetanilides with their chlorine constituents may not be compatible with the starch xanthide process (Foley and Wax, 1980). Plaster of Paris tablet encapsulated alachlor has been shown to successfully protect ornamental flowering plants and to control weeds (Smith and Verma, 1977). Pine kraft lignin when combined with 2,4-D ester has been shown to successfully control the release of 2 4-D at a zero-order rate which would allow for the use of 2,4-D as an effective plant growth regulator (DelliColli 1980).

Problems could arise from the use of both covalently linked and entrapment type polymers. Polyvinyl type polymers for pendant substituted herbicides would be prohibitively expensive for conventional agronomic use. Furthermore, pendant formulations would require enormous sums of money to test for toxicological effects of all potential by-products that contained the herbicide molecule or a portion of it. Polysaccharides such as starch and amylose are potential, readily available, carbon sources for soil microorganisms and so may not be very effective as good barriers to prevent microbial decomposition of the entrapped herbicide. In addition, the presence of polysaccharides in holding tanks or spray tanks could risk fermentation with resulting by-products which could clog lines and nozzles; this potential problem would be best overcome by the use of granular formulation approach. Plaster of Paris-encapsulated herbicides would be more suitable for ornamentals and perennial horticultural crops rather than agronomic field crops because of the potential for long lasting persistence of these tablets.

III. CANDIDATE HERBICIDES AND THEIR USES

The user may benefit from the technology that will reduce herbicide mobility in the soil and injury to crop plants. One herbicide which could benefit from the use of a controlled release agent such as kraft lignin is metribuzin. This compound is typically used in soybean [*Glycine max* (L.) Merr.] production as a preemergence herbicide, but may injure soybeans because of its mobility in soils (Ladlie *et al.*, 1976a,b). Metribuzin has a molecular weight of 214.3, has a water solubility of approximately 1200 ppm at 20°C, and is relatively stable when exposed to ultraviolet radiation (Weed Science Society of America, 1983). The compound inhibits plant growth by blocking electron transport in photosystem II at the B protein complex (Ashton and Crafts, 1981; Pfister and Arntzen, 1979). This in turn can result in the formation of free radicals which can cause the disruption of the chloroplast membrane and necrosis of the surrounding cell. Many soybean varieties are able to metabolize metribuzin to a nonphytotoxic compound called deaminated diketo metribuzin (DADK) and then conjugate the DADK with glycosides (Smith and Wilkinson, 1974). In a recent report by Frear *et al.* (1985) the metabolism in soybean was shown to include a homoglutathione conjugate of sulfoxidized metribuzin as a major metabolite. The half-life of metribuzin in soils has been reported to be from 5 to 20 days (Banks and Robinson, 1982; Bouchard *et al.*, 1982; Hyzak and Zimdahl, 1974; Savage, 1976). In addition, lower half-life values were obtained when the soil was warmed from 20°C to 35°C, which would indicate the importance of microbial activity (Hyzak and Zimdahl, 1974; Savage, 1976).

Metribuzin mobility in soil is primarily dependent on soil organic matter content and soil pH. The compound is predominately adsorbed by organic matter rather than by clay (Savage, 1976). Maximum adsorption of triazines usually occurs in soils with low pH levels (Ladlie *et al.*, 1976a; Weber *et al.*, 1969). Protonation of the metribuzin tertiary amine group at low pH levels may give the compound a positive charge which could then be attracted to negatively charged organic matter and/or clay colloids (Ladlie *et al.*, 1976a; Sullivan and Felbeck, 1968). As the pH of the soil is increased, metribuzin becomes more mobile (Ladlie *et al.*, 1976a), and in the case of soybean production this could result in an increase in phytotoxicity (Ladlie *et al.*, 1976b). Addition of organic matter with a low pH to greenhouse-grown potted soybeans resulted in a reduction of metribuzin phytotoxicity to soybeans as well as a reduction in giant foxtail (*Setaria faberi* Herrm.) control (Coble and Schrader, 1973). These results demonstrate that metribuzin is very mobile in neutral and alkaline soils, especially those soils with sandy textures and low organic matter content.

Another herbicide whose use might benefit from greater positioned selectivity is chloramben. It has a water solubility of 700 ppm at 25°C and moves readily in sandy soils (Weed Science Society of America, 1983). Soil organic matter and soil pH have been reported as the primary factors controlling adsorption to soil

(Wildung *et al.*, 1968). Formation of the methyl ester of the acid increased soil adsorption and decreased water solubility to 120 ppm at 25°C (McLane and Parkins, 1966). However, instability of the chloramben methyl ester in soils resulted in less than desired prevention of leaching and protection of sensitive crops (McLane and Parkins, 1966). Raboy and Hopen (1982) obtained limited success in controlling the release of chloramben methyl ester with starch xanthide.

Other herbicides potentially amenable to the use of this technology are the chloroacetanilides, such as alachlor and metolachlor. Formulation of alachlor as a microencapsulation package is a step in this direction. The list of herbicides mentioned thus far is by no means inclusive but is intended to illustrate the potential for the application of controlled release technology.

IV. CASE HISTORY: LIGNIN AS A CONTROLLED RELEASE AGENT

A. Background

The efficacy of a large number of commercial preplant incorporated and preemergence herbicides can be related to the organic matter content of the soil, hence the significance of K_{oc} (adsorption coefficient for soil organic carbon) values. Soil organic matter, which is rich in humic and fulvic acid, has been compared to lignin. Since lignin is an abundant by-product of the pulp and paper industry considerable interest has been focused on using lignins as controlled release agents specifically for herbicides readily bound to or partitioned into soil organic matter.

Lignin is relatively cheap compared to polyvinyl polymers and is not rapidly degraded as compared to polysaccharides. Because of the multifunctionality (Coring, 1971) and the adsorptive capacity of kraft lignin (Dunigan and MacIntosh, 1971), it should be capable of readily adsorbing a compound such as metribuzin. Lignin has been referred to as the most abundant aromatic source in nature (DelliColli, 1980), and is produced in the cell walls of higher plants via the shikimic acid pathway (Schubert, 1965).

The function of lignin in the plant appears to be to provide both structural and turgo-stress counteracting strength as well as serving as an antimicrobial agent (Smith and Wilkinson, 1974). The antimicrobial properties are thought to be due to the multihydroxyl functional groups that are abundant in lignin (Swain, 1979). The natural structure has been defined by milling woody tissue (Marton, 1971). In addition, there are differences in lignin structure that vary according to the location in the cell wall and middle lamella (Marton, 1971). How lignin is associated with cellulose and hemicellulose is not clear, although some research has described this association as "snake cages" in which polymers of lignin are wrapped around the cellulose and hemicellulose fibers which in turn imparts strength to the cell wall (Goring, 1971).

Lignin is persistent and remains in the soil for long periods of time because of the presence of multifunctional groups (Schubert, 1965). It is slowly degraded by white rot fungi, *Basidiomycetes* sp. (Schubert, 1965), and is believed to be the source of aromatic groups used in the formation of soil organic matter components such as humic and fulvic acid (Stevenson, 1982).

Lignin is very similar in structure to humic acid and both have been reported to have a high adsorptive capacity for atrazine (Dunigan and MacIntosh, 1971). Atrazine adsorption to humic acid has been found to be mainly due to hydrogen bonding with carboxylic and hydroxyl groups of the humic acids binding to amine groups of the triazine as measured by infrared (IR) spectrometry (Sullivan and Felbeck, 1968). Because of the similarity in structure between lignin and humic acid, the results suggest that lignin would adsorb triazine herbicides in a similar manner.

The concept of using natural unaltered lignin for a lignin–herbicide complex or as a controlled release mixture is impractical since unaltered lignin can only be obtained under laboratory conditions and only in small amounts. A more plentiful and realistic source is industrial grade lignin, which is a by-product of the paper pulping industry.

The two major types of industrial lignin are sulfite and alkaline kraft lignin. The source of these lignins is typically from conifer trees and the differences between the two types of lignin depend on differences in the chemical pulping processes. Sulfite pulping includes sulfonation and hydrolytic reactions producing sulfite salts alcohols, aldehydes, and carbohydrates as potential by-products (Glennie, 1971). Sulfite lignin is not suitable as a controlled release agent for water-soluble herbicides because of the water solubility factor and the accompanying impurities. Alkaline pulping is a more complex method which utilizes sodium sulfite and sodium hydroxide to solubilize and strip lignin away from wood chips. The lignin is then precipitated out of the pulping liquors by acidification. The most popular form of alkaline pulping is the kraft process, which uses sodium polysulfide, sodium thiosulfate, and sodium carbonate, in addition to sodium sulfide and sodium hydroxide (Marton, 1971). The result is a relatively pure methoxylated water-insoluble lignin which contains approximately 1.5% sulfur (Marton, 1971).

B. Characteristics of Lignin

The exact physical nature of kraft lignin is not fully known because of the distribution of functional groups and the cross linkage of the basic carbon six to carbon three (C_6–C_3) lignin units. The major C_6–C_3 units in pine lignin are sinapyl alcohol (Pearl, 1964). Methoxyl groups make up approximately 14% of pine kraft lignin and for every 100 C_6–C_3 units there are approximately 120 OH groups and 16 COOH groups. The average pine kraft lignin is thought to have 20 or more C_6–C_3 units for an average molecular weight of 1800. Some researchers have found molecular weights as high as 50,000 or more. These large molecular weight fractions consist of smaller units held together by secondary linkages to form larger units (DelliColli, 1980). Viscosity analysis suggests spherical shapes but other research suggests small platelike structures (DelliColli, 1980). There are free unattached low molecular weight catechol derivatives such as vanillin and protocatechuic acid in the lignin and these are produced by the time-controlled precipitation in the kraft process; the exact percentage composition of these water-soluble compounds is not known, but estimates are about as much as 1% of the total weight (Marton, 1971).

C. Methodology

1. Soil Thin-Layer Chromatography

Riggle and Penner (1987) evaluated an experimental kraft lignin, PC940C (Westvaco), and a commercial kraft lignin, INDULIN AG (Westvaco), for controlled release properties with metribuzin as measured by the soil thin-layer chromatography (STLC) technique (Helling, 1971; Helling and Turner, 1968; Ladlie *et al.*, 1976a) using carbonyl^{14}C-labeled metribuzin and ^{3}H-labeled PC94OC.

The STLC analysis was done in the following manner. Glass plates were coated with a slurry of sieved Spinks sandy loam soil using a plate spreader and then allowed to dry. The soil had a pH of 6.5 and an organic matter content of 0.8%. A 1-ml sample of 1220 ppm metribuzin stock solution was mixed with 10 mg of PC940C solid material and shaken for 15 sec; 250 μl was then transferred to a vial containing 5 μl of [^{14}C]metribuzin (spec. act. 21.9 mCi/mmol) in methanol. This vial was shaken for 1 hr and an aliquot was spotted onto a plate. The procedure was the same for INDULIN AG except that, after the 1 hr shaking time, the vial was stored for 24 hr before samples were spotted onto a plate. In addition, 2 ml of distilled water plus 19.2 mg of PC940C were mixed together and shaken for 15 sec and then 100 μl was transferred to a vial which contained 0.3 mg of [^{3}H]PC940C (spec. act. 0.73 mCi/mg). The vial was shaken for 1 hr and then samples were spotted onto a plate. The brown colored lignin was easily detected on the soil plates. Soil plates were developed in holding tanks with distilled water to an approximate height of 12.5 cm and then dried. Soil plates were then scanned using a Berthold 2D-TLC-scanner LB276 with methane as a carrier gas. Disintegrations were integrated to determine how much of the radiolabeled material remained at the point of application and how much had migrated with the water solvent front. All studies had two replications and were repeated.

2. Soil Column Chromatography

Riggle and Penner (1988) also determined the controlled release properties of PC940C with metribuzin as measured by the soil column chromatography (SCC) technique (Abernathy and Davidson, 1971; Davidson and McDougal, 1973; Weber and Whitacre, 1982).

The SCC analysis was done in the following manner. Soil columns were prepared using polyvinyl chloride tubing (4.6 cm wide, 36 cm long) and the columns were packed with air dried sieved Wando sandy loam soil. The soil had a pH of 7.1 and an organic matter content of 1.7%. Tubes were split lengthwise and taped back together and the bottom end of the columns were covered with cheesecloth. The same solutions and radiolabel were used as described for STLC.

A study was done to evaluate the potential of PC940C to control the release of metribuzin. PC940C was used at zero-, two-, and fourfold levels and these were multiples of 2.6 mg PC940C/ml of metribuzin stock solution. The corresponding amount of PC940C, depending on the rate tested, was weighed out and mixed with 1 ml of metribuzin stock solution. This was shaken for 15 sec and 200 μl were transferred to a vial containing 5 μl [^{14}C]metribuzin; the vial was shaken for 1 hr and its entire contents were transferred to the soil surface of a column. Two leaching treatments were used. The first, a single leaching of 51 ml (equivalent of 2.54 cm) of distilled water, was uniformly applied to the soil surface over a 2-hr time period. The second was identical to the first but with a 6-day drying period followed by an additional 51 ml of water which was also uniformly applied to the soil surface over a 2-hr time period. For the drying period, the leached columns were placed under sodium halide lamps at 1000 μE/m^2/sec for 15 hr illumination periods.

Another study was designed to test for possible competitive effects between metribuzin and alachlor for the controlled release properties of PC940C. This test was performed because metribuzin is typically tank-mixed with other preemergence herbicides, such as alachlor. At rates of zero-, two-, and fourfold, 5.2 mg of PC940C added to either 1 ml of metribuzin stock solution plus 1 ml of 240 ppm alachlor stock solution or 2 ml of metribuzin stock solution. Mixtures were also shaken for 15 sec and 200 μl was transferred to a vial containing 5 μl [^{14}C]metribuzin. The vial was shaken for 1 hr and its entire contents were applied to the soil surface of a column. Only the single leaching treatment as described above was used.

An additional study was designed to test the mobility of PC940C with or without metribuzin. A twofold rate of nonlabeled PC940C was used which corresponded to 5.2 mg PC940C/2 ml of either distilled water or metribuzin stock solution. These mixtures were shaken for 15 sec as previously described and 100 μl samples were transferred to vials which contained 0.1 mg [^{3}H]PC940C. Vials were shaken for 1 hr and a 10 μl sample was transferred from the vial to the soil surface. Only the single leaching treatment was used for the [^{3}H]PC940C study. All studies had two replications and were done two times.

Radiolabeled material distribution in the soil columns was determined in the following manner. The columns were split open after leaching and the moist soil was removed at 2.54 cm increments with the first increment split in 1.27 cm portions. The water front was typically 15 cm from the soil surface. The entire segments of collected soil were transferred to long-necked 250 ml Erlenmeyer ground glass flasks. Fifty milliliters of extraction solvent was added and the flasks were shaken for 2 hr using a tabletop wrist action shaker at 200 reciprocations/min. Samples from each flask were transferred to 50 ml glass centrifuge tubes and centrifuged at 1000 rpm for 10 min. Between 1 and 5 ml of collected supernatant was removed by using a 5 ml disposable glass pipette and added to a 20 ml glass vial which contained 10 ml of liquid scintillation radioassay fluid. Samples were radioassayed with a liquid scintillation spectrometer.

3. Greenhouse and Field Studies

Riggle (1985) conducted greenhouse studies in 1984 and 1985 at East Lansing, Michigan. The soil type used was a sandy loam with a pH of 6.5 and an organic matter content of 0.8%. Three soybean seeds ('Corsoy 79') were planted in 1-liter plastic pots with drainage holes. Pots were placed in 250-ml aluminum pans to allow for daily subirrigation with water. A combination of metribuzin at rates of 0 and 0.64 kg/ha and PC940C at 0, 1.15, and 2.30 liters/ha was applied preemergence. Materials were mixed together in plastic 1-liter bottles, shaken, and allowed to stand for 1 hr prior to application. Materials were applied by a belt sprayer with a pressure of 104 kPa and a spray volume of 235 liters/ha. All treatments were arranged in the greenhouse in a randomized design and six replications were used. The experiment was repeated. Container-grown plants were maintained at 2 to 30°C with natural illumination and sodium halide supplemental lighting at 500 $\mu E/m^2/sec$. Plants were rated for injury when they were in the third trifoliar growth stage [approximately 35 days after emergence (D.A.E.)].

A field study was conducted by Riggle (1985) at East Lansing, Michigan, in 1984. The soil type was a fine loam with a pH of 7.8 and an organic matter content of 2.6%. The experimental design had three levels of metribuzin at 0, 0.42, and 0.84 kg/ha and four levels of PC940C at 0, 0.38, 0.77, and 1.15 liters/ha. Metribuzin and PC940C were mixed together 4 hr prior to preemergence application with water as a carrier at a rate of 215 liters/ha and delivered through 100-mesh screen at a pressure of 207 kPa. The plots had four rows of 'Corsoy 79' soybeans and were 9.1 by 3 m with 0.76 m row width spacings. Planting was done on June 22, 1985, and soybean injury ratings and weed control ratings were measured 32 and 46 D.A.E. The predominate weeds in the plots were redroot pigweed (*Amaranthus retroflexus* L.) and common lamb's-quarter (*Chenopodium album* L.). Plots were periodically irrigated to ensure that metribuzin was mobilized into the soil profile.

D. Results and Discussion

1. Soil Thin-Layer Chromatography

Riggle and Penner (1987) found that of the two lignins tested, only the experimental material, PC940C, was able to retain metribuzin at the point of application on the soil plates after each development (Table I). INDULIN AC did not appear to alter metribuzin movement and was no different from the metribuzin control.

Tritiated PC940C did not appear mobile since most of the radiolabel remained at the point of application on the plates. After two 12.5 cm developments, only 3 to 5% of the ^3H moved from the point of application. There is the possibility that some ^3H exchange took place between the lignin and the water or water-soluble materials in the soil. If so, this would mean that the percentage of [^3H]PC940C measured at the point of application was underestimated.

The results of this STLC research conducted by Riggle and Penner (1987) suggest that PC940C is a non-water-soluble material which has a strong affinity for water-soluble metribuzin. These characteristics allow for a reduction in the mobility of metribuzin when combined with PC940C in an aqueous mixture. Because INDULIN AG did not restrict metribuzin mobility, this would indicate that the two lignins had different physical and/or chemical properties.

2. Soil Column Chromatography

Riggle and Penner (1988) reported that PC940C controlled the release of [^{14}C]metribuzin in the soil columns by keeping more of the label in the top portion of the columns (Table II). As a consequence less [^{14}C]metribuzin was

Table I Effect of Lignins on the Percentage of [^{14}C]Metribuzin Retained at the Point of Origin on Soil TLC Plates after Three Developments[a]

Treatment	Developments[b] 1	2	3
		% of Total[c]	
PC94OC	47.4 a	29.2 c	16.2 d
Indulin AC	0.0 e	0.0 e	0.0 e
Metribuzin control	0.0 e	0.0 e	0.0 e

[a] Adapted from Riggle (1985).

[b] Developments are multiples of approximately 12.5 cm.

[c] Numbers followed by the same letter are not significantly different according to Duncan's multiple range test at the 0.05 level.

Table II The Effect of PC94OC Lignin on [^{14}C]Metribuzin Distribution in a Wando Sandy Loam Soil Column following One or Two Leaching Periods with 2.54 cm of Water[a]

Sample depth (cm)	Treatments[b,c]					
	Metribuzin		Metribuzin plus PC940C (2×)		Metribuzin plus PC940C (4×)	
	1[d]	2	1	2	1	2
	% of total					
0–2.5	8.8 Ccd	8.4 Cc	16.3 Bc	12.1 BCbc	24.2 Aab	17.5 Bb
2.5–5.1	11.4 Ac	11.1 Ac	14.3 Ac	13.5 Ab	15.4 Ab	15.6 Abc
5.1–7.6	31.1 ABb	28.2 Bb	35.1 Aa	30.1 ABa	31.3 ABa	26.5 Ba
7.6–10.1	44.0 Aa	38.5 ABa	32.1 BCb	33.7 BCa	28.1 Ca	30.2 Ca
10.1–12.7	4.6 BCd	13.4 Ac	2.3 Cd	10.6 ABbc	1.0 Cc	10.2 ABc

[a] Adapted from Riggle (1985).

[b] Numbers followed by the same capital letter in the same row are not significantly different according to Duncan's multiple range test at the 0.05 level.

[c] Numbers followed by the same lowercase letter in the same column are not significantly different according to Duncan's multiple rang test at the 0.05 level.

[d] Leaching rates are as follows: (1) 2.54 cm of water and (2) 2.54 cm of water separated by a 6-day drying period.

leached to the lower portion of the columns. The effect occurred for both single- and double-leaching treatments and appeared to be linear as the concentration of PC940C increased. The double-leaching treatment also resulted in a decrease of ^{14}C-label percentages for the top segments of the columns.

The combination of alachlor and metribuzin did not interfere with [^{14}C]metribuzin retention to and release from PC940C (Table III). These results suggest that alachlor did not compete with ^{14}C-labeled and ^{12}C metribuzin for the binding sites of PC940C at the twofold rate. In addition, an interaction between alachlor and metribuzin reduced the mobility of [^{14}C]metribuzin. This would suggest that an interaction took place between the two compounds; hydrogen bonding may have occurred between the two, possibly between the primary amine group of metribuzin and the carbonyl group of alachlor. This would result in a larger and presumably less mobile molecule.

PC940C was not mobile in the soil columns when leached, as determined by the [^{3}H]PC940C leaching experiments. This was also supported by visual observations since the material remained on the soil surface after repeated leaching treatments.

A combination of metribuzin and [^{3}H]PC940C retained more [^{3}H]PC940C at the soil surface than did the [^{3}H]PC940C treatment alone. However, this difference is not very large when compared to the total amount of material applied. Penner and Riggle (1988) suggested that the portion of ^{3}H-label which was leached through the columns was either a water-soluble fraction of PC940C or a result of tritium exchange with water or a water-soluble material.

Table III Percentage of [^{14}C]Metribuzin Distribution in a Wando Sandy Loam Soil Column following Application of a Metribuzin–Alachlor Mixture Leached with 2.54 cm Water over 2 hr[a]

Sample depth (cm)	Treatment[b,c]					
	Metribuzin	Metribuzin plus alachlor	Metribuzin plus PC940C (2×)	Metribuzin plus alachlor plus PC940C (2×)	Metribuzin plus PC940C (4×)	Metribuzin plus alachlor plus PC940C (4×)
			(% of total)			
0–2.5	8.8 Cc	11.8 Cc	16.3 Bc	19.0 Bb	24.2 Ab	23.2 Ab
2.5–5.1	11.4 Dc	19.2 ABb	14.3 CDc	21.4 Ab	15.4 BCc	18.0 ABCc
5.1–7.6	31.1 Db	52.7 Aa	35.1 CDa	39.2 BCa	31.3 Da	40.5 Ba
7.6–10.1	44.0 Aa	16.3 Db	32.1 Bb	20.2 CDb	28.1 BCab	18.0 Dc
10.1–12.7	4.6 Ad	0.1 Cd	2.3 Bd	0.1 Cc	1.0 BCd	0.2 Cd

[a] Adapted from Riggle (1985).

[b] Numbers followed by the same capital letter in the same row are not significantly different according to Duncan's multiple range test at the 0.05 level.

[c] Numbers followed by the same lowercase letter in the same column are not significantly different according to Duncan's multiple range test at the 0.05 level.

3. Greenhouse and Field Studies

With the greenhouse study, Riggle (1985) observed reduction in soybean injury from metribuzin from 22% to 10 and 11% with PC940C at rates of 1.15 and 2.30 liters/ha, respectively. Since there appeared to be no concentration effect of PC940C in terms of injury reduction, it can be assumed that a saturation point had been reached at the 1.15 liters/ha rate.

Riggle (1985) observed that the lignin successfully reduced metribuzin related phytotoxicity by one-half or more to field grown soybeans evaluated 32 and 46 days after treatment. The 0.77 liter/ha rate of PC940C appeared to produce the maximum protective response.

The similarity in the degree of phytotoxicity for both the 0.42 and the 0.84 kg/ha rate of metribuzin may have been due to the high pH of the soil (7.7). Past research has shown that as the soil pH increases, the degree of metribuzin related injury to soybeans also increases (Ladlie *et al.*, 1976b).

PC940C did not appear to injure plants. This was apparent from the lack of injury to both weed species and soybeans which grew in plots that had been treated with PC940C alone. These results suggest that PC940C did not have any herbicidal activity.

PC940C did not have any effect on metribuzin-related control of both redroot pigweed and common lamb's-quarters. Both rates of metribuzin gave excellent control of both weed types with or without PC940C.

V. CONCLUDING REMARKS

In conclusion, the lignin PC940C reduced the mobility of metribuzin as measured by STLC and SCC and reduced metribuzin related phytotoxicity for both greenhouse- and field-grown soybeans. In addition, tritiated PC940C appeared to be immobile in the STLC and SCC studies; this characteristic is essential for a controlled release agent with soil-applied water-mobilized pre-emergence herbicides.

Successful utilization of lignins as protective agents for crop plants from herbicide injury by a controlled release strategy will be dependent on the careful selection of the appropriate lignin for a specific herbicide and the proper formulation technology. The successful outcome would be an adequate release of the herbicide into the soil for weed control but in an amount and at a rate that would not be injurious to the crop plants.

References

Abernathy, J. R., and Davidson, J. M. (1971). Effect of calcium chloride on prometryne and fluometuron adsorption in soil. *Weed Sci.* **19**, 517–521.

Allan, G. G., Chopra, C. S., Neogi, A. N., and Wilkins, R. M. (1971). Design and synthesis of controlled release pesticide–polymer combinations. *Nature (London)* **234**, 349–351.

Allan, G. G., Chopra, C. S., and Russell, R. M. (1972). Controlled release pesticides. III. Selective suppression of weeds and deciduous brush in the presence of conifers. *Int. Pest Control* **14**, 15–20.

Allan, G. G., Beer, J. W., and Cousin, M. J. (1977). Controlled release of herbicides from biodegradable substrates. *In* "Controlled Release Pesticides" (H. B. Scher, ed.), pp. 94–101. Am. Chem. Soc., Washington, D. C.

Ashton, F. M., and Crafts, A. S. (1981). "Mode of Action of Herbicides." Wiley (Interscience), New York.

Banks, P. A., and Robinson, E. L. (1982). The influence of straw mulch on the soil reception and persistence of metribuzin. *Weed Sci.* **30**, 164–165.

Bouchard, D. C., Lavy, T. L., and Marx, D. B. (1982). Fate of metribuzin, metolachlor, and fluometuron in soil. *Weed Sci.* **30**, 629–632.

Coble, H. D., and Schrader, J. W. (1973). Soybean tolerance to metribuzin. *Weed Sci.* **21**, 308–309.

Collins, R. L., and Doglia, S. (1973). Concentration of pesticides slowly released by diffusion. *Weed Sci.* **21**, 343–349.

Collins, R. L., Doglia, S., Mazak, R. A., and Samulski, E. T. (1973). Controlled release of herbicides theory. *Weed Sci.* **21**, 1–5.

Davidson, J. M., and McDougal, J. R. (1973). Experimental and predicted movement of three herbicides in a water-saturated soil. *J. Environ. Qual.* **2**, 428–433.

DelliColli, H. T. (1980). Pine kraft lignin as a possible delivery system. *In* "Controlled Release Technologies: Methods, Theory, and Application" (A. F. Kydonieus, ed.), Vol. 2, pp. 225–234. CRC Press, Boca Raton, Florida.

Dunigan, E. P., and MacIntosh, T. (1971). Atrazine soil organic matter interaction. *Weed Sci.* **19**, 279–282.

Foley, M. E., and Wax, L. M. (1980). Effects of starch xanthate and sludge polymer on the initial activity with delayed incorporation, residual activity, and crop safety of several herbicides. *Weed Sci.* **28**, 626–632.

Frear, D. S., Swanson, H. R., and Mansager, E. R. (1985). Alternative pathways of metribuzin metabolism in soybean: Formation of *N*-glucoside and homoglutathione conjugates. *Pestic. Biochem. Physiol.* **23**, 56–65.

Glennie, D. W. (1971). Reactions in sulfite pulping. *In* "Lignins: Occurrence, Formation Structure, and Reactions" (K. V. Sarkanen and C. H. Ludwig, eds.), pp. 597–631. Wiley (Interscience), New York.

Goring, D. A.I. (1971). Polymer properties of lignin and lignin derivatives. *In* "Lignins: Occurrence, Formation Structure, and Reactions" (K. V. Sarkanen and C. H. Ludwig, eds.), pp. 697–761. Wiley (Interscience), New York.

Harris, F. W. (1980). Polymers containing pendent pesticide substituents. *In* "Controlled Release Technologies: Methods, Theory, and Applications" (A. F. Kydonieus, ed.), Vol. 2, pp. 63–82. CRC Press, Boca Raton, Florida.

Hartmann, M., Bauer, H., and Wermann, K. (1985a). Polymeric herbicides biocide polymers. 13. + synthesis and hydrolysis of modified poly (acrylic acids) containing 3-amino-1,2,4-triazole. *Polym. Bull.* **13**, 195–200.

Hartmann, M., Kohrs, D., and Wermann, K. (1985b). Biocide polymers X. Synthese und hydrolyseverhalten von polymeren mit seitenstanding gebundenem 3-amino-1,2,4-triazol. *Acta Polym.* **36**, 185–187.

Helling, C. S. (1971). Pesticide mobility in soils. I. Parameters of thin-layer chromatography. *Soil Sci. Soc. Am. Proc.* **35**, 731–737.

Helling, C. S., and Turner, B. C. (1968). Pesticide mobility: Determination by soil thin-layer chromatography. *Science* **162**, 562–563.

Hyzak, D. L., and Zimdahl, R. L. (1974). Rate of degradation of metribuzin and two analogs in soil. *Weed Sci.* **22**, 75–79.

Ladlie, J. S., Meggitt, W. F., and Penner, D. (1976a). Effect of soil pH on microbial degradation, adsorption, and mobility of metribuzin. *Weed Sci.* **24**, 477–481.

Ladlie, J. S., Meggitt, W. F., and Penner, D. (1976b). Effect of pH on metribuzin activity in the soil. *Weed Sci.* **24**, 505–507.

McCormick, C. L., and Fooladi, M. M. (1980). Controlled activity polymers with labile bonds to pendent metribuzin. *In* "Controlled Release of Bioactive Materials" (W. Richard, ed.), pp. 317–330. Academic Press, New York.

McCormick, C. L., Anderson, K. W., Pelezo, J. A., and Lichatowich, D. K. (1981). Controlled release of metribuzin, 2,4-D, and model aromatic amines from polysaccharides and poly (vinyl alcohol). *In* "Controlled Release of Pesticides and Pharmaceuticals" (D.H Lewis, ed.), pp. 147–158. Plenum, New York.

McLane, S. R., and Parkins, M. D. (1966). Biological and physical attributes of several amiben derivatives. *Proc. Br. Weed Control Conf.* **8**, 283–290.

Marton, J. (1971). Reactions in alkaline pulping. *In* "Lignins: Occurrence, Formation Structure, and Reactions" (K. V. Sarkane and C. H. Ludwig, eds.), pp. 639–689. Wiley (Interscience), New York.

Mehltretter, C. L., Roth, W. B., Weakly, F. B., McGuire, T. A., and Russell, C. R. (1974). Potential controlled-release herbicides from 2,4-D esters of starch. *Weed Sci.* **22**, 415–418.

Pearl, I. A. (1964). Lignin chemistry. *Chem. Eng. News* July 6, pp. 81–92.

Pfister, K., and Arntzen, C. J. (1979). The mode of action of photosystem II specific inhibitors in herbicide-resistant weed biotypes. *Z. Naturforsch. C: Biosci.* **34C**, 966–1009.

Raboy, V., and Hopen, H. J. (1982). Effectiveness of starch xanthide formulations of chloramben for weed control in pumpkin (*Cucurbita moschata*). *Weed Sci.* **30**, 169–174.

Riggle, B.D, (1985). The controlled release of water-soluble herbicides. Ph.D. Dissertation, Michigan State University, East Lansing.

Riggle, B. D., and Penner, D. (1988). Controlled release of three herbicides with the kraft lignin PC94OC. *Weed Sci.* **36**, 131–136.

Riggle, B. D., and Penner, D. (1987). Evaluation of pine kraft lignins for controlled release of alachlor and metribuzin. *Weed Sci.* **35**, 243–246.

Savage, K. E. (1976). Adsorption and mobility of metribuzin in soil. *Weed Sci.* **24**, 525–528.

Schacht, E. H., Uesmarets, G. E., Coethals, E. J., and St. Pierre, T. (1978). Synthesis and characterization of 2,6-dichlorobenzaldehyde-generating polymers. *In* "Polymeric Drugs" (L. G. Donaruma and O. Vogl, eds.), p. 331. Academic Press, New York.

Schubert, W. J. (1965). "Lignin Biochemistry." Academic Press, New York.

Smith, A. E., and Verma, B. P. (1977). Weed control in nursery stock by controlled release of alachlor. *Weed Sci.* **25**, 175–178.

Smith, A. E., and Wilkinson, R. E. (1974). Differential adsorption, translocation, and metabolism of metribuzin 4-amino-6-tert-butyl-3-(methyl-thio)-as-triazin-5(4*H*)-one by soybean cultivars. *Plant Physiol.* **32**, 253–257.

Stevenson, F. J. (1982). "Humus Chemistry Genesis, Composition, Reactions." Wiley, New York.

Sullivan, J. D., Jr., and Felbeck, G. T., Jr. (1968). A study of the interaction of *s*-triazine herbicides with humic acids from three different soils. *Soil Sci.* **106**, 42–52.

Swain, T. (1979). Tannins and lignins. *In* "Herbivores: Their Interaction with Secondary Plant Metabolites" (G. A. Rosenthal and D. H. Janzen, eds.), pp. 674–681. Academic Press, New York.

Weber, J. B., and Whitacre, D. M. (1982). Mobility of herbicides in soil columns under saturated and unsaturated conditions. *Weed Sci.* **30**, 579–584.

Weber, J. B., Weed, S. B., and Ward, T. M. (1969). Adsorption of *s*-triazines by soil organic matter. *Weed Sci.* **17**, 417–421.

Weed Science Society of America (1983). "Herbicide Handbook," 5th ed." WSSA, Champaign, Illinois.

Wildung, R. E., Chesters, G., and Armstrong, D. E. (1968). Chloramben (amiben) degradation in soil. *Weed Res.* **8**, 213–225.

Chapter 13

Growth Regulators, Fungicides, and Other Agrochemicals as Herbicide Safeners

Sharad C. Phatak
Department of Horticulture
Coastal Plain Experiment Station
University of Georgia
Tifton, Georgia 31793

Charles S. Vavrina
Department of Horticulture
Cooperative Extension Service
University of Georgia
Statesboro, Georgia 30460

I. INTRODUCTION

Initial interest in herbicide safeners was stimulated by an accidental discovery involving 2,4-D and related compounds. In 1947 Hoffman (1978) noticed that tomato (*Lycopersicon esculentum* Mill.) plants treated with 2,4,6-T were protected from the injurious 2,4-D vapors. That initial observation led to an extensive screening of compounds for activity as antidotes. Herbicide safening involves an interaction between two (or more) chemicals and because many pesticides used in crop production are often applied tank-mixed or sequentially, the chances of detecting antidote activity are ever-increasing. Several growth-regulating substances, fungicides, and other agrochemicals have exhibited herbicide safening.

This chapter reviews the potential of selected growth regulators, fungicides, and other agrochemicals as crop safeners against injury from specific herbicides. A case history evaluating the efficacy of the growth retardants triapenthenol and BAS 140 810 as safeners against the herbicide metribuzin on soybeans [*Glycine max* (L.) Merr.] is discussed in greater detail.

II. GROWTH REGULATORS AS HERBICIDE SAFENERS

A. Auxins

As mentioned earlier, the potential safening activity of auxin analogs as crop safeners against herbicide injury was first realized by Hoffman (1978). In an unvented greenhouse, tomato plants treated with 2,4,6-T were able to withstand fumes of 2,4-D, whereas untreated or 2,4-D-treated tomatoes showed considerable epinasty. By 1960, Hoffman *et al.* had documented their findings with 2,4-D and 2,4,6-T and extended these observations to include 2,4-D, 2,4,6-T, and MCPA as antidotes against barban foliar injury on oats (*Avena sativa* L.). Hoffman (1978) believed initially that a "lock and key" mode of action was involved in the safening, but he later reversed his thinking to suggest antagonistic modes of action were at play as barban slows growth and 2,4-D accelerates growth. Hoffman (1978) also noted that none of the phenoxyacetic acid herbicides acted as an antidote of barban on oats when used as a seed treatment. Subsequent research by Hoffman (1962) led to the discovery of 4′-chloro-2-hydroxyimino acetanilide (S-449) as a seed-applied antidote for barban injury on wheat (*Triticum aestivum* L.).

B. Gibberellins

Symptoms of EPTC injury to corn (*Zea mays* L.) were noted to be very similar to those observed in corn treated with growth retardants (Donald and Fawcett, 1975; Harvey *et al.*, 1975). The stunting effect of EPTC on corn shoots was prevented by exogenous application of gibberellic acid (GA_3). However, GA_3 application failed to reverse the leaf twisting and rolling caused by EPTC (Harvey *et al.*, 1975). Wilkinson (1985) reported that reversal of a betacyanin efflux response in sugar beets (*Beta vulgaris* L.) was dependent on GA_3 and EPTC concentrations. GA_3 failed to reverse the effect of EPTC at 10^{-3} M concentrations. EPTC inhibits gibberellin biosynthesis in wheat (Wilkinson and Ashley, 1979) and corn (Donald *et al.*, 1979). The effects of herbicides and safeners on endogenous gibberellins are discussed in greater detail by Wilkinson in Chapter 10, this volume.

C. Cytokinins

Sullivan (1979) has shown improved sugar beet seedling emergence and growth in fields treated preemergence with the herbicides ethofumesate and diclofop-methyl when beet seed was pretreated with benzyladenine (BA). A significant increase in sugar beet root weight and recoverable sugar was the result of this interaction.

The inhibitory effects of fluridone on chlorophyll and carotenoid content of etiolated cucumber (*Cucumis sativus* L.) cotyledons were reversed by pretreatment with BA (Fletcher *et al.*, 1984). Many herbicides are known to have cytokinin-like activity (Bendixen, 1975; Hoffman, 1978; van Overbeek, 1964). Some of these herbicides and other cytokinins may prove to be antagonistic and become important as herbicide safeners in the future.

D. Growth Retardants

Growth retardants have shown promising potential as crop safeners against herbicide injury. The efficacy of the growth retardants cycocel (CCC), maleic hydrazide and daminozide in protecting selected crops from herbicide injury is reviewed briefly in this section. The potential of the recently introduced growth retardants triapenthenol (also known as RSW 0411) and BAS 140 810 [*N*-allyl-*N*-2-(2,4,6-trichlorophenoxy)-ethyl-piperidinium-bromide] in protecting soybeans against the herbicide metribuzin is discussed thoroughly in Section V.

1. CCC

In early reports by Cyanamid International (1966), CCC, a plant growth retardant, was shown to increase spinach (*Spinacia oleracea* L.) seedling resistance to herbicide injury (no specific herbicides mentioned). Kirkland (1973) reported that seed treatment with CCC provided winter wheat protection against injury caused by atrazine residues in soil. Ohali *et al.* (1979) have shown the involvement of CCC in increasing cotton (*Gossypium hirsutum* L.) resistance to fluometuron applied preemergence. Seed treatment with CCC at 100 ppm significantly reduced fluometuron injury to the foliage. The mode of the CCC safening action involved a reduced uptake of fluometuron resulting in reduced accumulation of the herbicide at its site of action. Ohali *et al.* (1979) reported that CCC reduced cotton transpiration; the relationship between fluometuron uptake and translocation and reduced transpiration was highly correlated. Fluometuron injury did not advance as the CCC growth-retarding effects diminished.

2. Maleic Hydrazide

Messier and Ashley (1981) reported that maleic hydrazide (MH) as a 0.2% foliar spray reduced metribuzin necrotic injury to tomato. Fresh weight of MH plus metribuzin-treated tomatoes was 83% of control as compared to 27% of control for metribuzin alone.

3. Daminozide

Phatak *et al.* (1985) have shown that daminozide provided protection to seedling potatoes (*Solanum tuberosum* L.) against injury caused by metribuzin applied postemergence. Daminozide at 2500 and 5000 ppm applied to the foliage 4 days prior to metribuzin application (0.56 kg/ha) protected 42-day-old potatoes. Of six cultivars screened, all received protection with 2500 ppm, although to varying degrees, suggesting that metribuzin cultivar tolerances remained intact. Overall, 25–42% of the control plants exhibited necrosis greater than 20%; 2–10% of plants treated with daminozide (2500 ppm) exhibited similar injury; and 0% of the daminozide-treated plants (5000 ppm) exhibited necrosis (Fig. 1). The daminozide-treated plants also had significantly higher total soluble solid concentrations (Fig. 2). The conjugation of metribuzin with sugars was suggested as a possible mode of action. Messier and Ashley (1981) reported that daminozide applied to the foliage of tomato offered little protection when metribuzin was added to nutrient solution at 2.36×10^{-5} M. Further work is needed in the field of herbicide/plant growth regulator interactions, as these results appear promising.

III. FUNGICIDES AS HERBICIDE SAFENERS

A. Diazosulfonates

The fungicide fenaminosulf (Dexon) was found to be strongly antagonistic to atrazine activity by Corbin and Sheets (1968). Cucumber and wheat dry weights

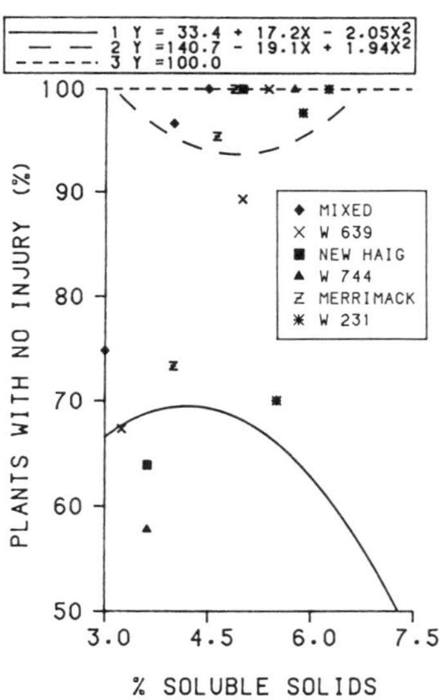

Fig. 1. Relationship of the percentage of potato plants with no injury versus the percentage of soluble solids in stem sap as influeced by daminozide treatments. Metribuzin was applied 4 days after daminozide treatments. Regression equations and regression lines represent daminozide treatments: (1) 0 ppm, (2) 2500 ppm, and (3) 5000 ppm. R^2 (Model) = 0.79. (From Phatak *et al.*, 1985.)

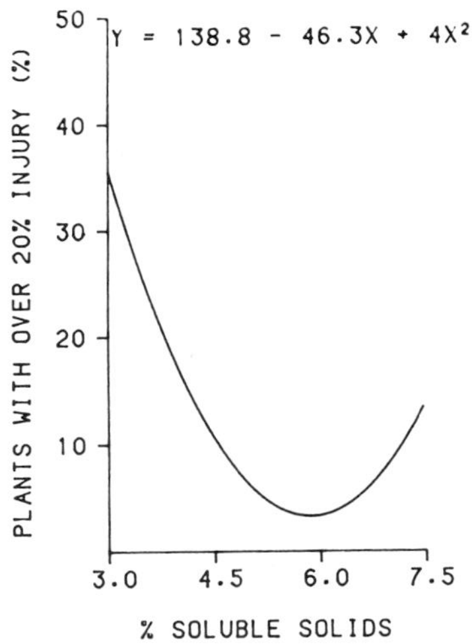

Fig. 2. Relationship of the percentage of potato plants with over 20% injury caused by metribuzin versus the percentage of soluble solids in the stem sap after applications of daminozide followed by metribuzin. R^2 (Model) = 0.38. (From Phatak *et al.*, 1985.)

were greater when atrazine and fenaminosulf were mixed than when atrazine was applied alone. Webster and Sheets (1969) reported that this safening response in the soil was due to a decrease in atrazine uptake induced by this fungicide. Fenaminosulf was found to alter root absorption and translocation of 14 *s*-triazines (Nash and Harris, 1969). In short-term (0–96 hr) laboratory studies with light- and dark-grown soybean seedlings, fenaminosulf was able to partially alleviate some alterations of phenylpropanoid metabolism induced by atrazine (Hoagland, 1981). These biochemical effects may be related to the antidotal activity of fenaminosulf against atrazine on soybeans.

Following this lead, Phillips and Bhagsari (1978) synthesized a series of diazosulfonates and screened them as safeners against residual levels of atrazine in soybean fields. Diazosulfonates tested were MBDS (*p*-methyl-benzenediazo sodium sulfonate), CBDS (*p*-chloro-benzenediazo sodium sulfonate), and BDS (benzenediazo sodium sulfonate). Fenaminosulf was included for comparisons. The results showed that fenaminosulf provided consistently better protection than other diazosulfonates tested. Compared to fenaminosulf, MBDS at 200 ppm provided only half the protection. No protection was provided by BDS. At low concentrations, CBDS increased atrazine injury to soybeans. According to Phillips and Bhagsari (1978), fenaminosulf at 400 ppm provided 43% protection to soybeans against 0.2 ppm atrazine after 66 days. The high concentrations of fenaminosulf needed to completely counteract residual atrazine toxicity on soybeans made it difficult to further exploit this fungicide as a practical crop safener. Other diazosulfonates may be more active than fenaminosulf and they remain to be tested.

B. Other Fungicides

Triadimefon, a systemic fungicide, exhibits plant growth-regulating activity. Low rates of triadimefon applied as a soil drench during tomato seedling emergence reduced the injury caused by postemergence application of metribuzin (Grier *et al.*, 1985). Carboxin, a seed-applied fungicide used for control of loose smut in wheat, has provided protection to wheat against the herbicide triallate (Miller and Nalewaja, 1980). Moderate protection was provided by the seed-applied fungicide HMI to rice (*Oryza sativa* L.) against injury caused by the herbicides simetryn, nitrofen, chlornitrofen, and propanil (Ogawa and Ota, 1976).

IV. OTHER AGROCHEMICALS AS HERBICIDE SAFENERS

Numerous antagonistic combinations have been observed in the literature but not fully exploited for their safener potential. Putnam and Penner (1974) and Hatzios and Penner (1985) have documented many antagonistic pesticide interactions. Urea and triazine herbicides interact antagonistically with dinitroaniline herbicides in soybeans (Ladlie *et al.*, 1977; Malefyt and Duke, 1981), buthidazole with acetanilide herbicides in corn (York and Slife, 1981), and 2,4-D with diclofop-methyl in oats (Kupatt and Ilnicki, 1985). Similarly, Phatak (1974) observed antagonistic interactions between the soil-applied insecticide chlorfenvinphos and the herbicides napropamide, alachlor, or DCPA on rutabaga (*Brassica napus* L.). Many more antagonistic interactions have been observed but not reported. Stephenson and Ezra (1983) have suggested that depending on the importance of the crops, herbicides, or weeds involved, many of these antagonistic combinations could become useful. The key is to find an antagonistic pesticide interaction which will selectively protect the crop but not the weeds.

V. CASE HISTORY: TRIAPENTHENOL AND BAS 140 810 AS SAFENERS AGAINST METRIBUZIN INJURY IN SOYBEANS

A. Triapenthenol/Metribuzin Interactions

Following the preliminary work of Vavrina and Phatak (1985) detailed investigations were made into the herbicide safening capacity of triapenthenol (Fig. 3) against metribuzin injury to soybeans (Vavrina, 1987). Studies were conducted to ascertain field application techniques, crop and weed response, and herbicide interactions in relation to the observed safening of soybeans from metribuzin injury by plant growth retardants. Triapenthenol (0.28 and 0.56 kg a.i./ha) was

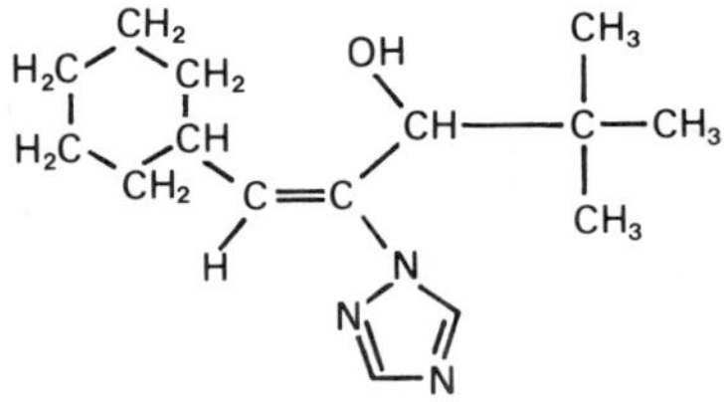

Fig. 3. Chemical structure of triapenthenol.

applied as a band treatment and as a tank mixture with (and without) metribuzin (0.56 or 1.12 kg a.i./ha) as a broadcast treatment. The metribuzin and triapenthenol band treatments were applied separately and each chemical application was followed by 0.6 cm irrigation water. The triapenthenol/metribuzin tank mix treatments also received 0.6 cm irrigation. Triapenthenol was not effective as a seed safener against metribuzin injury to soybeans. When applied as a seed treatment, triapenthenol inhibited the germination of soybean seeds.

1. Triapenthenol Banded/Metribuzin Broadcast over Soybeans

Triapenthenol banded prior to soybean germination reduced subsequent metribuzin-induced injury to 'Braxton' leaf tissue. Plants treated with metribuzin alone exhibited reduced fresh weights and dry weights at 30 days after planting. Triapenthenol-treated plants showed no difference in plant weight when compared to control plants. Metribuzin in combination with triapenthenol may or may not cause soybean weight loss dependent on rate, planting date, and growing conditions. However, in 1984 the triapenthenol/metribuzin treatment curbed metribuzin-induced weight loss. This result supported preliminary field trials of Vavrina and Phatak (1985).

Triapenthenol-treated soybean exhibited a typical antigibberellin height reduction. This was consistent with greenhouse and preliminary field data. Metribuzin alone caused a height reduction as a direct result of phytotoxicity. By 60 days after planting, the triapenthenol-induced reduction in height had disappeared. The 1.12 kg/ha rate of metribuzin (a rate 3.5 times greater than that recommended for coastal plain soils) stunted plants wherever used, regardless of whether triapenthenol was present or not.

Metribuzin at 1.12 kg a.i./ha applied in combination with triapenthenol injured soybeans and caused a yield loss. Triapenthenol alone did not directly affect yield. Soybean yield was negatively correlated with initial injury by metribuzin. Hagood *et al.* (1980) have suggested that early herbicide-induced injury in soybean is not a good indicator of yield potential. The length of time required for soybean maturity is sufficient to overcome early season herbicide injury. The level of injury achieved in this study correlated more directly with a population reduction and hence resulted in a highly correlated injury/yield parameter.

2. Triapenthenol/Metribuzin Tank-mixed over Soybeans

Results obtained with triapenthenol/metribuzin tank mix applications were quite similar to those of the study with the band-applied triapenthenol (Vavrina, 1987). Planting date significantly affected all parameters measured in the study, except initial injury, which remained consistent across dates. Except for reduced height, triapenthenol-treated plants behaved essentially as control plants throughout the study (Vavrina, 1987).

The triapenthenol/metribuzin tank mix reduced metribuzin-induced injury to soybeans particularly at 14 and 21 days after planting. Subsequently 30 days after planting, the triapenthenol/metribuzin tank mixture reduced plant weight loss associated with metribuzin injury alone. The triapenthenol/metribuzin tank mixture did not affect yield when compared to the control at the 0.56 kg/ha level of metribuzin. At the 1.12 kg/ha rate of metribuzin, yields of soybeans treated with the triapenthenol/metribuzin combination were greater than yields of plants treated with metribuzin alone, but not significantly so.

3. Triapenthenol Safening and Other Agrochemicals

The possible safening of soybeans against metribuzin injury by other agrochemicals, especially trifluralin (Ladlie *et al.*, 1977; Moomaw and Martin, 1978), was also investigated in this study. Compounds tested included the herbicides trifluralin and metolachlor and the insecticide/nematicide phenamiphos applied in all possible combinations.

Neither metolachlor nor phenamiphos alone or in combination with each other, trifluralin, and triapenthenol appeared to provide any safening to soybeans over and above those levels attained by the metribuzin/triapenthenol treatment alone. The efficacy of safening by trifluralin in comparison to that by triapenthenol was evaluated by measurements of soybean fresh weight at 30 days after planting. It was found that trifluralin alone did not provide the safening realized by triapenthenol alone. Trifluralin may have contributed additional safening in the metribuzin/trifluralin/triapenthenol combination, but not significantly greater than safening realized by triapenthenol alone.

4. Safening of Weed Species by Triapenthenol

The effect of the triapenthenol/metribuzin application on weed species was also studied. Metribuzin, applied at recommended rates, will control Florida beggarweed [*Desmodium tortuosum* (Sw.) DC] and sicklepod (*Cassia obtusifolia* L.). Triapenthenol did not reduce the herbicidal activity of metribuzin on either of these two weed species.

Metribuzin at 1.12 kg/ha caused a weight and population loss in Texas panicum (*Panicum texanum* Buckl.). Triapenthenol in combination with metribuzin did not change these reductions. Metribuzin will not control tall morning glory [*Ipomoea purpurea* (L.) Roth] when applied preemergence. Triapenthenol alone and in combination with metribuzin caused a population loss in tall morning glory. Furthermore, seedlings of tall morning glory treated with triapenthenol were stunted and prone to splashing of metribuzin-treated soil onto the foliage which caused phytotoxicity and plant death.

5. Triapenthenol and Soybean Cultivars

Experiments were conducted to assay the effect of the triapenthenol/metribuzin (at 1.12 kg a.i./ha) antagonistic interaction on selected soybean cultivars (Vavrina and Phatak, 1988). Following the methods of previous experimentation, six soybean cultivars, exhibiting differential tolerance to metribuzin, were assayed for their response to the safening effects of triapenthenol against metribuzin and included 'Essex' (tolerant), 'Braxton' (tolerant), 'Delta Pine 105' (midtolerant), 'Centennial' (midtolerant), 'Semmes' (susceptible), and 'Coker 156' (susceptible).

Metribuzin injury segregated along genetically defined lines as was expected (Fig. 4). Metribuzin alone at 1.12 kg a.i./ha caused a high incidence of stand necrosis (both 14 and 21 days after planting), significantly reduced final plant stand, and reduced plant fresh weight and dry weight 30 days after planting.

The application of triapenthenol at 0.56 kg a.i./ha by itself caused no visual injury to soybean cultivars. Triapenthenol reduced plant height and increased chlorophyll content in a manner similar to previously published laboratory and field reports (Vavrina and Phatak, 1985; Vavrina, 1987). In general, triapenthe-

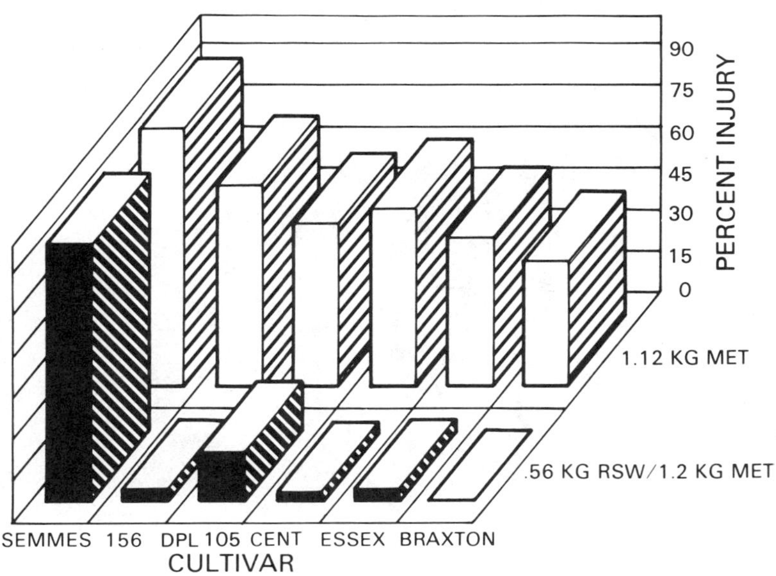

Fig. 4. Metribuzin-induced injury to soybean cultivars and safening conveyed by triapenthenol plant growth regulator applied as a broadcast treatment at planting.

nol appeared to enhance the survival of metribuzin-treated soybean seedlings 30 days after planting. Treatment with triapenthenol showed a consistent reduction of metribuzin-induced injury on all soybean cultivars tested (the 'Coker 156' data were surprising and bear further investigation). This effect was more dramatic when viewed within individual cultivars, sometimes reducing metribuzin-induced injury by 80–100% (Fig. 4). The safening conveyed by triapenthenol extended across all genetically defined metribuzin-tolerant lines by significantly safening all cultivars except 'Semmes'.

The addition of triapenthenol to the metribuzin treatment reduced the weight loss exhibited in soybean seedlings 30 days after planting from metribuzin treatment alone. At 60 days after planting, dry weights of soybeans treated with the metribuzin/triapenthenol combination approached and often exceeded the dry weight of seedlings exposed to triapenthenol treatment alone (Fig. 5). These data resemble the results of earlier cultivar trials (Vavrina and Phatak, 1985).

Plant population at emergence was unaffected by treatment across cultivars, though differences between cultivars occurred. The metribuzin-by-cultivar interaction showed a soybean stand loss across cultivar at both dates. Stand loss was greatest in 'Semmes', but all cultivars suffered stand reduction. The triapenthenol-by-cultivar interaction indicated that triapenthenol treatment could not reduce metribuzin-induced stand reduction to susceptible plants within cultivars.

Treatments with triapenthenol altered yield compared to the control in an inconsistent manner, increasing yield in some cultivars while reducing yield in

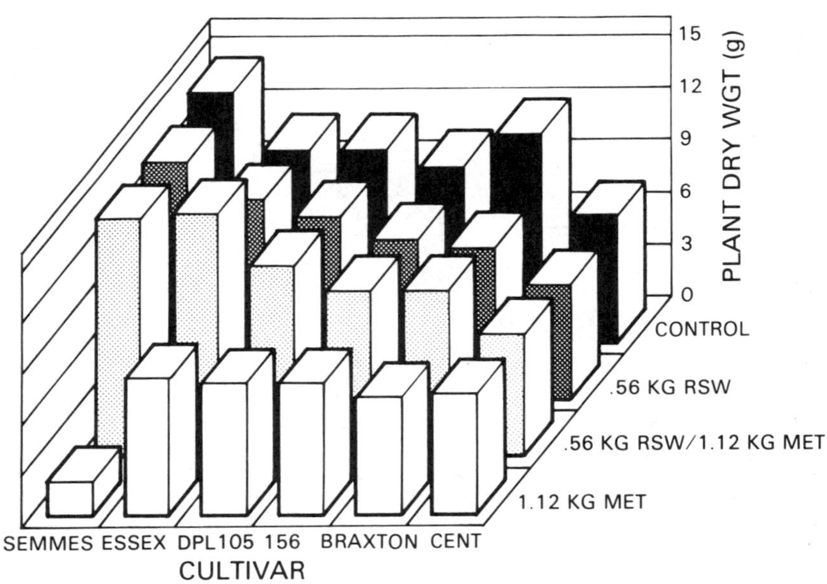

Fig. 5. Effect of triapenthenol plant growth regulator, metribuzin, and triapenthenol plus metribuzin treatment applied to soybean cultivars at planting on 60-day plant dry weight.

others. Metribuzin treatment at 1.12 kg a.i./ha consistently reduced cultivar yields, however. The triapenthenol/metribuzin treatment could not stabilize the yield loss caused by the 1.12 kg/ha metribuzin rate in most cultivars. Planting date-by-herbicide-by-safener interactions were inconsistent and appeared to be date dependent.

Cultivar yield was negatively correlated with injury ($p = 0.01$) and positively correlated ($p = 0.01$) with population and fresh weight/dry weight at 30 days after planting. It appears that triapenthenol may reduce metribuzin-induced necrosis, but it may not alter the overall metribuzin effect on yield reduction at the 1.12 kg a.i./ha rate. Cultivar itself appears to maintain the overriding control on soybean yield with respect to metribuzin susceptibility or tolerance.

B. BAS 140 810/Metribuzin Interactions on Soybeans

BAS 140 810 (Fig. 6) is an experimental piperidinium compound structurally similar to mepiquat chloride. BAS 140 810 possesses a positively charged ammonium moiety and is believed to act as an inhibitor of gibberellin biosynthesis between geranylgeranyl pyrophosphate (GGPP) and *ent*-kaurene (Rademacher and Jung, 1986).

Field experiments similar to those conducted with triapenthenol were carried out with BAS 140 810 at two locations on similar soils in 1984 and 1985 (Vavrina, 1987). Planting dates for the two sites were separated by approximately one month (May 21 and June 24), which provided for different growing regimes. The soybean cultivar used throughout the study was 'Braxton', a metribuzin-tolerant cultivar. The results of the BAS 140 810 trials were very similar to those described for triapenthenol.

BAS 140 810 was applied as a seed treatment at two rates, 0.005 and 0.009 kg a.i./45.35 kg seed with and without metribuzin at 0.56 and 1.12 kg/ha. Data taken during the study included visual injury ratings (as percentage of necrosis) 14, 21, and 28 days after planting, plant population at emergence and 30 and 60 days after planting, plant fresh weight and dry weight based on 1 m of row 30 and 60 days after planting, and yield.

BAS 140 180 caused no visual injury other than height reduction regardless of rate. BAS 140 810 reduced soybean height 30 days after planting, but the height reduction was absent 60 days after planting. BAS 140 810 reduced grain yield

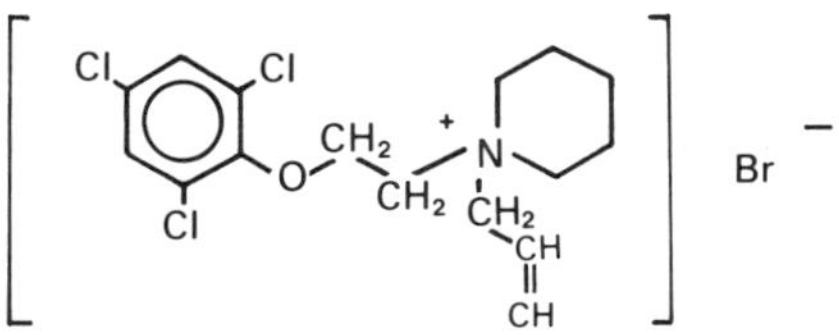

Fig. 6. Chemical structure of BAS 140 810.

slightly. Metribuzin at 1.12 kg/ha caused a high incidence of necrotic injury and reduced plant height, plant weight, and yield.

The BAS 140 810/metribuzin combination reduced necrotic injury, reduced plant height, reduced metribuzin-induced weight loss 30 days after planting, and produced yields commensurate with the controls. These data indicated that BAS 140 810 treatment could curb soybean yield loss associated with higher than normal rates of metribuzin.

C. Mode of Action of Triapenthenol and BAS 140 810 as Safeners Against Metribuzin Injury in Soybeans

A floating leaf disk assay (Truelove *et al.*, 1974) confirmed the safening activity of triapenthenol and BAS 140 810 against metribuzin-induced injury on soybean leaf tissue (Vavrina *et al.*, 1988). Figure 7 depicts the response of soybean leaf tissue treated with the growth retardants and metribuzin at 10^{-6} *M* using the floating leaf disk assay. Both triapenthenol- and BAS 140 810-treated tissues required longer time periods before oxygen evolution ceased and the tissue subsequently sank.

Differing area to weight ratios may be partially involved in this phenomenon. Other differences in metabolism resulting from growth regulator treatments (Vavrina, 1987) may also play a role. For example, the stoichiometry of herbicide molecule to photosynthetic unit may be altered. Plants treated with these retardants exhibited increased chlorophyll contents (Vavrina, 1987). This may result in less damage from the same amount of herbicide.

An additional study was conducted to characterize the uptake, translocation, and metabolite partitioning in [^{14}C]metribuzin-treated soybean seedlings previ-

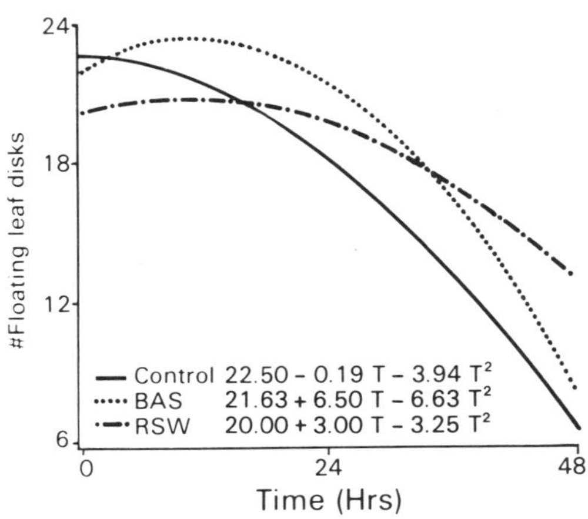

Fig. 7. The effect of plant growth-regulating chemicals applied preemergence on floating leaf disks of soybean trifoliate leaves in a solution of $10^{-6}M$ metribuzin.

ously exposed to the two growth retardants. 'Braxton' soybean was used to determine possible alterations in these physiological processes.

Braxton tolerance to metribuzin was enhanced by treatment with the two growth retardants. Time-course autoradiographs indicated a greater restriction of [^{14}C]metribuzin and metabolites in the stems and veins of soybean treated with the growth retardants (Vavrina *et al.*, 1988). These data confirm the results of Falb and Smith (1984), who showed that herbicide restriction in the stems and veins of metribuzin-tolerant soybeans contributes to the selective action of this herbicide. Frear *et al.* (1985) also indicated that vascular localization of metribuzin played a primary role in influencing soybean tolerance.

The accumulation of ^{14}C in stems of soybean plants treated with triapenthenol and BAS 140 810 generally remained significantly higher throughout the time-course study. The antigibberellin effect of these growth retardants increased the vascular sequestering of metribuzin and the metabolite moieties within 'Braxton' soybean (Vavrina *et al.*, 1988).

The polar metabolite pool (nontoxic metribuzin moieties) of stem and root portions from soybeans treated with the growth retardants was significantly greater than in nontreated plants. The differential rate of formation of polar metabolites has been cited by Falb and Smith (1984) as a major factor contributing to the differential tolerance of selected soybean cultivars to metribuzin. The significance of increased levels of polar metabolites in the stem and root, the major sites of metribuzin detoxication, further contributes to plant growth regulator safening capacity. The nonpolar metabolite pools varied little among treatments. An increase in the nonpolar fraction in stems of soybeans treated with growth retardants may imply continued translocation and suggests a stronger role for vascular localization in reducing metribuzin injury in the foliage.

Analysis of the degradation products of metribuzin by thin-layer chromatography (TLC) revealed changes in metabolite partitioning induced by the two growth retardants (Fig. 8). Plants treated with the two growth retardants produced more conjugated metabolites and less deaminated (DA), diketo (DK), and deaminated-diketo (DADK) moieties than plants treated with metribuzin alone. Frear *et al.* (1985) have shown that the primary metabolites of metribuzin in soybean are a homoglutathione conjugate and an *N*-glucoside conjugate. In our studies, the characterization of the predominate conjugates of metribuzin in soybeans was difficult since metabolite analysis was based only on TLC procedures. However, R_f values of the conjugates coincided with previously published reports of a possible conjugate or "unknown" (Falb and Smith, 1984; Mangeot *et al.*, 1979). Vascular localization would increase the probability of conjugation and the "washing out" phenomenon (Hess, 1985). Nevertheless, the reduction in DA, DK, and DADK production can not be answered at present.

In general, four mechanisms of action have been suggested for chemical safeners (Hatzios, 1983): (1) interference with herbicide uptake and/or translocation, (2) competitive inhibition at the site of herbicide action, (3) stimulation

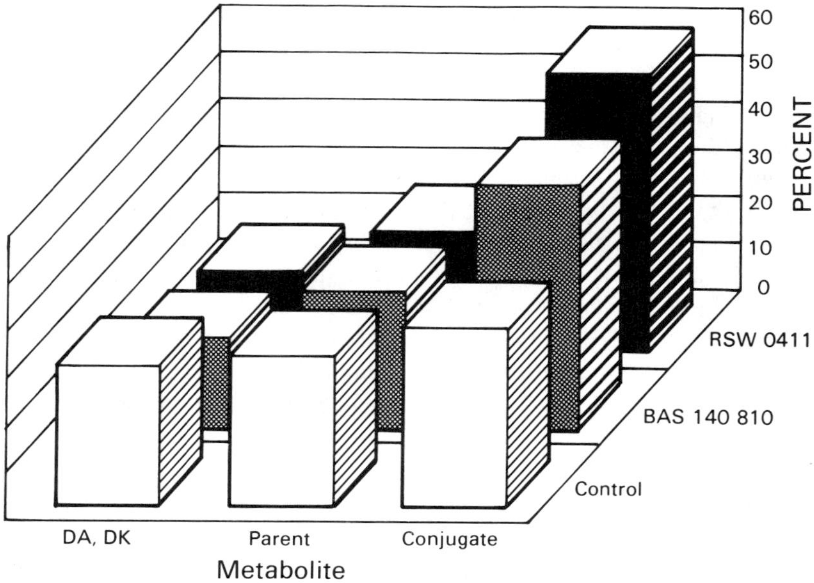

Fig. 8. Metabolite partitioning of [^{14}C]metribuzin in soybean tissue 96 hr after treatment in 21-day-old plants treated preemergence with plant growth-regulating substances.

of herbicide metabolism, and (4) a combination of these mechanisms. The safening action of triapenthenol and BAS 140 810 described here seems to be the result of a combination of mechanisms involving interference with herbicide translocation and stimulation of herbicide metabolism. Tolerance and susceptibility to metribuzin are governed by detoxication and immobilization of the herbicidal moiety (Frear *et al.*, 1983, 1985). The amount of parent compound reaching the chloroplast determines the level of injury (Mangeot *et al.*, 1979).

Apart from the effects of triapenthenol and BAS 140 810 on metribuzin detoxication, the physiological effects of these two growth retardants on soybeans may also be important for their action as safeners against metribuzin. Smaller, thicker leaves formed in response to treatments with these retardants (Vavrina, 1987) may provide increased barriers to herbicide movement. Increased chlorophyll and carotenoid levels observed in retardant-treated plants (Vavrina, 1987) could aid in quenching free radicals generated during photosynthetic electron transport or during herbicide-induced lipid peroxidation. Furthermore, the stoichiometry of herbicide molecule to photosynthetic unit is altered by the increased chlorophyll levels in plants treated with these two growth retardants and may lead to less damage from the same amount of herbicide.

Water use modifications induced by plant growth retardants may influence metribuzin uptake into soybeans (Vavrina, 1987). Reduced water use by soybean seedlings will result in longer residence of metribuzin in the soil/water phase, where microbial and physical factors can contribute to its chemical decomposition. In turn, this process may contribute to greater survival of the soybean seedlings to metribuzin.

VI. CONCLUDING REMARKS

Daminozide, triapenthenol, and BAS 140 810 have protected potatoes and soybeans from injury caused by metribuzin, a photosynthesis-inhibiting herbicide. According to Hatzios (1983), all previous attempts to chemically safen photosynthesis-inhibiting herbicides have failed or have been only partially successful. Our success with the use of growth retardants to protect potatoes (Phatak *et al.*, 1985) and soybeans (Vavrina, 1987) from metribuzin injury warrants further investigations on the safening effects of these and related compounds with urea, uracil, and triazine herbicides on economically important crops.

Plant growth regulation with retardants could protect soybeans against the photosynthesis-inhibiting herbicide metribuzin through: (1) increased vascular localization of the parent herbicide and/or its metabolites, (2) enhanced formation of polar metabolites, (3) altered metabolite pools favoring metribuzin conjugates, (4) modified stoichiometric relationships between herbicide molecules and the photosynthetic unit, and (5) altered water use associated parameters.

Regardless of the economic impact of antagonistic plant growth regulator/herbicide combinations, consideration of plant growth regulator/herbicide combinations in the future may provide us with interesting challenges and leads for the development of new marketable crop safeners for herbicides.

References

Bendixen, L. E. (1975). Cytokinin effects induced in purple nutsedge by perfluidone. *Weed Sci.* **23**, 445–447.

Corbin, F. T., and Sheets, T. J. (1968). Antagonistic effects of certain pesticide combinations on plants. *Abstr. Weed Sci. Soc. Am.* **8**, 22.

Cyanamid International (1966). "Cyocel Plant Growth Regulator," Tech. Inf. Bull. Cyanamid Int., Princeton, New Jersey.

Donald, W. W., and Fawcett, R. S. (1975). Prevention of EPTC injury to hybrid corn with antidote R-29148 and exogenous GA_3. *Proc.—North Cent. Weed Control Conf.* **30**, 28–29.

Donald, W. W., Fawcett, R. S., and Harvey, R. G. (1979). EPTC effects on corn (*Zea mays*) growth and endogenous gibberellins. *Weed Sci.* **27**. 122–127.

Falb, L. N., and Smith, A. E., Jr. (1984). Metribuzin metabolites in soybeans. Characterization of the intraspecific differential tolerance. *J. Agric. Food Chem.* **32**, 1425–1428.

Fletcher, R. A., Mero, V. S. S., and Bhardwaj, S. N. (1984). Reversal of the inhibitor action of fluridone by benzyladenine and potasium in greening cucumber cotyledons. *Abstr. Weed Sci. Soc. Am.* **24**, 93.

Frear, D. S., Mansager, E. R., Swanson, H. R., and Tanaka, F. S. (1983). Metribuzin metabolism in tomato: Isolation and identification of *N*-glucoside conjugates. *Pestic. Biochem. Physiol.* **19**, 270–281.

Frear, D. S., Swanson, H. R., and Mansager, E. R. (1985). Alternate pathways of metribuzin metabolism in soybeans: Formation of *N*-glucoside and homoglutathione conjugates. *Pestic. Biochem. Physiol.* **23**, 56–65.

Grier, R., Fletcher, R. A., and Souza Machado, V. (1985). Triadimenfon, a potential protectant for metribuzin in direct-seeded tomato. *Abstr. Weed Sci. Soc. Am.* **25**, 72.

Hagood, E. S., Williams, J. L., and Bauman, T. J. (1980). Influence of herbicide injury on yield potential of soybean (*Glycine max.*). *Weed Sci.* **28**, 40–45.

Harvey, B. M. R., Chang, F. Y., and Fletcher, R. A. (1975). Relationship between *S*-ethyl dipropylthiocarbamate injury and peroxidase activity in corn seedlings. *Can. J. Bot.* **53**, 225–230.

Hatzios, K. K. (1983). Herbicide antidotes: Development, chemistry and mode of action. *Adv. Agron.* **36**, 265–316.

Hatzios, K. K., and Penner, D. (1985). Interactions of herbicides with other agrochemicals in higher plants. *Rev. Weed Sci.* **1**, 1–63.

Hess, F. D. (1985). Herbicide absorption and translocation and their relationship to plant tolerances and susceptibility. *In* "Weed Physiology" (S. O. Duke, ed.), Vol. 2, pp. 191–214. CRC Press, Boca Raton, Florida.

Hoagland, R. E. (1981). Atrazine and lesan effects on soybean growth and phenolic metabolism. *Proc., South. Weed Sci. Soc.* **34**, 262.

Hoffman, O. L. (1962). Chemical seed treatments as herbicide antidotes. *Weeds* **10**, 322–323.

Hoffman, O. L. (1978). Herbicide antidotes: From concept to practice. *In* "Chemistry and Action of Herbicide Antidotes" (F. M. Pallos and J. E. Casida, eds.), pp. 1–13. Academic Press, New York.

Hoffman, O. L., Gull, P. W., Zeisig, H. C., and Epperly, J. R. (1960). Factors influencing wild oat control with barban. *Proc.—North. Cent. Weed Control Conf.* **17**, 20.

Kirkland, K. (1973). Reduction in atrazine toxicity in winter wheat by growth retardant CCC. *Diss. Abstr.* **13**, 4066-4067.

Kupatt, C. C., and Ilnicki, R. D. (1985). Preemergence pyridazinones reduce atrazine injury to soybeans. *Abstr. Weed Sci. Soc. Am.* **25**, 21.

Ladlie, J. S., Meggitt, W. F., and Penner, D. (1977). Effect of trifluralin and metribuzin combinations on soybean tolerance to metribuzin. *Weed Sci.* **25**, 88–93.

Malefyt, T., and Duke, W. B. (1981). Soybean protection from triazine injury by pendimethalin. *Proc. Annu. Meet. Northeast. Weed Sci. Soc.* **35**, 59.

Mangeot, B. L., Slife, F. E., and Rieck, C. E. (1979). Differential metabolism of metribuzin by two soybean cultivars. *Weed Sci* **27**, 267–269.

Messier, B. B., and Ashley, R. A. (1981). Effect of trifluralin, MH, and daminozide on tomato tolerance to metribuzin. *Proc. Annu. Meet. Northeast. Weed Sci. Soc.* **35**, 163.

Miller, S. D., and Nalewaja, J. D. (1980). Herbicide antidote with triallate. *Agron. J.* **72**, 662–664.

Moomaw, R. S., and Martin, A. R. (1978). Interaction of metribuzin and trifluralin with soil type on soybean (*Glycine max*) growth. *Weed Sci.* **26**, 327–331.

Nash, R. G., and Harris, W. G. (1969). Dexon fungicide antagonism toward herbicidal activity of *s*-triazines. *Abstr. Weed Sci. Soc. Am.* **9**, 240.

Ogawa, M., and Ota, Y. (1976). Effects of 3-hydroxy-5-methylisoxazole on reduction of herbicide injury to rice seedlings. *Proc. Asian-Pac. Weed Sci. Soc. Conf., 5th, 1975*, pp. 303–306.

Ohali, Y., Eshel, Y., Marani, A., and Rubin, B. (1979). Increasing cotton resistance to fluometuron by treating seed with chlormequat. *Phytoparasitica* **7**, 139.

Phatak, S. C. (1974). Effect of herbicide–insecticide combinations on rutabaga emergence. *Can. J. Plant Sci.* **54**, 517-520.

Phatak, S. C., Jaworski, C. A., and Ghate, S. R. (1985). Daminozide protects potato seedlings from metribuzin phytotoxicity. *HortScience* **20**, 690–691.

Phillips, B. A., and Bhagsari, A. S. (1978). Diazosulfonates as protectants of atrazine toxicity to soybeans. *In* "Chemistry and Action of Herbicide Antidotes" (F. M. Pallos and J. E. Casida, eds.), pp. 21–34. Academic Press, New York.

Putnam, A. R., and Penner, D. (1974). Pesticide interactions in higher plants. *Residue Rev.* **50**, 73–110.

Rademacher, W., and Jung, J. (1986). GA biosynthesis inhibitors—An update. *Proc. Plant Growth Regul. Soc. Am.* **13**, 102–114.

Stephenson, G. R., and Ezra, G. (1983). Herbicide antidotes: A new era in selective chemical weed control. *In* "Plant Growth Regulating Chemicals" (L. G. Nickell, ed.), pp. 193–212. CRC Press, Boca Raton, Florida.

Sullivan, E. F. (1979). Herbicide/regulant mixtures for sugarbeets. *Plant Growth Regul. Bull.* **7**, 12–13.

Truelove, B., Davis, D. E., and Jones, L. R. (1974). A new method for detecting photosynthesis inhibitors. *Weed Sci.* **22**, 15–18.

van Overbeek, J. (1964). Survey of mechanisms of herbicide action. *In* "Physiology and Biochemistry of Herbicides" (L. J. Audus, ed.), pp. 387–400. Academic Press, New York.

Vavrina, C. S. (1987). Plant growth regulators as herbicide safeners for metribuzin induced injury to soybeans (*Glycine max*). Ph.D. Dissertation, University of Georgia, Athens.

Vavrina, C. S., and Phatak, S. C. (1985). Use of plant growth regulators to protect soybean [*Glycine max* (L.) Merr.] seedlings from metribuzin injury. *Proc. Plant Growth Regul. Soc. Am.* **12**, 18–36.

Vavrina, C. S., and Phatak, S. C. (1988). Plant growth regulators protect soybean (*Glycine max*) cultivars from herbicide injury. *J. Plant Growth Regul.* **7**, (2), 85–94.

Vavrina, C. S., Phatak, S. C., and Smith, A. E. (1988). Uptake, translocation, and metabolite partitioning of ^{14}C-labeled metribuzin in plant growth regulated soybeans (*Glycine max*). *J. Plant Growth Regul.* **7**, (2), 77–84.

Webster, H. L., and Sheets, T. J. (1969). A physiological explanation of the atrazine–dexon interaction. *Abstr. Weed Sci. Soc. Am.* **9**, 239.

Wilkinson, R. E. (1985). Gibberellin influence on membrane disruption by thiocarbamate herbicides. *Pestic. Biochem. Physiol.* **24**, 103–111.

Wilkinson, R. E., and Ashley, D. (1979). EPTC induced modification of gibberellin biosynthesis. *Weed Sci.* **27**, 270–274.

York, A. C., and Slife, F. W. (1981). Interaction of buthidazole and acetanilide herbicides. *Weed Sci.* **29**, 461–468.

Chapter 14

Herbicide Prosafeners: Chemistry, Safening Activity, and Mode of Action

Baruch Rubin
Department of Field and Vegetable Crops
Faculty of Agriculture
The Hebrew University of Jerusalem
Rehovot 76100, Israel

Osamu Kirino
Pesticide Research Laboratory
Takarazuka Research Center
Sumitomo Chemical Company Ltd.
4-2-1 Takatsukasa, Takarazuka
Hyogo 665, Japan

I. INTRODUCTION

The site of action of pesticides is generally different from the site of their application. Generally a pesticide penetrates the cuticle and is translocated to the site(s) of action in the organism before it can exert its pesticidal action. During such transportation processes, the chemical structure of the pesticide is often altered by chemical and/or biological reactions. Therefore, many pesticides appear to be propesticides which need to be activated to their ultimate toxicant form in the environment and/or in the organism (Casida, 1983). Activation processes may involve hydrolysis, oxidation, reduction, and rearrangement reactions (Casida and Ruzo, 1986).

Pesticides must be sufficiently stable and soluble for convenient manufacturing and storage and formulated efficiently for optimal use. Introduction of chemically or biologically degradable groups into a toxicant changes its solubility and partitioning properties. Degradable groups also allow the stabilization of a reactive, photolabile, or volatile toxicant. Furthermore, propesticides may be used to improve selective toxicity by exploiting species differences in the balance of activation and detoxication reactions. For example, the phenoxybutanoic acid herbicides 2,4-DB and MCPB are activated to their corresponding phenoxyacetic acids by β-oxidation in susceptible plants. Since the activity of the oxidation system is different in various plant species, the phenoxybutanoic acid proherbicides exhibit higher selectivity than the ultimate herbicides, phenoxyacetic acids (Loos, 1975; Wain, 1958).

Herbicide safeners or antidotes are compounds which selectively reduce or eliminate plant injury caused by herbicides, thus improving crop selectivity. All herbicide safeners that are presently marketed can be applied either as seed dressings or as tank mixtures with herbicides to avoid any extra operation. To be effective in the field, herbicide safeners applied as prepackaged tank mixtures with herbicides, should not be more soluble than the herbicides to avoid leaching of the safener from the zone of protection during heavy rainfall or irrigation. Such herbicide safeners should also remain in the protection zone as long as the crop is susceptible to the respective herbicides (Buzio and Burt, 1980; Donald and Fawcett, 1976).

Plant thiols including glutathione (GSH) are involved in both herbicide detoxication and the action of herbicide safeners (Corbett *et al.*, 1984; Fedtke, 1981, 1982; Hatzios, 1983; Hatzios and Penner, 1982). The thiocarbamate herbicide EPTC is activated by mixed function oxidases to its sulfoxide and sulfone and the activated derivatives are efficient carbamoylating agents for GSH in corn (*Zea mays* L.) (Casida *et al.*, 1974; Horváth and Pulay, 1980; Hubbell and Casida, 1977). Metribuzin is metabolized in part by S-demethylation involving an intermediate sulfoxidation step which predisposes this herbicide for conjugation with GSH in mouse liver (Bleeke and Casida, 1984; Bleeke *et al.*, 1985). The herbicide safener dichlormid enhances sulfate metabolism and elevates the GSH content in several plant systems (Adams *et al.*, 1983; Ezra and Gressel, 1982; Lay and Casida, 1976; Lay *et al.*, 1975; Rennenberg *et al.*, 1982). Similar effects have been recently reported for the new safener MG-191 (2-dichloromethyl-2-methyl-1,3-dioxolane) (Dutka and Komives, 1987). Dichlormid also increases the GSH-dependent metabolism of alachlor in corn (Leavitt and Penner, 1979; Mozer *et al.*, 1983) and of EPTC via its sulfoxide (Lay and Casida, 1976). The protective action of dichlormid extends to several other herbicides, regardless of whether they are detoxified by conjugation with GSH in plants or not (Hatzios, 1983; Pallos and Casida, 1978; Rubin and Casida, 1985; Sweetser, 1985).

OTC (L-2-oxothiazolidine-4-carboxylic acid) was reported to increase GSH levels in mouse liver by enzymatic degradation to *S*-carboxy-L-cysteine which undergoes spontaneous decarboxylation to produce L-cysteine. This L-cysteine serves as a source for GSH biosynthesis (Williamson *et al.*, 1982). Hilton and Pillai (1986) have found that OTC circumvented, at least partially, the growth inhibition of wheat (*Triticum aestivum* L.), grain sorghum [*Sorghum bicolor* (L.) Moench] and corn caused by tridiphane. OTC also partially circumvented the tridiphane-induced reduction in the extractable amount of GSH and glutathione-*S*-transferase activity in treated corn roots.

Some of the herbicides which react with GSH may exert their herbicidal activity through analogous derivatization of critical enzymes or membrane thiols. Thus, chemicals which elevate or deplete GSH and other thiols in plants may serve as herbicide safeners or herbicides, respectively, and chemicals which require activation processes to exert their activity may serve as herbicide prosafeners or proherbicides. The scope of this chapter is to review the chemistry, safening properties, and mode of action of *N*-phenylmaleamic acids and their progenitors which act as prosafeners protecting sorghum against chloroacetanilide herbicides.

II. *N*-PHENYLMALEAMIC ACIDS AND THEIR PROGENITORS AS HERBICIDE SAFENERS

Maleimides like *N*-ethylmaleimide (NEM) and *N*-(4-chlorophenyl)maleimide (CPMI) are known thiol reagents (Adams *et al.*, 1983; Augustin *et al.*, 1978) and related N-aryl cyclic imides possess herbicidal (Ohta *et al.*, 1976) or fungicidal (Fujinami *et al.*, 1972) activity. Alachlor and some other chloroacetanilides provide selective control of grasses and certain broad-leaved weeds in soybean [*Glycine max* (L.) Merr.] and corn. Their selective action is extended to sorghum with suitable herbicide safeners, including flurazole (Brinker *et al.*, 1982; Schafer *et al.*, 1980, 1981) and appropriate oxime ether derivatives such as cyometrinil and oxabetrinil (CGA-92194) (Hatzios, 1984; Ketchersid and Merkle, 1981, 1982; Rufener *et al.*, 1982).

Alachlor and metolachlor are detoxified in part by GSH conjugation which is facilitated by the action of several safeners (Fuerst and Gronwald, 1986; Hatzios and Penner, 1982; Leavitt and Penner, 1979; Mozer *et al.*, 1983; Stephenson and Ezra, 1982; Stephenson *et al.*, 1983; Zama and Hatzios, 1986). Compounds that alter GSH levels may therefore influence the phytotoxicity of herbicides or be herbicidal themselves.

A series of *N*-phenylmaleimides and related isomaleimides and maleamic acids were synthesized and screened for phytotoxicity and herbicide safening activity. The structure of representative compounds from these chemical classes such as CPMI, CPIM [*N*-(4-chlorophenyl)isomaleimide] and CPMA [*N*-(4-chlorophenyl)maleamic acid] are illustrated in Fig. 1. It was found that the *N*-phenylmaleimides and *N*-phenylisomaleimides are highly selective herbicide safeners rather than herbicides, protecting sorghum from alachlor injury (Kirino *et al.*, 1985; Rubin *et al.*, 1985).

The *N*-phenylmaleimides and *N*-phenylisomaleimides were hydrolyzed rapidly to their corresponding *N*-phenylmaleamic acids. Therefore *N*-phenylmaleimides

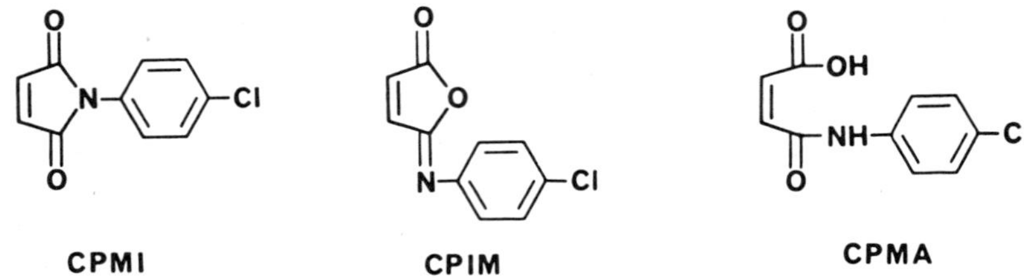

Fig. 1. Three types of herbicide safeners, *N*-(4-chlorophenyl)maleimide (CPMI), *N*-(4-chlorophenyl)isomaleimide (CPIM), and *N*-(4-chlorophenyl)maleamic acid (CPMA).

and *N*-phenylisomaleimides appear to be herbicide prosafeners and *N*-phenylmaleamic acids are the actual herbicide safeners.

A. Chemistry

The chemistry of cyclic carboxylic monoimides is well established and has been briefly reviewed (Hargreaves *et al.*, 1970). In general, cyclic carboxylic monoimides are synthesized by the reaction of dicarboxylic acids with amines. The most important step is cyclization and this may be achieved by heating to a temperature of not more than 200°C. It has been suggested that in some cases the reaction of the acids may proceed via the formation of the acid anhydride. The reaction of maleic anhydride with amines in an inert solvent gives maleamic acids in quantitative yield (Fig. 2). The maleimides are then synthesized by

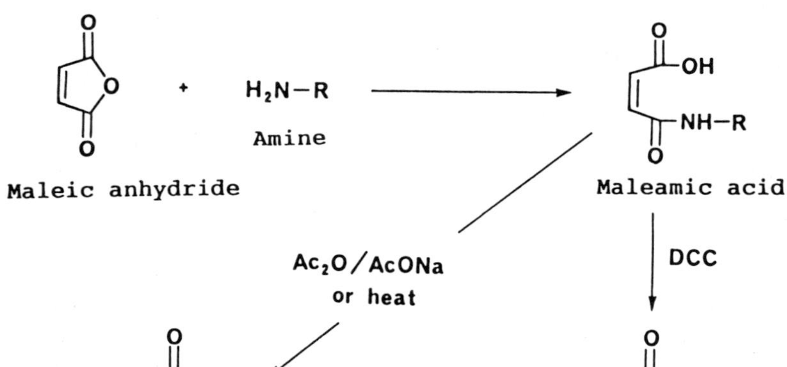

Fig. 2. Synthesis of maleamic acid, maleimide, and isomaleimide. DCC = dicyclohexylcarbodiimide and Ac_2O/AcONa = acetic anhydride/sodium acetate mixture.

dehydration of the maleamic acids with acetic anhydride and sodium acetate or by direct heating.

The isomaleimides requiring a nucleophilic displacement by a neighboring amide group on an activated carboxyl function are obtained by dehydration of the maleamic acids with dicyclohexylcarbodiimide (DCC). The isomaleimides are generally unstable with respect to the corresponding maleimides (Cotter *et al.*, 1961; Roderick, 1957; Sauers, 1969). It has been suggested that the isomaleimides are the primary products of the dehydration reactions using an acetic anhydride–sodium acetate mixture at moderate temperatures and that the maleimides are formed by the rearrangement of the isomaleimides. Fusion or prolonged heating in an inert solvent converts isomaleimides to their maleimide isomers.

Hydrolysis of maleimides and isomaleimides with hot 1 *N* sodium hydroxide for about 15 min gives the same product, maleamic acids (Roderick, 1957). The alkaline saponification of maleimides as well as isomaleimides is a simple and quantitative process comprising the attack of hydroxide ion on the carbonyl carbon (Fig. 3). The substitution of the N-hydrogen by phenyl group in maleimide causes a striking increase in the rate of ring opening by the hydroxide ion. This occurs because the electron-withdrawing effect of the phenyl group makes the imide ring less stable because of a conjugation of the lone-pair electrons of nitrogen with the phenyl ring. Both maleimides and isomaleimides are also hydrolyzed by acids. The reaction mechanism involves nucleophilic attack of water on the carbonyl carbon of the unstable conjugated acid of the imide (Fig. 3).

B. Safening Activity

In laboratory experiments, CPMI did not injure sorghum or barnyard grass [*Echinochloa crus-galli* (L.) Beauv.] and selectively protected sorghum from

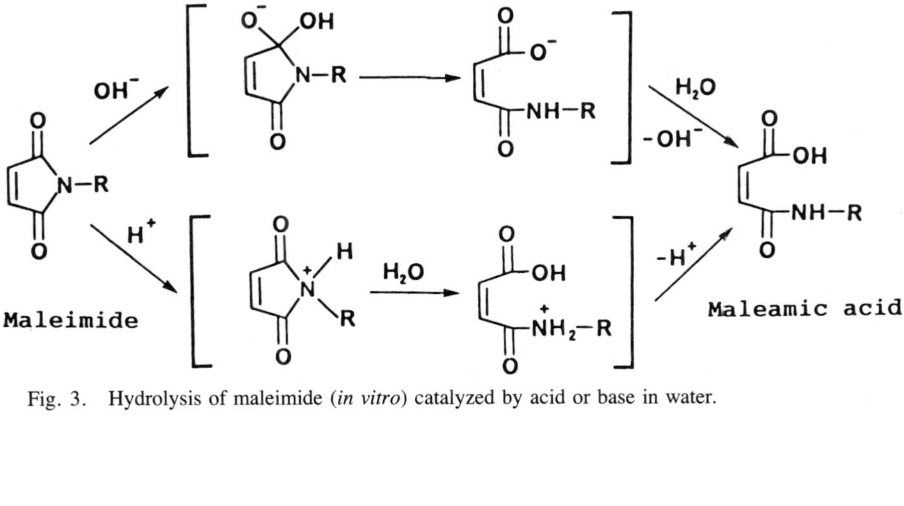

Fig. 3. Hydrolysis of maleimide (*in vitro*) catalyzed by acid or base in water.

alachlor injury without reducing its herbicidal activity against barnyard grass (Table I). Other *N*-(4-chlorophenyl)imides and -amides were also active as herbicides or herbicide safeners. The compounds 3,4,5,6-tetrahydrophthalimide, 3,4-dimethylmaleimide, and acetamide were phytotoxic and none of them had useful herbicide safening activity.

Substitution of the maleimide ring with one or two methyl groups or a tetramethylene moiety resulted in compounds that are herbicides (see also Ohta *et al.*, 1976) rather than herbicide safeners, indicating the importance of the hydrophobicity around the imide ring for the herbicidal activity.

This study established that the unsubstituted maleimide ring is favorable but not essential for herbicide safening activity. It also indicated the importance of an interaction between the unsaturated bond and a specific site critical for the herbicide safening effect. In contrast, the succinimide is not phytotoxic and shows selective herbicide safening activity, although it is not as effective as CPMI.

Crotonamide, phthalamide, 3-methylmaleimide, and *N*-alkylmaleimides including NEM are of much less interest as phytotoxicants or herbicide safeners. The *N*-alkylmaleimides were neither phytotoxic nor alachlor safeners, suggesting

Table I Effect of Preemergence Application of Maleimides and Related Compounds and Alachlor on Growth of Sorghum and Barnyard Grass[a]

	Plant injury rating[b]			
	Alachlor (ppm)			
	0	1	0	1
Safener (100 ppm)	Sorghum		Barnyard grass	
Controls				
None	0	3	0	3
Flurazole	0	0	0	2
N-(4-Chlorophenyl) derivatives				
Maleimide (CPMI)	0	0	0	3
3-Methylmaleimide	1	3	1	3
3,4-Dimethylmaleimide	2	3	1	3
Tetrahydrophthalimide	3	3	3	3
Phthalimide	1	2	1	3
Succinimide	0	1	0	3
Acetamide	1	3	2	3
Crotonamide	1	2	1	3
N-Alkylmaleimide[c]	0	3	0	3

[a]From Rubin *et al.* (1985), with permission.

[b]Rating: graded response from 0 = no or slight injury to 3 = severe injury as determined 8 days after sowing.

[c]*N*-alkylmaleimides = R-N-C(O)-CR_1 = CR_2-C(O). The analogs examined were (R,R_1,R_2) = Me,H,H; Et,H,H; Et,Me,H; Et,Me,Me; *n*-Pr,H,H; *i*-Pr,H,H; *i*-Bu,H,H. The second, third, and seventh analogs were slightly phytotoxic to sorghum (rated 1).

that they may be too unstable. The herbicide safener flurazole, used as a reference standard, slightly retarded sorghum growth at 100 ppm and offered complete protection against alachlor injury. However, in contrast to CPMI, it also slightly protected barnyard grass from the herbicidal action (Table I).

C. Effect of Phenyl Substituents

The three superior herbicide safeners in each of the *N*-phenylmaleimide (MI), *N*-phenylisomaleimide (IM), and *N*-phenylmaleamic acid (MA) series have a 4-Cl, 4-Me, or 4-F substituent (Table II). In general, substitution at the 4-position provided a greater herbicide safening activity than substitution of the 2- or 3-position. Disubstitution at 2,4-, 3,4-, or 2,6-positions reduced the safening activity (Rubin *et al.*, 1985). Therefore the effect of phenyl substituents at the 4-position was analyzed quantitatively by using the adaptive least-squares (ALS) method (Moriguchi and Komatsu, 1981) because the activity is represented in a form of activity classes.

According to the ALS method, the discriminant score for classification, S can be represented by

Table II Effect of Phenyl Substituents at the 4-Position of *N*-Phenylmaleimides (MI), *N*-Phenylisomaleimides (IM), and *N*-Phenylmaleamic Acids (MA) at 100 ppm as Herbicide Safeners for Alachlor Injury at 1 ppm in Sorghum

		Herbicide safening activity[a]					
		Safener type					
		MI		IM		MA	
4-Phenyl substituent	L[b]	Obsd.	Calc.[c]	Obsd.	Calc.[d]	Obsd.	Calc.[e]
H	2.06	2	2	2	2	2	3
F	2.65	3	3	2	2	3	3
Me	3.00	3	3	3	3	3	3
NO_2	3.44	2	3	2	3	2	2
Cl	3.52	3	3	3	3	3	2
OMe	3.98	3	2	2	2	2	2
COMe	4.06	2	2	1	2	1	1
Et	4.11	1	2	2	2	1	1
CN	4.23	1	2	1	2	1	1
OEt	4.92	1	1	2	2	1	1
n-Bu	6.17	1	1	1	1	1	1

[a]Discriminant scores: 3 = excellent herbicide safening activity, 2 = moderate, and 1 = poor.
[b] Sterimol length substituent constant.
[c] Determined by Eq. (2).
[d] Determined by Eq. (3).
[e] Determined by Eq. (4).

$$S = W_0\,\mathrm{hr} + W_1X_1 + W_2X_2 + \ldots + W_pX_p \quad (1)$$

where

$$X_{\mathrm{k}} \quad (k = 1, 2, \ldots, p)$$

is the *k*th descriptor for the structure and

$$W_{\mathrm{k}} \quad (k = 1, 2, \ldots, p)$$

is the weight coefficient. In this case, $S = 3$ means excellent herbicide safening activity, $2 =$ moderate, and $1 =$ poor.

The accuracy of the correlation is confirmed by the leave-one-out prediction. In preliminary analyses, the length of the substituents was found to be important for variations in safening activity against alachlor. The best discriminant functions formulated from 30 times of iterative calculations are Eqs. (2), (3), and (4) for MIs, IMs, and MAs, respectively.

MIs:

$$S = -0.277L^2 + 1.591L + 0.370 \quad (2)$$
$$n = 11 \text{ (three classes)}, R_s = 0.766$$
$$n_{\mathrm{mis}} = 4\ (0),\ n_{\mathrm{mis}} \text{ (after leave-one-out)} = 7\ (0)$$

IMs:

$$S = -0.263L^2 + 1.690\mathrm{L} - 0.182 \quad (3)$$
$$n = 11 \text{ (three classes)}, R_s = 0.789$$
$$n_{\mathrm{mis}} = 3\ (0),\ n_{\mathrm{mis}} \text{ (after leave-one-out)} = 8\ (0)$$

MAs:

$$S = -0.944L + 5.332 \quad (4)$$
$$n = 11 \text{ (three classes)}, R_s = 0.915$$
$$n_{\mathrm{mis}} = 2\ (0),\ n_{\mathrm{mis}} \text{ (after leave-one-out)} = 5\ (0)$$

In these equations, n is the number of compounds included in the correlation, R_s is the Spearman's rank correlation coefficient (see Morigushi and Kumatsu, 1981), and n_{mis} is the number of misclassified compounds. The figure in parentheses after the value of n_{mis} is the number of compounds misclassified into the next class but one. L is the Sterimol length substituent constant which expresses the length of the substituent (in Å) along the axis connecting the aromatic carbon where the substituents are located with the α-atom of substituents. The Sterimol parameters are a set of values representing the steric shape of substituents (Verloop *et al.*, 1976).

According to Eqs. (2) and (3), the herbicide safening activity of MIs and IMs is related parabolically to the length of the substituents and reaches a maximum when L is 2.87 Å for MIs and L is 3.21 Å for IMs. The L^2 term is lacking in Eq. (4) for MAs and the negative sign of the L term means that the shorter the substituent the higher the activity. Considering the fact that the optimal length of the substituents for MIs and IMs is relatively small, the structural requirements may be similar in the three MI, IM, and MA series. The significance of the short length of substituents indicates that the steric interaction between the substituents and the specific site critical for the activity is important in determining variations in safening activity.

When compared on a quantitative basis, the superior herbicide safeners (i.e., 4-Cl, 4-Me, and 4-F substituents) are of similar activity in MI, IM, and MA series (Table III). The safening effectiveness is considerably lower at 10 than at 50 ppm in contrast to flurazole and dichlormid. However, flurazole even at 10 ppm inhibited sorghum growth. The trans isomer of 4-chlorophenylmaleamic

Table III Effect of Preemergence Application of *N*-Phenyl-maleimides, -isomaleimides, -maleamic Acids, and Related Compounds as Safeners for Alachlor Injury on Sorghum[a]

	Shoot fresh weight (% of control)[b]		
		Alachlor at 3 ppm with safener at indicated ppm	
Candidate safener	Safener alone at 50 ppm	10	50
None (control)	100	20	20
Maleimides			
4-Cl	99	55	82
4-Me	84	38	75
4-F	94	42	72
Isomaleimides			
4-Cl	99	50	77
4-Me	99	48	61
4-F	82	43	71
Maleamic acids			
4-Cl	91	49	77
4-Me	98[c]	43	60
4-F	106	50[c]	81
4-Cl fumaramic acid	100	38	41[c]
4-Cl succinamic acid	110[c]	37	50
Dichlormid	85 (98)[d]	65	79
Flurazole	73 (75)[d]	68	74

[a]From Rubin *et al.* (1985), with permission.

[b]Untreated control: 72 mg/plant 10 days after sowing. Similar relationships are evident with plant height as the criterion.

[c]SE values of 13-18%. In all other cases SE values of up to 12%.

[d]Safener alone at 10 ppm.

acid (fumaramic acid) and the saturated derivative (succinamic acid) possess much lower herbicide safening activity than the corresponding MA (Table III).

D. Hydrolytic Stability

Maleimides and isomaleimides are spontaneously hydrolyzed in aqueous solutions as noted in Section II, A. CPMI and the corresponding isomaleimide (CPIM) were confirmed to be hydrolyzed rapidly to the maleamic acid (CPMA) in 0.2 *M* phosphate buffer, pH 7.4, at 25°C (Fig. 4) with a half-life of about 2 hr. Similar results were obtained at pH 6.8. CPMA was stable under these conditions and no hydrolysis was observed over a period of 15 days. Since a week or longer incubation period is required for the bioassay of the herbicide safening effect, maleimides and isomaleimides must be hydrolyzed to the maleamic acids during the test period. It therefore appears that maleimides and isomaleimides may be prosafeners undergoing activation by hydrolytic conversion to the actual herbicide safeners, maleamic acids. The similarity in structural requirements for phenyl substituents in MI, IM, and MA series supports our suggestion that a conversion of the maleimides and isomaleimides to the maleamic acids is involved in the mechanism of herbicide safening action.

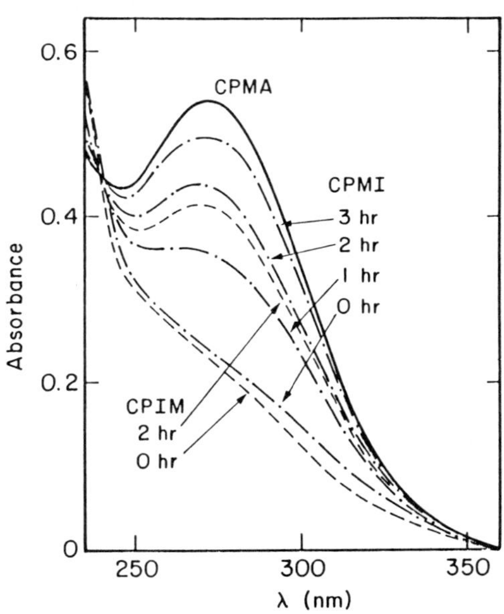

Fig. 4. Ultraviolet spectra illustrating hydrolytic conversion of a representative maleimide (CPMI) and isomaleimide (CPIM) to the corresponding maleamic acid (CPMA) [Reprinted with permission from Rubin *et al.* (1985). Copyright 1985 American Chemical Society.]

III. SAFENING PROPERTIES OF *N*-PHENYLMALEAMIC ACIDS AND THEIR PROGENITORS

Several environmental factors such as temperature, light, soil moisture, and soil type have been reported to influence the efficacy of herbicide safeners (Hatzios, 1983). In addition, the timing of the herbicide safener and herbicide applications is important (Donald and Fawcett, 1976). The varietal differences of plants may also change the efficacy of herbicide safeners (Lay and Niland, 1985; Rubin and Casida, 1985; Sagaral and Foy, 1982; Schafer *et al.*, 1980).

A. Effect of Application Time

Alachlor under the conditions used was more phytotoxic to sorghum seedlings when applied at 2 rather than at 0 days after sowing (Table IV), possibly because of a greater root and/or coleoptile exposure to the herbicide. Applied at 7 days after sowing, alachlor caused no crop injury. The activity of CPMI was dependent on proper timing, since simultaneous treatment of CPMI at 100 ppm with alachlor at 2 ppm immediately after sowing showed the highest protection, as compared to delayed application of any combination of the two compounds. Timing was less critical with the more potent and stable flurazole, which at 10 ppm provided good and consistent protection at all application times. The herbicide safening effect of dichlormid at 10 ppm was similar to that of CPMI at 100 ppm. All herbicide safeners at any application time showed little or no phytotoxicity.

Table IV Effect of Application Time on the Action of Three Safeners for Alachlor Injury to Sorghum[a]

	Shoot fresh weight (% of control)[b]				
			Safener with alachlor at 2 ppm on indicated day after sowing		
Safener (ppm)	**Treatment day**	**Safener alone**	**0**	**2**	**4**
None (control)		100	53	29	58
CPMI (100)	0	96	85	61[c]	70[c]
	2	97		53	70
	4	92			80[c]
Flurazole (10)	0	86	81	77	77
	2	91[c]		82	83
	4	90			84
Dichlormid (10)	0	99	93	40	62
	2	100		47	74
	4	101			61

[a] From Rubin *et al.* (1985), with permission.

[b] Untreated control: 80- mg/plant 14 days after sowing.

[c] SE values of 14–16%. In all other cases SE values of up to 12%.

B. Safening Spectrum

The herbicide safening effectiveness of CPMI, flurazole, and dichlormid against alachlor, EPTC, and chlorsulfuron injury was compared on six major crops (Table V). The herbicide safeners alone were not phytotoxic except for CPMI on soybeans. Alachlor at 2 ppm severely damaged sorghum, rice (*Oryza sativa* L.), and wheat with lesser injury to cotton (*Gossypium hirsutum* L.). Protection against alachlor injury was provided by all three herbicide safeners only in sorghum and by CPMI and dichlormid in wheat. EPTC at 10 ppm caused severe injury to all crops but corn, and the use of herbicide safeners did not alleviate its phytotoxicity in sensitive crops. Chlorsulfuron at 20 ppb caused significant injury to soybean, cotton, and rice, but not to wheat. This herbicide did not greatly decrease the fresh weight of sorghum and corn shoots but almost completely inhibited their root elongation. Dichlormid offered some protection to corn against chlorsulfuron injury in which the root growth inhibition was considerably alleviated. In this study, none of the herbicide safeners at 10 or 100 ppm was

Table V Spectrum of Safening Action with Various Crops and Herbicides Applied Preemergence[a]

Herbicide and safener (ppm)	Shoot fresh weight (% of control)[b]					
	Sorghum	**Corn**	**Rice**	**Wheat**	**Cotton**	**Soybean**
No herbicide						
CPMI (100)	97[c]	86[c]	89	108	95	58
Flurazole (10)	86[c]	103	90	92	96	97
Dichlormid (10)	94[c]	97	93	95	99	100
Alachlor (2 ppm)						
None	41[c]	107	0	45	76	91
CPMI (100)	80	76	0	67[c]	63	52
Flurazole (10)	78	99	14	48	77	91
Dichlormid (10)	73	92[c]	0	75	72	82
EPTC (10 ppm)						
None	0	101	0	0	0	46
CPMI (100)	0	81	0	0	0	38
Flurazole (10)	0	97	0	0	10	52
Dichlormid (10)	0	102	0	0	0	44
Chlorsulfuron (20ppb)						
None	102[d]	81[e]	61	94	65	41
CPMI (100)	87[d]	66[e]	77	108[c]	67	17
Flurazole (10)	100[c]	84[c,d]	86	100	57	37
Dichlormid (10)	94	91[d]	79	106[c]	56	35

[a] From Rubin *et al.* (1985), with permission.
[b] Untreated controls (mg/plant): sorghum 65, corn 457, rice 35, wheat 114, cotton 435, and soybean 456.
[c] SE values of 13–19%. In all other cases SE values of up to 12%.
[d] Slight inhibition of root growth.
[e] Severe inhibition of root growth.

phytotoxic to barnyard grass, johnsongrass [*Sorghum halepense* (L.) Pers.], or green foxtail [*Setaria viridis* (L.) Beauv.], whereas alachlor at 2 ppm completely killed these weeds. CPMI and dichlormid did not reduce the herbicidal activity of alachlor but flurazole completely protected johnsongrass from this herbicide.

The safening activity of flurazole and dichlormid seems to be more general and less selective (Hatzios, 1983, 1984; Schafer *et al.*, 1980; Stephenson and Ezra, 1982) than that of the maleimide CPMI, relative to both the herbicide and the plant. CPMI acts only with alachlor in sorghum and less so in wheat. This may be an advantage for CPMI since the action of flurazole extends to protecting the weed johnsongrass from alachlor (Schafer *et al.*, 1981).

IV. MODE OF SAFENING ACTION OF *N*-PHENYLMALEAMIC ACIDS

A. Effect on Herbicide Uptake, Distribution, and Metabolism

The exact mechanisms of currently available herbicide safeners are not fully understood. There are some crop protectants that work outside the plant by interfering mainly with herbicide absorption. The most successful example is activated carbon which exhibits a great absorptive capacity and acts as a physical barrier to herbicide uptake (Blair *et al.*, 1976; Clapp, 1974, 1975). However, the majority of chemical herbicide safeners counteract herbicide injury by acting inside the plant. The mode of action of herbicide safeners is considered to involve several mechanisms (see Chapter 3, this volume, for an overview). The following mechanisms were examined as potential explanations of the safening action of *N*-phenylmaleamic acids:

1. Alteration of Alachlor Uptake and/or Translocation in Safened Sorghum Plants

Reduction in herbicide uptake due to a safener treatment has been proposed for naphthalic anhydride (NA) in sorghum (Ahrens and Davis, 1978), for dichlormid in corn cell culture (Ezra *et al.*, 1982), and for cyometrinil in sorghum (Ebert, 1982; Ketchersid *et al.*, 1982). However, other studies have shown that the safeners rather increased herbicide uptake and translocation (Ezra *et al.*, 1982; Fuerst and Gronwald, 1986; Zama and Hatzios, 1986).

Using [^{14}C]CPMA uniformly labeled at the phenyl ring and autoradiography, we found that CPMA applied preemergence accumulated mainly in the roots and lower shoots of both 4- and 10-day-old sorghum plants, with slight translocation to the shoot and some accumulation in the coleoptile and first node (Rubin *et al.*, 1985).

[^{14}C]Alachlor (methoxy labeled) applied preemergence was readily absorbed by sorghum seedlings. Immediately following emergence (4 days after sowing), the ^{14}C was mainly found in the seeds and roots and CPMA or flurazole significantly increased alachlor uptake without altering its distribution patterns (Table VI). Seven days after treatment the total amount absorbed increased by two- to fourfold. The only safener-related differences found were an increased alachlor uptake with flurazole and reduced alachlor levels in the shoot with both herbicide safeners. Autoradiography revealed that at 7 days the ^{14}C is distributed throughout the plant. These data rule out the possible involvement of differential uptake and translocation of alachlor in the safening mechanism of CPMA and flurazole.

2. Counteraction of Herbicide Phytotoxicity through a Competitive Inhibition

Most of the herbicide safeners have been shown to counteract the thiocarbamate and chloroacetanilide inhibition of plant lipid metabolism (see Hatzios, 1983, for comprehensive review). Since no attempt was made by the authors to clarify this mechanism regarding CPMA, we cannot rule it out as a possible mechanism. However, Warmund *et al.* (1985) have found only slight effects of flurazole on total lipid synthesis and fatty acid composition in germinating grain sorghum.

3. Stimulation of Herbicide Metabolic Detoxication by the Protected Plant

Enhancement of the degradation of thiocarbamate and chloroacetanilide herbicides as a result of treatment with various safeners has been proposed in corn and grain sorghum (Carringer *et al.*, 1978a,b; Dutka and Komives, 1987; Fuerst and Gronwald, 1986; Horvath and Polay, 1980; Hubbel and Casida, 1977; Lay and Casida, 1976; Lay and Niland, 1985; Lay *et al.*, 1975; Leavitt and Penner, 1979; Zama and Hatzios, 1986). Similar results have been reported for the herbicide

Table VI Effect of Safeners on Uptake and Distribution of Radioactivity from [^{14}C]Alachlor in Sorghum Plants[a]

		^{14}C distribution (% of total)		
Safener	**Uptake**[b]	**Root**	**Seed**	**Shoot**
		4 days after treatment		
None	100 ± 20[c]	47 ± 10	41 ± 8	12 ± 5
CPMA	218 ± 38	51 ± 7	37 ± 7	12 ± 9
Flurazole	209 ± 15	49 ± 8	39 ± 3	12 ± 6
		7 days after treatment		
None	100 ± 11[c]	73 ± 8	15 ± 5	12 ± 2
CPMA	96 ± 12	80 ± 7	12 ± 4	8 ± 4
Flurazole	145 ± 9	76 ± 1	17 ± 4	7 ± 2

[a]From Rubin *et al.* (1985), with permission.
[b]6110 dpm/plant at 4 days and 26340 dpm/plant at 7 days.
[c]SE values.

chlorsulfuron (Sweetser, 1985). In our studies (Kirino *et al.*, 1985; Rubin *et al.*, 1985), [^{14}C]alachlor metabolism was not affected by herbicide safeners. Analyses of sorghum seedlings, 7 days after treatment, have shown that ethanol extraction recovered 75–80% of the total ^{14}C in both roots and shoots with no differences between alachlor alone and alachlor plus CPMA treatments. No parent compound was found in the roots or shoot extracts following separation on thin layer chromatography (TLC). All metabolites were confined to the origin (R_f 0.00) versus alachlor at R_f 0.68 in a benzene–methanol (19:1, v/v) solvent system. TLC analysis with *n*-butanol–acetic acid–water (6:3:2, v/v/v) revealed four major root metabolites (R_f's of 0.34, 0.45, 0.48, and 0.63; alachlor R_f was 0.95) and two minor metabolites (R_f's of 0.4 and 0.6). Neither qualitative nor quantitative differences were found between the alachlor alone and alachlor plus CPMA treatments. Shoot extracts had three major products (R_f of 0.33, 0.50, and 0.61) in the latter solvent system and an additional major spot in the solvent front which contained large amounts of extracted chlorophyll. The lack of safener induced differences in alachlor metabolism was also evident with shoot extracts. Thus, we concluded that CPMA did not exert its safening activity by altering alachlor degradation in sorghum.

B. Effect on Glutathione Content and Enzyme Activity

It is well established that herbicide safeners such as dichlormid, OTC, and MG-191 and several herbicides elevate GSH content in treated plants (Adams *et al.*, 1983; Carringer *et al.*, 1978b; Dutka and Komives, 1987; Ezra and Gressel, 1982; Hilton and Pillai, 1986; Lay and Casida, 1976; Lay and Niland, 1985; Rennenberg *et al.*, 1982; Rubin and Casida, 1985; Stephenson *et al.*, 1983). Several reports have shown that in treated plants, herbicide safeners caused an increase in enzyme activity, e.g., glutathione-*S*-transferase (Lay and Casida, 1976; Lay and Niland, 1985; Mozer *et al.*, 1983), mixed function oxidase (Hatzios, 1983), and acetohydroxyacid synthase (Rubin and Casida, 1985).

1. Effect on Glutathione Content

Under *in vitro* conditions, the maleimides NEM and CPMI reacted rapidly with GSH in 0.2 *M* phosphate buffer, pH 7.4, at 5°C and depleted the GSH levels within 15 min, whereas CPMA, which does not have a maleimide ring, and alachlor did not. Thus, the unreacted GSH was 100% in the control, 24 ± 4% for NEM, 36 ± 5% for CPMI, and 93 ± 6% for both CPMA and alachlor. Similar results were found when the experiments were conducted at ratios of 10:1 and 100:1 of thiol reagent:GSH. However, when applied preemergence, alachlor gave 15% and the herbicide safeners (CPMI, CPMA, flurazole and dichlormid) gave 60–130% increase in the GSH content (determined as free thiol) in sorghum roots 4 and 7 days after sowing and treatment (Table VII). The elevation ratio in the GSH level was maintained or increased between days 4 and

Table VII Effect of Preemergence Application of Various Safeners on Glutathione (GSH) Content in Sorghum Roots[a]

Safener	Alachlor (ppm)	GSH (mol/g fresh wt)		Ratio relative to control	
		4 days	7 days	4 days	7 days
None	0	0.87 ± 0.01[b]	0.47 ± 0.04	1.00	1.00
	1	1.00 ± 0.06	0.55 ± 0.04	1.15	1.17
CPMI	0	1.38 ± 0.05	0.80 ± 0.10	1.59	1.70
(100 ppm)	1	1.46 ± 0.05	0.78 ± 0.03	1.68	1.66
CPMA	0	1.42 ± 0.05	0.77 ± 0.04	1.63	1.64
(100 ppm)	1	1.49 ± 0.08	0.78 ± 0.04	1.71	1.66
Flurazole	0	1.69 ± 0.02	1.18 ± 0.13	1.94	2.51
(10 ppm)	1	1.73 ± 0.02	1.09 ± 0.07	1.99	2.32
Dichlormid	0	1.59 ± 0.01	0.95 ± 0.03	1.83	2.02
(10 ppm)	1	1.59 ± 0.01	1.05 ± 0.06	1.83	2.23

[a]From Rubin *et al.* (1985), with permission.
[b]SE values.

7 even though the absolute content declined. This was probably due to dilution of the thiol pool by plant growth, depletion by utilization, and exhaustion of stored reserves (Hilton and Pillai, 1986).

Simultaneous treatment of herbicide safeners with alachlor resulted in an increase of GSH content in the same pattern obtained with the safeners alone. A more detailed analysis of the CPMA-induced changes in GSH content in 4-day-old sorghum is shown in Fig. 5. The elevated GSH content was already evident at 8–48 hr after application of CPMA with or without alachlor.

Barnyard grass root contained lower levels of GSH (0.51 ± 0.02 and 0.28 ± 0.05 mol/g fresh wt at 4 and 7 days after sowing, respectively) but CPMA (which does not protect this weed from alachlor injury) did not induce any increase in GSH levels. Thus, the initial antidotal effect of CPMI could conceivably be due to its strong reaction with thiols (Augustin *et al.*, 1978), but this is not consistent with either the observed elevated GSH level *in vitro* or its similarity in action to CPMA which did not react (*in vitro*) with thiols.

V. CONCLUDING REMARKS

Substituted *N*-phenylmaleamic acids are a new class of herbicide safeners offering protection to grain sorghum against chloroacetanilide herbicide injury. These safeners are also useful as probes in examining species specificity and the factors controlling GSH synthesis and functions in plants. The studies reported in this chapter provide a hint for the development of new herbicide safeners for practical agricultural uses.

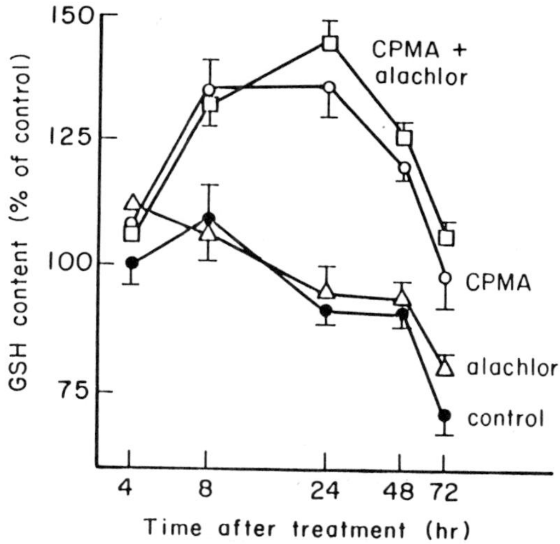

Fig. 5. Effect of *N*-(4-chlorophenyl)maleamic acid (CPMA) and alachlor alone and in combination on the glutathione (GSH) content of sorghum roots. GSH content is relative to untreated control at 4 hr. Bars indicate SE. [Reprinted with permission from Rubin *et al.* (1985). Copyright 1985 American Chemical Society.]

References

Adams, C. A., Blee, E., and Casida, J. E. (1983). Dichloroacetamide herbicide antidotes enhance sulfate metabolism in corn roots. *Pestic. Biochem. Physiol.* **19**, 350–360.

Ahrens, W. H., and Davis, D. E. (1978). Seed protectant effects on metolachlor absorption and translocation. *Proc., South. Weed Sci. Soc.* **31**, 249.

Augustin, M., Werndl, R., Kohler, M., and Ruttinger, H.-H. (1978). Die thiolphilie arylsubstituierter maleinimidderivate. *Pharmazie* **33**, 191–192.

Blair, A. M., Parker, C., and Kassasian, L. (1976). Herbicide protectants and antidotes—A review. *Pestic. Artic. News Summ.* **22**, 65–74.

Bleeke, M. S., and Casida, J. E. (1984). Metribuzin metabolism in mammals and liver microsomal oxidase systems: Identification, synthesis, and reactions. *J. Agric. Food Chem.* **32**, 749–755.

Bleeke, M. S., Smith, M. T., and Casida, J. E. (1985). Metabolism and toxicity of metribuzin in mouse liver. *Pestic. Biochem. Physiol.* **23**, 123–130.

Brinker, R., Schafer, D., Radke, R., Boeken, G., and Frazier, H. (1982). The effectiveness of MON-4606 as seed safener against alachlor and acetochlor in grain sorghum. *Proc. Br. Crop Prot. Conf.—Weeds* **2**, 469–473.

Buzio, C. A., and Burt, G. W. (1980). Leaching of EPTC and R-25788 in soil. *Weed Sci.* **28**, 241–245.

Carringer, R. D., Rieck, C. E., and Bush, L. P. (1978a). Metabolism of EPTC in corn (*Zea mays*). *Weed Sci.* **26**, 157–160.

Carringer, R. D., Rieck, C. E., and Bush, L. P. (1978b). Effect of R-25788 on EPTC metabolism in corn (*Zea mays*). *Weed Sci.* **26**, 167–171.

Casida, J. E. (1983). Propesticides: Bioactivation in pesticide design and toxicological evaluation. *In* "Pesticide Chemistry: Human Welfare and the Environment" (J. Miyamoto and P. C. Kearney, eds.), Vol. 3, pp. 239–244. Pergamon, Oxford.

Casida, J. E., and Ruzo, L. O. (1986). Reactive intermediates in pesticide metabolism: Peracid oxidation as possible biomimetic models. *Xenobiotica* **16**, 1003–1015.

Casida, J. E., Gray, R. A., and Tilles, H. (1974). Thiocarbamate sulfoxides: potent, selective and biodegradable herbicides. *Science* **184**, 573–574.

Clapp, K. E. (1974). A progress report on agricultural uses of activated charcoal. *Proc. West. Soc. Weed Sci.* **27**, 40–42.

Clapp, K. E. (1975). Methods of applying activated charcoal. *Proc. West. Soc. Weed Sci.* **28**, 31–34.

Corbett, J. R., Wright, K., and Baillie, A. C. (1984). "The Biochemical Mode of Action of Pesticides," 2nd ed. Academic Press, New York.

Cotter, R. J., Sauers, C. K., and Whelan, J. M. (1961). The synthesis of N-substituted isomaleimides. *J. Org. Chem.* **26**, 10–15.

Donald, W. W., and Fawcett, R. S. (1976). Prevention of EPTC injury in corn: A developmental and anatomical study. *Abstr. Weed Sci. Soc. Am.* **16**, 70.

Dutka, F., and Komives, T. (1987). MG-191: A new selective herbicide antidote. *In* "Pesticide Science and Biotechnology" (R. Greenhalgh and T. R. Roberts, eds.), pp. 201–204. Blackwell, Oxford.

Ebert, E. (1982). The role of waxes in the uptake of metolachlor into sorghum in relation to the protectant CGA 43089. *Weed Res.* **22**, 305–311.

Ezra, G., and Gressel, J. (1982). Rapid effects of thiocarbamate herbicide and its dichloroacetamide protectant on macromolecular syntheses and glutathione levels in maize cell culture. *Pestic. Biochem. Physiol.* **17**, 48–58.

Ezra, G., Krochmal, E., and Gressel, J. (1982). Competition between a thiocarbamate herbicide and herbicide protectants at the level of uptake into maize cell culture. *Pestic. Biochem. Physiol.* **18**, 107–112.

Fedtke, C. (1981). Wirkung von Herbiziden und Herbizid-Antidots auf den Glutathion-Gehalt in Mais- und Soja- Bohnenwurzlen. *Z. Pflanzenkrank. (Pflanzenpathol.) Pflanzenschutz, Sonderh.* **9**, 141–146.

Fedtke, C. (1982). "Biochemistry and Physiology of Herbicide Action." Springer-Verlag, Berlin and New York.

Fuerst, E. P., and Gronwald, J. W. (1986). Induction of rapid metabolism of metolachlor in sorghum (*Sorghum bicolor*) shoots by CGA-92194 and other antidotes. *Weed Sci.* **34**, 354–361.

Fujinami, A., Ozaki, T., and Nodera, K. (1972). Studies on biological activity of cyclic imides compounds. Part 2. Antimicrobial activity of 1-phenylpyrrolidine-2,5-diones and related compounds. *Agric. Biol. Chem.* **36**, 318–323.

Hargreaves, M. K., Pritchard, J. G., and Dave, H. R. (1970). Cyclic carboxylic monoimides. *Chem. Rev.* **70**, 439–469.

Hatzios, K. K. (1983). Herbicide antidotes: Development, chemistry and mode of action. *Adv. Agron.* **36**, 265–316.

Hatzios, K. K. (1984). Interaction between selected herbicides and protectants on corn (*Zea mays*). *Weed Sci.* **32**, 51–58.

Hatzios, K. K., and Penner, D. (1982). "Metabolism of Herbicides in Higher Plants," Burgess, Minneapolis, Minnesota.

Hilton, J. L., and Pillai, P. (1986). L-2-Oxothiazolidine-4-carboxylic acid protection against tridiphane toxicity. *Weed Sci.* **34**, 669–675.

Horváth, L., and Pulay, Á. (1980). Metabolism of EPTC in germinating corn: Sulfone as the true carbamoylating agent. *Pestic. Biochem. Physiol.* **14**, 265–270.

Hubbell, J. P., and Casida, J. E. (1977). Metabolic fate of the *N,N*-dialkylcarbamoyl moiety of thiocarbamate herbicides in rat and corn. *J. Agric. Food. Chem.* **25**, 404–413.

Ketchersid, M. L., and Merkle, M. G. (1981). The effect of CGA-43089 on adsorption and metabolism of metolachlor in grain sorghum. *Abstr. Weed Sci. Soc. Am.* **21**, 100.

Ketchersid, M. L., and Merkle, M. G. (1982). CGA-43089 (concep) and MON-4606 (screen) as protectants in sorghum. *Proc., South. Weed Sci. Soc.* **35**, 368.

Ketchersid, M. L., Vietor, D. M., and Merkle, M. G. (1982). CGA-43089 effects on metolachlor uptake and membrane permeability in grain sorghum (*Sorghum bicolor*). *J. Plant Growth Regul.* **1**, 285–294.

Kirino, O., Rubin, B., and Casida, J. E. (1985). *N*-Phenyl-maleimides, -isomaleimides and -maleamic acids as selective herbicide antidotes. *Agric. Biol. Chem.* **49**, 267–268.

Lay, M.-M., and Casida, J. E. (1976). Dichloroacetamide antidotes enhance thiocarbamate sulfoxide detoxification by elevating corn root glutathione content and glutathione *S*-transferase activity. *Pestic. Biochem. Physiol.* **6**, 442–456.

Lay, M.-M., and Niland, A. N. (1985). Biochemical response of inbred and hybrid corn (*Zea mays*) to R-25788 and its distribution with EPTC in corn seedlings. *Pestic. Biochem. Physiol.* **23**, 131–140.

Lay, M.-M., Hubbell, J. P., and Casida, J. E. (1975). Dichloroacetamide antidotes for thiocarbamate herbicides: Mode of action. *Science* **189**, 287–289.

Leavitt, J. R. C., and Penner, D. (1979). *In vitro* conjugation of glutathione and other thiols with acetanilide herbicides and EPTC sulfoxide and the action of the herbicide antidote R-25788. *J. Agric. Food. Chem.* **27**, 533–536.

Loos, M. A. (1975). Phenoxyalkanoic acids. *In* "Herbicide: Chemistry Degradation and Mode of Action" (P. C. Kearney and D. D. Kaufman, eds.), Vol. 1, pp. 1–128. Dekker, New York.

Moriguchi, I., and Komatsu, K. (1981). Structure–activity studies of withaferin analogues using the adaptive least-squares methods. *Eur. J. Med. Chem.* **16**, 19–23.

Mozer, T. J., Tiemeier, D. C., and Jaworski, E. G. (1983). Purification and characterization of corn glutathione-*S*-transferase. *Biochemistry* **22**, 1068–1072.

Ohta, H., Suzuki, S., Watanabe, H., Jikihara, T., Matsuya, K., and Wakabayashi, K. (1976). Structure–activity relationship of cyclic imide herbicides. I. N-Substituted phenyl-3,4,5,6-tetrahydrophtalimides and related compounds. *Agric. Biol. Chem.* **40**, 745–751.

Pallos, F. M., and Casida, J. E., eds. (1978). "Chemistry and Action of Herbicide Antidotes." Academic Press, New York.

Rennenberg, H., Birk, C., and Schaer, B. (1982). Effect of *N-N*-diallyl-2,2-dichloroacetamide (R-25788) on efflux and synthesis of glutathione in tobacco suspension cultures. *Phytochemistry* **21**, 5–8.

Roderick, W. R. (1957). The "isomerism" of N-substituted maleimides. *J. Am. Chem. Soc.* **79**, 1710–1712.

Rubin, B., and Casida, J. E. (1985). R-25788 effects on chlorsulfuron injury and acetohydroxyacid synthase activity. *Weed Sci.* **33**, 462–468.

Rubin, B., Kirino, O., and Casida, J. E. (1985). Chemistry and action of *N*-phenylmaleamic acids and their progenitors as selective herbicide antidotes. *J. Agric. Food. Chem.* **33**, 489–494.

Rufener, J., Nyfeler, A., and Peek, J. W. (1982). CGA 92194, a new safener to protect sorghum from injurious effects of metolachlor. *Proc. Br. Crop Prot. Conf.—Weeds* **2**, 461–467.

Sagaral, E. G., and Foy, C. L. (1982). Response of several corn (*Zea mays*) cultivars and weed species to EPTC with and without the antidote R-25788. *Weed Sci.* **30**, 64–69.

Sauers, C. K. (1969). The dehydration of *N*-arylmaleamic acids with acetic anhydride. *J. Org. Chem.* **34**, 2275–2279.

Schafer, D. E., Brinker, R. J., and Radke, R. O. (1980). Introduction of a new safening agent for alachlor on grain sorghum. *Proc.—North Cent. Weed Control Conf.* **35**, 67–71.

Schafer, D. E., Brinker, R. J., and Radke, R. O. (1981). Evaluation of the behavior and performance of MON-4606, a new safener for alachlor in grain sorghum. *Abstr. Weed Sci. Soc. Am.* **21**, 26.

Stephenson, G. R., and Ezra, G. (1982). The mode of action of herbicide safeners. *Proc. Br. Crop Prot. Conf.—Weeds* **2**, 451–459.

Stephenson, G. R., Ali, A., and Ashton, F. M. (1983). Influence of herbicides and antidotes on glutathione levels in maize seedlings. *In* "Pesticide Chemistry: Human Welfare and the Environment" (J. Miyamoto and P. C. Kearney, eds.), Vol. 3, pp. 219–224. Pergamon, Oxford.

Sweetser, P. B. (1985). Safening of sulfonylurea herbicides to cereal crops: Mode of herbicide antidote action. *Proc. Br. Crop Prot. Conf.—Weeds*. **3**, 1147–1154.

Verloop, A., Hoogenstraaten, W., and Tipker, J. (1976). Development and application of new steric substituent parameters in drug design. *Drug Des.*, **7**, 165–207.

Wain, R. L. (1958). Relation of chemical structure to activity for 2,4-D-type herbicide and plant growth regulator. *Adv. Pest Control Res.*, **2**, 263–305.

Warmund, M. R., Kerr, H. D., and Peters, E. J. (1985). Lipid metabolism in grain sorghum (*Sorghum bicolor*) treated with alachlor plus flurazole. *Weed Sci.* **33**, 25–28.

Williamson, J. M., Boettecher, B., and Meister, A. (1982). Intracellular cysteine delivery system that protects against toxicity by promoting glutathione synthesis. *Proc. Natl. Acad. Sci. U.S.A.* **79**, 6246–6249.

Zama, P., and Hatzios, K. K. (1986). Effects of CGA-92194 on the chemical reactivity of metolachlor with glutathione and metabolism of metolachlor in grain sorghum (*Sorghum bicolor*). *Weed Sci.* **34**, 834–841.

Chapter 15

Microbial Herbicide Safeners

Jeffrey S. Karns
Pesticide Degradation Laboratory
USDA, Agricultural Research Service
Beltsville Agricultural Research Center–West
Beltsville, Maryland 20705

I. INTRODUCTION

The use of pesticide-transforming microorganisms to protect or "safen" a particular crop plant from the action of a herbicide while leaving other plants susceptible is, at present, little more than a concept. In lieu of any actual research to review in the area of microbial safeners, this chapter will briefly describe what is known about the microbial degradation of some important classes of herbicides and then speculate on how the degradative properties of selected microorganisms might be exploited to broaden the useful spectrum of these herbicides.

All herbicides probably undergo some type of degradation due to soil microbes. This degradation can be as extensive as complete mineralization of the compound to CO_2 and water, or as simple as the hydrolysis or oxidation of a single chemical bond. Most of the compounds presently used as herbicides can be completely inactivated by the cleavage or modification of a single chemical linkage.

II. MICROBIAL DEGRADATION OF MAJOR CLASSES OF HERBICIDES

A. Phenoxyalkanoic Acids

The chlorinated phenoxyalkanoic acids were among the first organic compounds to be employed as herbicides. Their selective killing of dicotyledonous plants made them extremely useful in the cultivation of such cereal crops as wheat (*Triticum aestivum* L.) and corn (*Zea mays* L.) which form the backbone of American agriculture. These herbicides are generally applied to the foliage of growing plants and are absorbed through the cuticle and transported to their site of action. They act as artificial auxins in susceptible plants and disrupt normal regulation of growth, thereby killing the plant.

Phenoxyalkanoic acid herbicides have the chemical structure shown in Fig. 1. The most commonly used members of this family include 2,4-D ($R_1=H$; $R_2=Cl$; $R_3=Cl$; $R_4=H$), MCPA ($R_1=H$; $R_2=CH_3$; $R_3=Cl$; $R_4=H$), 2,4,5-T ($R_1=H$; $R_2=Cl$; $R_3=Cl$; $R_4=Cl$), and Silvex ($R_1=CH_3$; $R_2=Cl$; $R_3=Cl$; $R_4=Cl$). Because of problems with dioxin contamination, 2,4,5-T and silvex are no longer used in the United States.

The microbial degradation of phenoxyalkanoic acids has received considerable attention. The pioneering work of Audus (1949) demonstrated that soil microbial populations responded to the repeated application of 2,4-D by developing the capability to rapidly degrade the herbicide. Subsequent investigations resulted in the isolation of pure cultures of microorganisms capable of degrading 2,4-D (Tiedje *et al.*, 1969) and/or MCPA (Gaunt and Evans, 1961). These studies determined the pathway by which these compounds were degraded by bacteria of the genus *Arthrobacter* and various *Pseudomonas*. The most important mechanism of degradation of these compounds was the cleavage of the ether linkage between the C_2 of the alkanoic acid side chain and C_1 of the aromatic portion of the molecule, resulting in the release of the corresponding substituted phenol (Evans *et al.*, 1971; Gaunt and Evans, 1971; Steenson and Walker, 1957). This cleavage completely destroys the herbicidal activity of the compound. An oxidative cleavage system was implicated in the cleavage of the ether linkage of 4-chloro-2-methylphenoxyacetic acid (Gamar and Gaunt, 1971). Chakrabarty and co-workers isolated a strain of *Pseudomonas cepacia* that was capable of

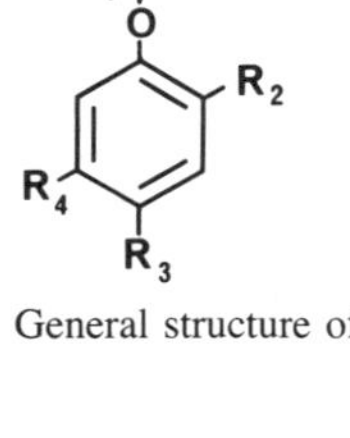

Fig. 1. General structure of phenoxyalkanoic acid herbicides.

rapidly and completely degrading 2,4,5-T (Kilbane *et al.*, 1982). Studies on the regulation of 2,4,5-T metabolism in this strain implicated a similar oxidative cleavage system as 2,4,5-trichlorophenol appeared to be a major metabolite (Karns *et al.*, 1983).

Genetic analysis has proven that the genes for the degradation of 2,4-D and MCPA in bacteria are plasmid-encoded (Don and Pemberton, 1981; Fisher *et al.*, 1978) and that these plasmids are often capable of conjugal transfer between various species of bacteria (see Fig. 2). Elegant studies using transposon mutagenesis and molecular cloning have localized the specific genes for the degradation of the chlorophenol metabolite (Don *et al.*, 1985). However, the gene(s) encoding the oxidative ether cleavage enzyme(s) were not isolated in this study.

B. Chlorinated Aliphatic Acids

Like the phenoxyalkanoic acids, the chlorinated aliphatic acids were among the earliest man-made herbicides. The two most common members of this class of herbicides are trichloroacetic acid (TCA) and dalapon (Fig. 3). They are usually used for the control of annual and perennial grasses, with TCA usually being soil incorporated and dalapon usually used as a foliar application. Partly because of their age and partly because of intense interest in the degradation of chlorinated compounds, there exists a wealth of literature on the biological degradation of these compounds in soils and pure cultures (Foy, 1975).

Among the more recent and complete works on the microbial degradation of chlorinated aliphatics are those of Slater's group in England and Kawasaki's group in Japan. Both groups isolated pure cultures of bacteria capable of utilizing chlorinated aliphatic acids as a sole source of carbon and energy (Kawasaki *et al.* 1981a; Senior *et al.*, 1976). In both cases the pathway of degradation was via hydrolytic dechlorination of the compound to yield a saturated aliphatic acid that is readily degraded by the organisms. The removal of the chlorine atoms from the α-carbon of TCA or dalapon completely destroys their herbicidal activity (Foy, 1975).

Genetic studies have revealed that the genes encoding the haloacetate halidohydrolases (EC 3.8.1.3) that allow *Moraxella* sp. strain B to utilize chloroacetates were encoded on a plasmid (Kawasaki *et al.*, 1981a), whereas the genes encoding the permeases and 2-haloaliphatic acid dehalogenases (EC 3.8.1.2) that allowed *Pseudomonas putida* PP3 to utilize dalapon and other halogenated aliphatics were contained on transposable DNA elements (transposon, see Fig. 2) located within the chromosome of the bacterium (Slater *et al.*, 1985). A recent report suggests that the dehalogenase genes in six new soil bacteria (isolated by their ability to utilize chlorinated aliphatic compounds as carbon and energy sources) were plasmid-borne and that at least one of the genes could move into the antibiotic resistant plasmid R68.45 (Hardman *et al.*, 1986), suggesting that it is on a transposon.

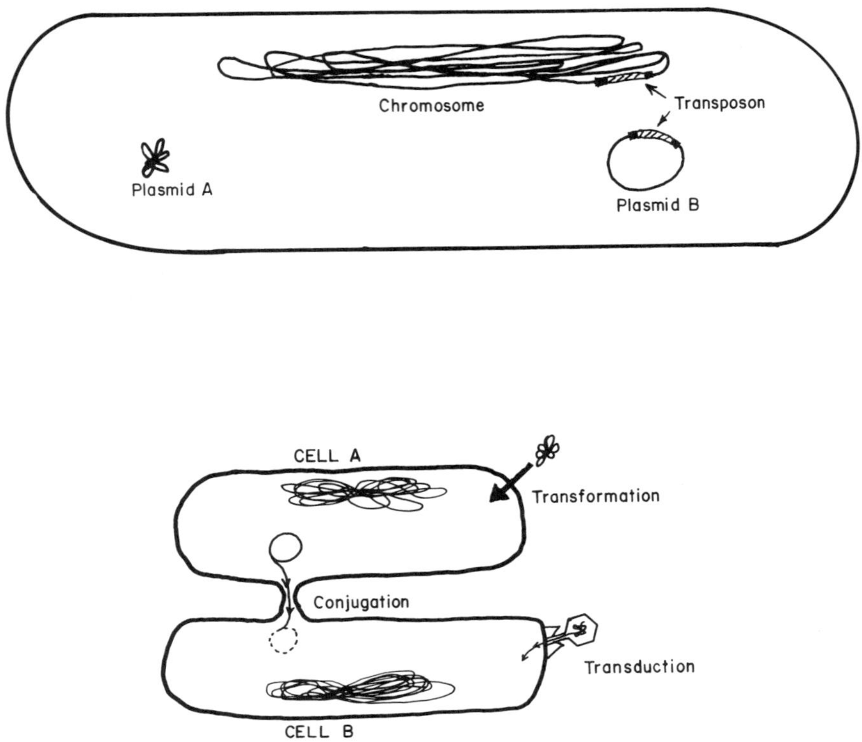

Fig. 2. A bacterial genetics primer. DNA encoding pesticide degrading enzymes can be exchanged by members of the soil microflora. The bacterial chromosome is an extremely large circle of DNA that exists as a very twisted (supercoiled) mass within the cell. Plasmids are smaller circles of DNA that replicate inside the cell independently of the chromosome. They, too, exist as supercoiled molecules. Some plasmids are present in one copy per cell, while others may exist in many copies per cell. Plasmids usually encode genes that are useful to the cell, but that are not required for normal cell maintenance. Transposons are discrete pieces of DNA that can move from location to location within a cell's DNA (jumping genes). Transposons cannot replicate by themselves, but can jump from plasmid to chromosome (and vice versa) using specialized enzymes encoded in the DNA at the ends of the transposon (enlarged solid lines). They can carry one or more genes in their coding region (hatched area).

There are three ways that DNA can be passed between bacteria under natural conditions. Conjugation is the direct transfer of DNA from one cell to another during a period of actual cell-to-cell contact. The plasmid replicates in cell A and the copy is transferred to cell B, where it recircularizes to form a replicating plasmid. Transformation is the uptake of naked DNA by a cell. Transduction is DNA transfer mediated by a bacterial virus (bacteriophage).

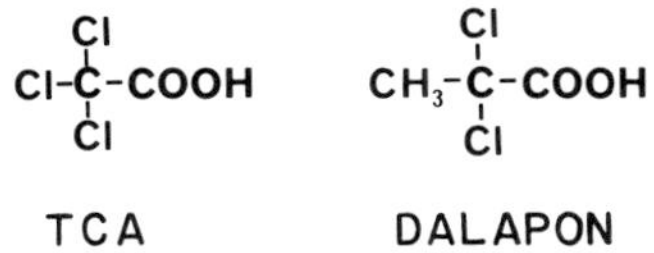

Fig. 3. Structure of two chlorinated aliphatic acid herbicides.

Several 2-haloaliphatic acid dehalogenases have been purified and characterized (Motosugi *et al.*, 1982; Weightman *et al.*, 1982). The enzymes differ in the stereospecificity of the substrates utilized and products released, indicating that different reaction mechanisms may exist for the hydrolytic removal of halogens from 2-haloaliphatic acids. Two different haloacetic acid halidohydrolases (differing in their ability to dehalogenate fluoroacetate) were isolated from *Moraxella* and *Pseudomonas* which were able to degrade chloroacetates (Kawasaki *et al.*, 1981b,c). It is not known why these organisms always seem to have two distinct dehalogenation enzymes.

C. N-Substituted Carbamates

Several different esters of carbamic acid are known to have herbicidal properties. The herbicidal N-substituted carbamates fall into two classes, the *N*-methylcarbamates and the *N*-phenylcarbamates or carbanilates. Their general structures are shown in Fig. 4. Few *N*-methylcarbamates are presently used as herbicides. Karbutilate is the only herbicide of this type listed in the "Herbicide Handbook" of the Weed Science Society of America (Beste, 1983). *N*-Methylcarbamates have found more extensive use as agricultural insecticides because of their ability to inhibit acetylcholinesterase (EC 3.1.1.7). Of the *N*-phenylcarbamates, propham, chlorpropham, and barban are still used as selective herbicides. The carbamate herbicides act through the roots of susceptible plants, so they are usually applied directly to soil.

The cleavage of the carbamate linkage of an N-substituted carbamate herbicide by hydrolytic or oxidative enzyme systems completely neutralizes the herbicidal activity of these compounds. Several researchers have reported the degradation of carbamate herbicides by microorganisms (Kaufman, 1967). Kearney (1965) isolated and purified an enzyme that was very effective in hydrolyzing a wide

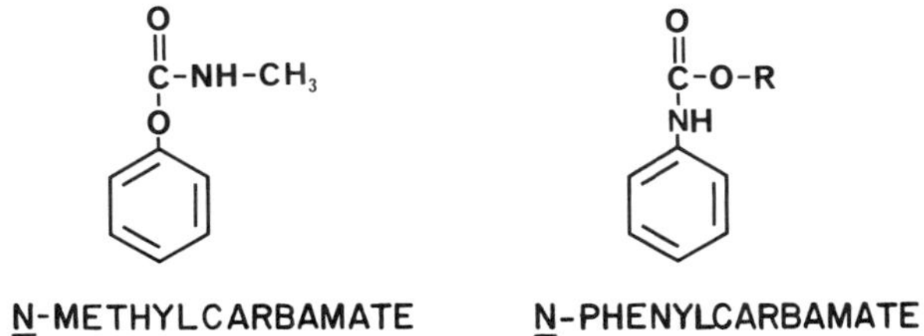

Fig. 4. General structure of carbamate herbicides.

range of *N*-phenylcarbamate herbicides (including both propham and chlorpropham) from a strain of *Pseudomonas striata* that could use chlorpropham as a sole carbon and energy source. An aryl acylamidase (EC 3.5.1.13) from *Bacillus subtilis* had some ability to hydrolyze propham, although propanil and urea herbicides were much better substrates (Engelhardt *et al.*, 1973). In my own laboratory we have isolated and purified a microbial hydrolase able to rapidly degrade a wide range of *N*-methylcarbamate compounds from a culture of *Achromobacter* that utilizes the insecticide carbofuran as a nitrogen source (Derbyshire *et al.*, 1987; Karns *et al.*, 1986). All these enzymes hydrolytically cleave the carbamate side chain from the aromatic portion of the molecule, acting at either the amide linkage (amidases; EC 3.5.1) or the ester linkage (esterases; EC 3.1.1). The products of the reaction are the same in either case as a result of the chemical instability of the reaction products (Fig. 5).

D. Thiocarbamates

The thiocarbamates are structurally similar to the other carbamate pesticides except that the ester oxygen is replaced by a sulfur atom. The general structure of the thiocarbamate herbicides is shown in Fig. 6, where R_1 is an alkyl or chlorinated alkyl and R_2 and R_3 are usually identical straight- or branched-chain alkyl groups. Because of their high volatility, the thiocarbamate herbicides are applied as preplant, soil incorporated herbicides.

For several years the thiocarbamate herbicides EPTC and butylate have been susceptible to a condition known as "problem" or "aggressive" soils. In these soils, the resident microbes degrade the herbicide so rapidly that it does not persist long enough to control weeds (Tuxhorn *et al.*, 1986; Wilson, 1984). While the biological degradation of thiocarbamates in soil has been well documented, very few reports document the isolation of organisms capable of degrading these compounds. Thus, little or nothing is known about the biochemical mechanisms of thiocarbamate degradation. The most likely route of degradation would be through the hydrolysis of the thiol ester to yield a mercaptan, CO_2, and a dialkyl substituted amine.

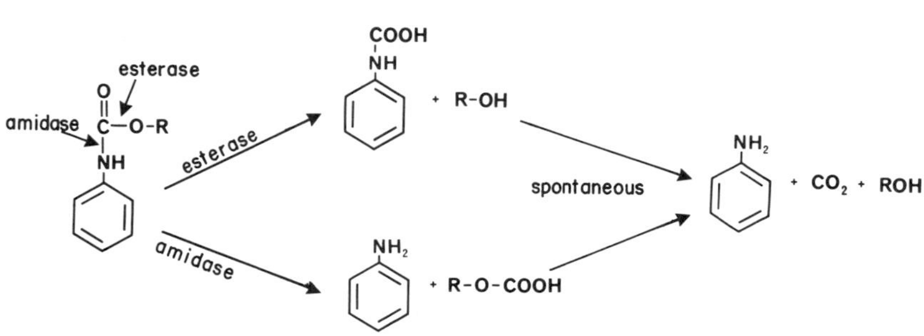

Fig. 5. Hydrolysis of *N*-phenylcarbamates by amidases or esterases.

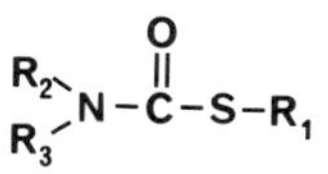

Fig. 6. General structure of thiocarbamate herbicides.

A very recent report has described the isolation of an *Arthrobacter* species capable of degrading EPTC that utilizes it as a source of carbon and energy (Tam *et al.*, 1987). A plasmid in this strain was shown to be required for EPTC degradation. The enzymatic mechanisms of EPTC degradation in this strain are unknown. Imai and Kuwatsuka (1986a,b) have reported the degradation of molinate by several bacteria and fungi isolated from a molinate perfused soil. Several reaction products were observed, with "hydrophilic substances" accounting for close to 50% of the added molinate. The direct product of the hydrolysis of the thiolcarbamate ester of molinate accounted for very little of the total reaction products. Little is known about the herbicidal activities of the molinate acid that accounted for 42% of the metabolites observed when the herbicide was incubated with cultured cells in solution. As is the case with the other carbamate herbicides, the direct hydrolysis of the carbamate ester linkage will entirely eliminate the herbicidal activity of the thiocarbamates.

E. Chloroacetanilides

The chloroacetanilides comprise a newer class of selective herbicides that are somewhat related to the *N*-phenylcarbamates. Alachlor and metolachlor are widely used to control weeds in corn and soybeans. Propachlor is also a member of this class of herbicides. Chloroacetanilides have the general structure shown in Fig. 7, where R_1 is an alkyl or oxyalkyl group. The aromatic ring may also have various substitutions; in alachlor R_2 and R_3 are both ethyl groups while in metolachlor R_2 is an ethyl group and R_3 is a methyl group.

Little is known about the microbial degradation of alachlor or metolachlor. The fungus *Chaetomium globosum* was shown to slowly degrade alachlor and metolachlor (McGahen and Tiedje, 1978; Tiedje and Hagedorn, 1975). The most direct route for the inactivation of these compounds is the cleavage of the amide linkage, a reaction that was demonstrated in both the aforementioned studies. Other authors (Krause *et al.*, 1985; Saxena *et al.*, 1987) have documented transformation of metolachlor by bacteria and fungi, however, these transforma-

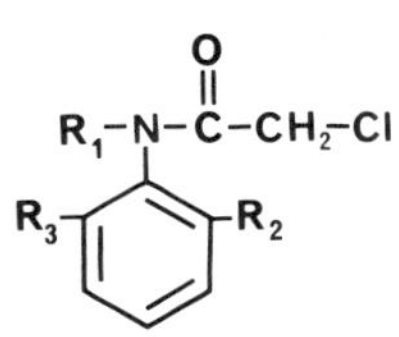

Fig. 7. General structure of chloroacetanilide herbicides.

tions only involved the oxidation of the side chains on the aromatic ring (R_2 and R_3) or the N-linked side chain (R_1). It is unclear whether these transformations would have any appreciable effect on the herbicidal activity of metolachlor. Novick *et al.* (1986) studied the degradation of alachlor and propachlor in pretreated soil samples and in aquifers and found that alachlor was not readily degraded whereas propachlor could be rapidly degraded by a mixture of two bacteria. The bacterial mixture mineralized over 50% of ring labeled propachlor in just over 2 days, indicating that the degradation was extensive as well as fast. *N*-Isopropylaniline was identified as an intermediate in the degradation of propachlor by this mixture of bacteria, demonstrating that cleavage of the amide linkage was probably a major mechanism of propachlor degradation.

F. Substituted Ureas

Related to the carbamates and chloroacetanilides, the urea herbicides are broad spectrum herbicides that have been used extensively since their discovery just after World War II. The general structure of these compounds is shown in Fig. 8, where R_1 is an aromatic moiety (usually halogenated) and R_2 and R_3 are alkyl or oxyalkyl groups. The most notable members of this class of herbicide are monuron, diuron, and linuron. More recently, a new and potent subclass of the urea herbicides, the sulfonylureas, has been developed. Chlorsulfuron, whose structure is given in Fig. 9, is a commercially developed herbicide in this class.

All the substituted urea herbicides have an amide bond which should be readily hydrolyzable by microbial enzymes. The aryl acylamidase from *Bacillus sphaericus* (Engelhardt *et al.*, 1973) was induced by linuron in the culture medium. Although the enzyme hydrolyzed the acylanilide herbicide propanil at a more rapid rate than it hydrolyzed linuron, it had a lower K_m for linuron, suggesting that the enzyme had a greater affinity for urea herbicides. There has been a report on the degradation of chlorsulfuron by soil fungi (Joshi *et al.*, 1985). Since the sulfonylureas are still somewhat new, I expect that more reports of their degradation will be forthcoming.

G. *s*-Triazines

The *s*-triazine herbicides include some of the most used compounds in American agriculture. Atrazine, the most used member of this class, has had great effectiveness for the control of weeds in corn. *s*-Triazines have the general

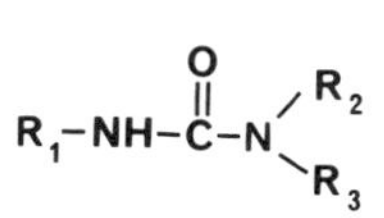

Fig. 8. General structure of substituted urea herbicides.

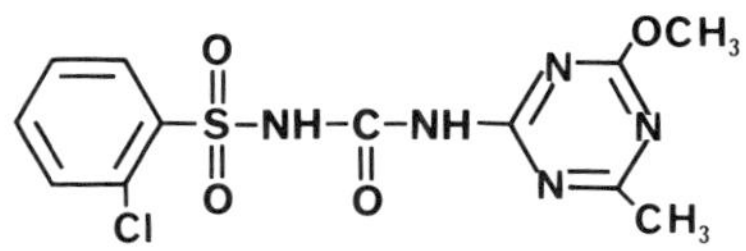

Fig. 9. Structure of chlorsulfuron, a sulfonylurea herbicide.

structure shown in Fig 10, where R_1 is a chlorine, thioalkyl, or methoxy group and R_2 and R_3 are alkyl groups. While removal or modification of the R_2 or R_3 can affect the biological activity of *s*-triazine herbicides somewhat, only the removal or modification of the R_1 substituent completely inactivates these compounds. Triazine herbicides are generally applied to soil as preemergence treatments.

Microorganisms have been shown to carry out various biochemical transformations of *s*-triazine compounds. The removal of the R_1 chlorine of atrazine has been shown to occur in sterile soils at a very slow rate, indicating that the reaction can occur nonbiologically (Harris, 1967). There have been reports of microbial degradation of *s*-triazines (Cook and Hutter, 1984; Giardina *et al.*, 1982; Kearney *et al.*, 1965; Wolf and Martin, 1975). One of the most recent reports describes the isolation of several *Pseudomonas* species capable of dealkylating and dechlorinating atrazine while utilizing it as a carbon source (Behki and Kahn, 1986). In all cases the rate of atrazine degradation was slow. Although it was not shown to be effective against atrazine itself, Cook and Hutter (1986) have assayed and partially purified a ring dechlorinating hydrolase (EC 3.8.1) from *Rhodococcus corallinus* that was capable of removing chlorine from several different substituted *s*-triazines. A crude extract was capable of completely dechlorinating a 0.5 m*M* solution of deethylsimazine in 5 hr. Such an enzyme would be a useful tool for inactivating *s*-triazine herbicides.

H. Miscellaneous Herbicides

Propanil is a herbicide used in the cultivation of rice. As the structure in Fig. 11 shows, it has an aliphatic acid moiety linked to an aniline group via an amide bond. The hydrolysis of propanil by microbial enzymes has been well documented. *Bacillus sphaericus* (Engelhardt *et al.*, 1973), *Fusarium oxysporum* (Blake and Kaufman, 1975), *Fusarium solanii* (Lanzilotta and Pramer, 1970), and *Aspergillus nidulans* (Pelsy *et al.*, 1987) have all been shown to produce

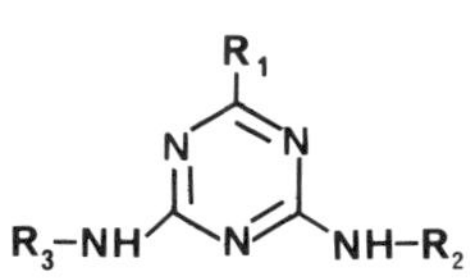

Fig. 10. General structure of *s*-triazine herbicides.

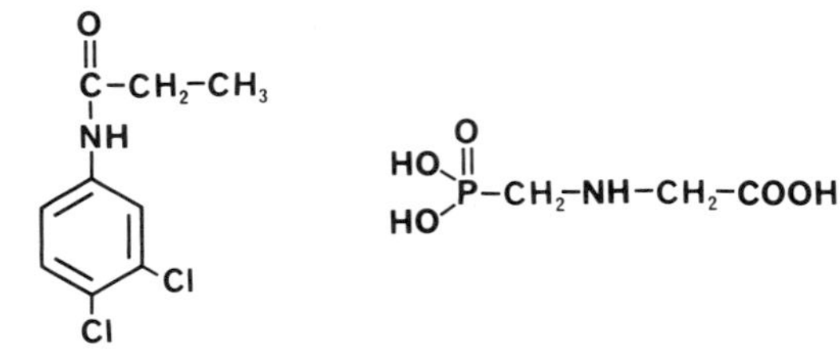

Fig. 11. Structure of the miscellaneous herbicides propanil and glyphosate.

aryl acylamidases capable of hydrolyzing propanil, yielding propionic acid and 3,4-dichloroaniline as the nonherbicidal products.

Glyphosate is a new, broad-spectrum herbicide that has found extensive use in minimum tillage systems, where its ability to kill any plant and its rapid dissipation make it useful in clearing land prior to planting. As seen in Fig. 11, glyphosate is a phosphonate compound (having a carbon–phosphorus bond). It blocks the synthesis of aromatic amino acids, thus killing the plant. For the same reason it is toxic to microbes. Degradation of glyphosate by soil microorganisms has been reported by several authors (Moore *et al.*, 1983; Nomura and Hilton, 1976; Talbot *et al.*, 1984). Shinabarger and Braymer (1986) studied the pathway of glyphosate metabolism in a *Pseudomonas* species and found that the first step was the cleavage of the phosphonate linkage. The product of that reaction (sarcosine) was then oxidatively cleaved to yield glycine and formaldehyde. The cleavage of the phosphonate linkage destroys the herbicidal activity of glyphosate.

III. PROSPECTS FOR THE USE OF MICROBES AS HERBICIDE SAFENERS

A. General Considerations

As the previous sections have shown, most herbicides are subject to some degree of microbial degradation. In some cases the herbicide molecule is completely degraded, whereas in others only one or two simple oxidation or hydrolysis steps are carried out. As stressed in the previous section, one or two simple hydrolytic or oxidative reactions can alter a herbicide molecule enough to eliminate its herbicidal activity. Thus, microorganisms producing hydrolases or oxygenases capable of transforming a particular herbicide might be used as effective safeners to protect crop plants from the action of that herbicide.

To assess the potential of microbes as safeners for a particular herbicide it is first necessary to note how that herbicide is used for crop protection. Herbicides can be generally classified as foliar- or soil-applied depending on whether they are applied to the leaves of existing plants to kill them or applied to soil to prevent the emergence of weed species. Considering the natural limitations to the density of microorganisms on leaf surfaces, it seems unlikely that microbial safeners will be developed for use with foliar-applied herbicides. Foliar-applied herbicides such as 2,4-D, glyphosate, or paraquat will be in too high a localized concentration and are too rapidly absorbed into the cuticle of the leaf surface for microbial safeners (based on the biodegradation properties described) to work effectively. The most likely use for microbial safeners of this type will be in conjunction with the soil-incorporated, preemergence herbicides such as the chloroacetanilides, triazines, thiocarbamates, carbamates, and ureas.

Natural examples of microbial safeners can be seen in the "problem soils" phenomenon mentioned earlier. The rate of metabolism of the herbicides in these soils is so great that the chemical cannot persist at the concentrations required to control weed pests. One could hypothesize that if these degrading microorganisms were specifically associated only with the crop plants so that the zones of degradation were limited to the soil around the seed or roots of the crop plant, the degrading microbes would be useful safeners. The "problem soils" phenomenon also helps to illustrate an inherent danger in the use of microbes as safeners. Unless the safeners used have tight specificities for the crop plants they are to protect or are limited in their range by some other criteria, they might survive in the soil and limit the useful life of the herbicide. Thus, certain safeguards will need to be built into microbial safeners to assure that neither the organisms themselves nor the genes which encode the herbicide degradative property can spread beyond the area and time of use.

B. Genetically Engineered Microbial Safeners

The suggestion that safeguards be "built-in" to microbial safeners implies a role for genetic engineering in their creation. Indeed, it is quite likely that the methods of DNA manipulation that have become commonly associated with the term genetic engineering will play a major role in the development of microbial safeners.

It is very likely that most useful herbicide degradation genes will be found in free-living soil bacteria and will need to be isolated and manipulated in the laboratory so that they can be inserted into those bacteria or fungi of the rhizosphere of the crop plant to be protected. Alternatively, the gene or genes encoding the pesticide-degrading enzymes could be manipulated so that the enzymes are expressed at high levels in organisms that are currently produced industrially. These organisms could then be used as killed or attenuated cells put down in large numbers at the time of planting or herbicide treatment. When used in this manner (as an active enzyme in a killed cell rather than in an actively dividing cell) they would offer safe, nonreproducing,and temporary protection for the germinating plant. It may also be possible to isolate the enzyme of interest from an industrial fermentation and fix it to some solid support such as glass or silica for use in the same manner. Hydrolases would be particularly amenable to such applications since they require no cofactors. As an example, a dehalogenase (EC 3.8.1) that removed the chlorine from the triazine herbicide atrazine might be of use to farmers rotating corn and soybeans. If cells containing such a dehalogenase, or a fixed enzyme preparation, were placed into the furrow when planting soybeans after corn, a farmer could be assured that residual atrazine remaining after corn cultivation would not affect his soybeans.

IV. CONCLUDING REMARKS

In summary, it seems reasonable to predict that herbicide degrading microorganisms may be developed into safeners to protect crop plants from the action of herbicides. Such technology could be used to protect crops from damage due to herbicide carryover in rotated crops or might be used to expand the number of crops on which a given herbicide can be used.

Because of my professional bias as a microbiologist studying the biodegradation of pesticides, I have ignored some other perfectly rational approaches to the use of microbes as herbicide safeners. Some laboratories have attempted to render plants resistant to glyphosate by amplifying the number of genes encoding the sensitive enzyme, 5-enolpyruvyl-shikimate-3-phosphate synthase (EPSP synthase; EC 2.5.1.19), as this was shown to render *Escherichia coli* resistant to glyphosate (Rogers *et al.*, 1983).

An alternative approach might be to develop root colonizing microorganisms which produce excess phenylalanine in the root zone. If the plant can take up and use the phenylalanine produced by the microbe, it might be more resistant to the action of glyphosate. Such an approach is not without precedent. In work that first suggested the possibility of using microbes as herbicide safeners, Sobieszczanski (1975) noted that cultures of a *Pseudomonas* could render some plants less sensitive to the action of the herbicide CIPC. In this case the protective activity was shown to reside in extracts of the culture supernatants. Thus, the bacterium seemed to produce a factor that helped the plant counteract the toxic effect of the herbicide.

I have also not done justice to the large contributions that can be made in the areas of plant–microbe interactions. In the previous section I stressed my belief that the microbes developed as safeners must have a close association with the specific crop plant to be protected. This is true whether the protective effect is due to degradation of herbicides or to the production of amino acids or other protective factors. This type of use of the plant–microbe interaction for novel purposes also is not without precedent. The genetically engineered bacterial insecticide developed by Monsanto Corp. (Obukowicz *et al.*, 1986) takes advantage of a *Pseudomonas fluorescens* that is a known corn root colonizer to maintain a concentration of *Bacillus thuringiensis* δ-endotoxin in the root zone that is sufficient for controlling soil borne insect pests of corn.

There do seem to be real possibilities for the development of microbial herbicide safeners. As more becomes known about the biochemical mechanisms of pesticide degradation, plant–microbe interactions, and the genetic engineering of soil bacteria, the potential for the development of microbial safeners will grow. The development of such safeners will require a unique interaction among soil microbiologists, biochemists, plant physiologists, and molecular biologists.

References

Audus, L. J. (1949). Biological detoxification of 2,4-dichlorophenoxyacetic acid in soil. *Plant Soil* **2,** 31–36.

Behki, R. M., and Khan, S. U.. (1986). Degradation of atrazine by *Pseudomonas* N-Dealkylation and dehalogenation of atrazine and its metabolites. *J. Agric. Food Chem.* **34,** 746–749.

Beste, C. E. (1983). "Herbicide Handbook," 5th ed. Weed Sci. Soc. of Am., Champaign, Illinois.

Blake, J., and Kaufman, D. D. (1975). Characterization of acylanilide-hydrolyzing enzymes from *Fusarium oxysporum* Schlect. *Pestic. Biochem. Physiol.* **5,** 305–313.

Cook, A. M., and Hutter, R. (1984). Deethylsimazine: bacterial dechlorination, deamination and complete degradation. *J. Agric. Food Chem.* **32,** 581–585.

Cook, A. M., and Hutter, R. (1986). Ring dechlorination of deethylsimazine by hydrolases from *Rhodococcus corallinus. FEMS Microbiol. Lett.* **34,** 335–338.

Derbyshire, M. K., Karns, J. S., Kearney, P. C., and Nelson, J. O. (1987). Purification and characterization of an *N*-methylcarbamate pesticide hydrolyzing enzyme. *J. Agric. Food Chem.* **35,** 871–877.

Don, R. H., and Pemberton, J. M. (1981). Properties of six pesticide degradation plasmids isolated from *Alcaligenes eutrophus* and *Alcaligenes paradoxus. J. Bacteriol.* **145,** 681–686.

Don, R. H., Weightman, A. J., Knackmuss, H. J., and Timmis, K. N. (1985). Transposon mutagenesis and cloning analysis of the pathways for 2,4-dichlorophenoxyacetic acid and 3-chlorobenzoate in *Alcaligenes eutrophus* JMP134 (pJP4). *J. Bacteriol.* **161,** 85–90.

Engelhardt, G., Wallnofer, P. R., and Plapp, R. (1973). Purification and properties of an aryl acylamidase of *Bacillus sphaericus* catalyzing the hydrolysis of various phenylamide herbicides and fungicides. *Appl. Microbiol.* **26,** 709–718.

Evans, W. C., Smith, B. S. W., Fernley, H. N., and Davies, J. J. (1971). Bacterial metabolism of 2,4-dichlorophenoxyacetate. *Biochem. J.* **122,** 543–550.

Fisher, P. R., Appelton, J., and Pemberton, J. M. (1978). Isolation and characterization of the pesticide degrading plasmid pJP1 from *Alcaligenes paradoxus. J. Bacteriol.* **135,** 798–804.

Foy, C. L. (1975). The chlorinated aliphatic acids. *In* "Herbicides: Chemistry, Degradation, and Mode of Action" (P. C. Kearney and D. D. Kaufman, eds.), Vol. I. pp. 399–452. Dekker, New York.

Gamar, Y., and Gaunt, J. K. (1971). Bacterial metabolism of 4-chloro-2-methylphenoxyacetate. Formation of glyoxalate by side chain cleavage. *Biochem. J.* **122,** 527–533.

Gaunt, J. K., and Evans, W. C. (1961). Metabolism of 4-chloro-2-methylphenoxyacetic acid by a soil microorganism. *Biochem. J.* **79,** 25P.

Gaunt, J. K., and Evans, W. C. (1971). Metabolism of 4-chloro-2-methylphenoxyacetic acid by a soil pseudomonad. Preliminary evidence for the metabolic pathway. *Biochem. J.* **122,** 533–537.

Giardina, M. C., Giardi, M. T., and Filacchione, G. (1982). Atrazine metabolism by *Nocardia*: Elucidation of initial pathway and synthesis of potential metabolites. *Agric. Biol. Chem.* **46,** 1439–1445.

Hardman, D. J., Gowland, P. C., and Slater, J. H. (1986). Large plasmids from soil bacteria enriched on halogenated alkanoic acids. *Appl. Environ. Microbiol.* **51,** 44–51.

Harris, C. I. (1967). Fate of 2-chloro-*s*-triazine herbicides in soil. *J. Agric. Food Chem.* **15,** 157–162.

Imai, Y., and Kuwatsuka, S. (1986a). Characteristics of microflora degrading the herbicide molinate in soil. *J. Pestic. Sci.* **11,** 57–63.

Imai, Y., and Kuwatsuka, S. (1986b). The mode of metabolism of the herbicide molinate by four strains of microorganisms isolated from soil. *J. Pestic. Sci.* **11,** 111–117.

Joshi, M. M., Brown, H. M., and Romesser, J. A. (1985). Degradation of chlorsulfuron by soil microorganisms. *Weed Sci.* **33,** 888–893.

Karns, J. S., Duttagupta, S., and Chakrabarty, A. M. (1983). Regulation of 2,4,5-trichlorophenoxyacetic acid and chlorophenol metabolism in *Pseudomonas cepacia* AC1100. *Appl. Environ. Microbiol.* **46,** 1182–1186.

Karns, J. S., Mulbry, W. W., Nelson, J. O., and Kearney, P. C. (1986). Metabolism of carbofuran by a pure bacterial culture. *Pestic. Biochem. Physiol.* **25,** 211–217.

Kaufman, D. D. (1967). Degradation of carbamate herbicides in soil. *J. Agric. Food Chem.* **15,** 582–591.

Kawasaki, H., Tone, N., and Tonomura, K. (1981a). Plasmid determined dehalogenation of haloacetates in *Moraxella* sp. *Agric. Biol. Chem.* **45,** 29–34.

Kawasaki, H., Tone, N., and Tonomura, K. (1981b). Purification and properties of haloacetate halidohydrolase specified by plasmid from *Moraxella* sp. strain B. *Agric. Biol. Chem.* **45,** 35–42.

Kawasaki, H., Miyoshi, K., and Tonomura, K. (1981c). Purification, crystallization and properties of haloacetate halidrohydrolase from *Pseudomonas* species. *Agric. Biol. Chem.* **45,** 543–544.

Kearney, P. C. (1965). Purification and properties of an enzyme responsible for hydrolyzing phenylcarbamates. *J. Agric. Food. Chem.* **13,** 561–564.

Kearney, P. C., Kaufman, D. D., and Sheets, T. J. (1965). Metabolites of simazine by *Aspergillus fumigatus*. *J. Agric. Food Chem.* **13,** 369–372.

Kilbane, J. J., Chatterjee, D. K., Karns, J. S., Kellogg, S. T., and Chakrabarty, A. M. (1982). Biodegradation of 2,4,5-trichlorophenoxyacetic acid by a pure culture of *Pseudomonas cepacia*. *Appl. Environ. Microbiol.* **44,** 72–78.

Krause, A., Hancock, W. G., Minard, R. D., Freyer, A. J., Honeycutt, R. C. LeBaron, H. M., Paulson, D. L., Liu, S.-Y., and Bollag, J.-M. (1985). Microbial transformation of the herbicide metolachlor by a soil actinomycete. *J. Agric. Food Chem.* **34,** 584–589.

Lanzilotta, R. P., and Pramer, D. (1970). Herbicide transformation II. Studies with an acylamidase of *Fusarium solanii*. *Appl. Microbiol.* **19,** 307–312.

McGahen, L. L., and Tiedje, J. M. (1978). Metabolism of two new acylanilide herbicides, Antor herbicide (H-22234) and Dual (metolachlor) by the soil fungus *Chaetomium globosum*. *J. Agric. Food Chem.* **26,** 414–423.

Moore, J. K., Braymer, H. D., and Larson, A. D. (1983). Isolation of a *Pseudomonas* sp. which utilizes the phosphonate herbicide glyphosate. *Appl. Environ. Microbiol.* **46,** 316–320.

Motosugi, K., Esaki, N., and Soda, K. (1982). Purification and properties of a new enzyme, DL-2-haloacid dehalogenase from *Pseudomonas* sp. *J. Bacteriol.* **150,** 522–527.

Nomura, N. S., and Hilton, H. W. (1976). The absorption and degradation of glyphosate in five Hawaiian sugarcane soils. *Weed Res.* **17,** 113–121.

Novick, N. J., Mukherjee, R., and Alexander, M. (1986). Metabolism of alachlor and propachlor in suspensions of pretreated soils and in samples from groundwater aquifers. *J. Agric. Food Chem.* **34,** 721–725.

Obukowicz, M. G., Perlak, F. J., Kusano-Kretzmer, K., Mayer, E. J., Bolten, S. L., and Watrud, L. S. (1986). Tn5-mediated integration of the delta-endotoxin gene from *Bacillus thuringiensis* into the chromosome of root colonizing pseudomonads. *J. Bacteriol.* **168,** 982–989.

Pelsy, F., Leroux, P., and Heslot, H. (1987). Properties of an *Aspergillus nidulans* propanil hydrolase. *Pestic. Biochem. Physiol.* **27,** 182–188.

Rogers, S. G., Brand, L. A., Holder, S. B., Sharps, E. S., and Brackin, M. J. (1983). Amplification of the *Aro*A gene from *Escherichia coli* results in tolerance to the herbicide glyphosate. *Appl. Environ. Microbiol.* **46,** 37–43.

Saxena, A., Zhang, R., and Bollag, J.-M. (1987). Microorganisms capable of metabolizing the herbicide metolachlor. *Appl. Environ. Microbiol.* **53,** 390–396.

Senior, E., Bull, A. T., and Slater, J. H. (1976). Enzyme evolution in a microbial community growing on the herbicide dalapon. *Nature (London)* **263,** 476–479.

Shinabarger, D. L., and Braymer, H. D. (1986). Glyphosate catabolism by *Pseudomonas* sp. strain PG2982. *J. Bacteriol.* **168,** 702–707.

Slater, J. H., Weightman, A. J., and Hall, B. G. (1985). Dehalogenase genes of *Pseudomonas putida* PP3 on chromosomally located transposable elements. *Mol. Biol. Evol.* **2,** 557–567.

Sobieszczanski, J. (1975). Microorganisms as a factor changing the action of herbicides on plants. *Rocz. Glebozn. T.* **26,** 79–89.

Steenson, T. I., and Walker, N. (1957). The pathway of breakdown of 2,4-dichloro- and 4-chloro-2-methylphenoxyacetic acid by bacteria. *J. Gen. Microbiol.* **16,** 146–152.

Talbot, H. W., Johnson, L. M., and Munnecke, D. M. (1984). Glyphosate utilization by *Pseudomonas* sp., and *Alcaligenes* sp. isolated from environmental sources. *Curr. Microbiol.* **10,** 255–260.

Tam, A. C., Behki, R. M., and Khan, S. U. (1987). Isolation and characterization of an *S*-ethyl-*N*,*N*-dipropylthiocarbamate degrading *Arthrobacter* strain and evidence for plasmid associated *S*-ethyl-*N*,*N*-dipropylthiocarbamate degradation. *Appl. Environ. Microbiol.* **53,** 1088–1093.

Tiedje, J. M., and Hagedorn, M. L. (1975). Degradation of alachlor by a soil fungus, *Chaetomium globosum*. *J. Agric. Food Chem.* **23,** 77–82.

Tiedje, J. M., Duxbury, J. M., Alexander, M., and Dawson, J. E.(1969). 2,4-D metabolism: Pathway of degradation of chlorocatechols by *Arthrobacter* sp. *J. Agric. Food Chem.* **17,** 1021–1026.

Tuxhorn, G. L., Roeth, F. W., Martin, A. R., and Wilson, R. G. (1986). Butylate persistence and activity in soils previously treated with thiocarbamates. *Weed Sci.* **34,** 961–965.

Weightman, A. J., Weightman, A. L., and Slater, J. H. (1982). Stereospecificity of 2-monochloropropionate dehalogenation by the two dehalogenases of *Pseudomonas putida* PP3: Evidence for two different dehalogenation mechanisms. *J. Gen. Microbiol.* **128,** 1755–1762.

Wilson, R. G. (1984). Accelerated degradation of thiocarbamate herbicides in soil with prior thiocarbamate herbicide exposure. *Weed Sci.* **32,** 264–268.

Wolf, D. C., and Martin, J. P. (1975). Microbial decomposition of ring-^{14}C atrazine, cyanuric acid, and 2-chloro-4,6-diamino-*s*-triazine. *J. Environ. Qual.* **4,** 134–139.

Part Four

Summary

Chapter 16

Herbicide Safeners: Progress and Prospects

Kriton K. Hatzios
Department of Plant Pathology, Physiology and Weed Science
Virginia Polytechnic Institute and State University
Blacksburg, Virginia 24061-0330

Robert E. Hoagland
USDA, ARS, Southern Weed Science Laboratory
Stoneville, Mississippi 38776

I. INTRODUCTION

The development and commercialization of crop safeners for herbicides has opened a new era in chemical weed control. Safening agents improve the usefulness of a herbicide by increasing the ratio of crop safety to weed control. Thus, a major benefit of the safener concept is making herbicides more selective.

The concept of chemical crop safening against herbicide injury is also appealing for other reasons. At the user's level, an appreciable increase in crop selectivity of herbicides allows the use of higher rates of specific chemicals without injury to crop or the use of more potent herbicides to obtain a wider spectrum of weed control, including weeds that are closely related to the crop. Furthermore, safeners provide the farmer with greater flexibility in rotating his crops since they could protect susceptible crops against injury caused by soil-applied, persistent herbicides used in an earlier season.

At the manufacturer's level, safeners add versatility to the current arsenal of proprietary herbicides by extending the patent life of reliable herbicides and by expanding their use into new markets. The cost of the development of safeners for an established herbicide compares favorably with the cost of development of new herbicides (Peek *et al.*, 1981). Thus, safeners may allow cheaper herbicides to replace expensive new ones designed for the control of similar weed problems.

At the level of basic research, safeners provide plant physiologists and biochemists with useful tools to probe the mechanisms of action of selected herbicides as well as the regulation of plant metabolic processes.

Safeners may be applied to the crop seed prior to planting ("seed dressings"), to transplants, to the soil or with the herbicide in a prepackaged formulated mixture. Applications of safeners together with the herbicide in a single formulation package are advantageous to a manufacturer because a single company controls both components of such formulations. This kind of safener application is also desirable to the farmer because it does not burden him with any extra operations.

In recent years, however, the use of crop seeds as carriers for chemicals (including safeners) has been viewed as an effective and economical delivery system which is gaining wider acceptance (Baughan *et al.*, 1985; Jeffs, 1978). The release of a safener from a seed system must be manipulated in such a way that it fits the location, timing, and concentration required for protecting the seed or emerging seedling of a particular crop against injury caused from specific soil-applied herbicides (Muller and Nyffeler, 1981). To facilitate their marketing, seed safeners are presently sold as treated seeds. This strategy is desirable to seed companies affiliated with manufacturers of herbicide safeners since safened seeds may increase considerably the sales of a particular crop cultivar or hybrid. Our present knowledge of seed technology and the expanding use of seeds as carriers for chemicals would be beneficial to the application of microbial herbicide safeners that might be engineered by means of recombinant DNA technology in the near future.

Regardless of how safeners are applied, they should protect a given crop from herbicide injury for the desired length of time, be compatible with other chemical treatments (e.g., fungicides, insecticides, fertilizers, etc.) or agronomic operations, and cause no phytotoxicity to the crop.

Several environmental factors such as light, temperature, soil moisture and soil type may affect the field performance of herbicide safeners and these effects need to be established during the evaluation of potential crop/herbicide/safener combinations. In addition, the timing of herbicide and safener applications as well as the differential intraspecific response of cultivars or hybrids of a crop to herbicide/safener combinations need to be established for optimum results.

II. EXPANSION OF THE SAFENER CONCEPT

By necessity, search for safeners has concentrated on the most significant crop/weed/herbicide combinations (Gray *et al.*, 1982; Hoffman, 1978; Peek *et al.*, 1981). To date, the range of crops protected by chemical safeners against herbicide injury is rather limited. Practical success has been realized primarily with large-seeded grass crops such as corn (*Zea mays* L.), grain sorghum [*Sorghum bicolor* (L.) Moench], and rice (*Oryza sativa* L.) which can be safened against chloroacetanilide and thiocarbamate herbicide injury with a number of chemical safeners (Hatzios, 1983; Parker, 1983). Other grass crops such as wheat (*Triticum aestivum* L.), oats (*Avena sativa* L.), barley (*Hordeum vulgare* L.), and teff [*Eragrostris tef* (Zucc.) Trotter] can also be protected against thiocarbamate and chloroacetanilide herbicides by means of chemical safeners, but such combinations have not been exploited commercially (Parker, 1983). Commercialized safeners are members of diverse chemical groups, including dichloroacetamides (e.g., dichlormid), naphthopyranones (e.g., NA), oxime ether derivatives (e.g., cyometrinil), substituted thiazoles (e.g., flurazole), dichloromethyl-dioxolans (e.g., MG-191), and substituted phenyl-pyrimidines (e.g., fenclorim).

The botanical and chemical specificities of herbicide safeners have often been cited as factors increasing the difficulty for the discovery of new safeners (Casida, 1978; Gray *et al.*, 1982; Peek *et al.*, 1981). Most of the currently marketed safeners are particularly effective against soil-applied, shoot-absorbed herbicides such as the thiocarbamates and the chloroacetanilides. Success has been very limited in developing chemical safeners for the protection of any crop from photosynthesis-inhibiting herbicides or broad spectrum weed killers such as glyphosate and paraquat.

The observation that crops protected by current chemical safeners against herbicide injury include primarily plant species that are damaged but not killed by these herbicides demonstrates further the botanical specificity and physiological complexity of the safener concept. The practicality of this observation, however, can be questioned in view of recent findings demonstrating acceptable protection of corn and other grass crops against injury caused by several diverse groups of herbicides, including the sulfonylureas (e.g., chlorsulfuron), imidazolinones (e.g., imazaquin), arylophenoxyalkanoic acids (e.g., diclofop-methyl), cyclohexenones (e.g., sethoxydim), and isoxazolidinones (e.g., clomazone or FMC-57020) (Barrett, Chapter 9, this volume; Hatzios, Chapter 1, this volume). Many of these herbicides are very phytotoxic to these grass crops and cause their death even at low concentrations.

So far, the interactions of broad-leaved crops with chemical safeners against herbicide injury have not been very promising. Adsorbents (e.g., activated charcoal and lignin by-products) and growth regulators of the triazole group (e.g., triapenthenol and triadimefon) offer a promising degree of protection to broad-leaved crops such as soybeans [*Glycine max* (L.) Merr.] against injury caused by selected herbicides including photosynthetic inhibitors such as metribuzin (Fletcher, 1985; Hoagland, Chapter 11, this volume; Phatak and Vavrina, Chapter 13, this volume; Riggle and Penner, Chapter 12, this volume).

III. MECHANISMS OF SAFENER ACTION: THE CHALLENGE CONTINUES

Although the literature on the mechanisms of action of herbicide safeners is getting more and more voluminous each year, there is still disagreement among researchers on how safeners work to protect crops against herbicide injury (Hatzios, Chapter 3, this volume).

At the physiological and biochemical level, two theories have attracted the most attention by proposing the following mechanisms for the action of herbicide safeners: (1) a safener-induced enhancement of herbicide detoxication in protected plants (Lay and Casida, 1978) and (2) a competitive antagonism of herbicide effects by safeners at a common site of action (Stephenson and Chang, 1978). The view that safeners may act by more than one mechanism has also been proposed (Ezra *et al.*, 1983). At the molecular level, the "gene activation" theory, which has been implicated in the action of natural or synthetic plant hormones (Theologis, 1986), also appears promising for explaining the protective action of herbicide safeners (Wiegand *et al.*, 1986).

In recent years, however, increasing evidence has accumulated to indicate that most, if not all, of the currently used safeners act primarily by enhancing the rate of the metabolic detoxication of selected herbicides. The safening of grass crops from the phytotoxic effects of chemically and biologically diverse groups of herbicides such as the thiocarbamates, chloroacetanilides, sulfonylureas, imidazolinones, arylophenoxyalkanoic acids, cyclohexenones, and isoxazolidinones on grass crops does not favor the "competitive antagonism" theory for the action of herbicide safeners. The aforementioned classes of herbicides exert their action by a variety of biochemical and physiological mechanisms which in some cases are well defined (e.g. sulfonylureas, imidazolinones, and isoxazolidinones) and in others just speculative (e.g. chloroacetanilides, thiocarbamates, cyclohexenones, and arylophenoxyalkanoic acids) (Fedtke and Trebst, 1987). The metabolic detoxication of these herbicides in higher plants is rather well characterized and an enhancement of their metabolism by safeners offers a more acceptable hypothesis for explaining the action of safeners which protect grass crops against injury caused by these unrelated herbicides.

Herbicide safeners may enhance the metabolism of selected herbicides in protected plants by inducing either the biosynthesis of specific cofactors (e.g., reduced glutathione, GSH) or the synthesis and activity of key metabolic enzymes involved in herbicide biotransformations such as the glutathione-*S*-transferases (GST) and oxidative enzymes (e.g. mixed-function oxidases, peroxidases, peroxygenases, etc.) (Breaux *et al.*, Chapter 7, this volume; Gronwald, Chapter 4, this volume; Sweetser, 1985).

On the basis of presently available evidence, the effects of herbicide safeners on glutathione biosynthesis correlate rather poorly with their ability to ameliorate herbicidal effects on grass crops (Gronwald, Chapter 4, this volume). However, the degree of protection of grass crops from chloroacetanilide and thiocarbamate herbicides seems to correlate rather strongly with the ability of safeners to enhance GST activity in crops such as corn and grain sorghum (Gronwald, Chapter 4, this volume).

In spite of the progress made during the past decade, we still do not have all the answers and facts to explain conclusively the mechanisms of action of the currently marketed crop safeners for herbicides. The work of unravelling the mechanisms of action of chemical regulators of crop tolerance to herbicides will continue to challenge the ingenuity of investigators in the future. Selected areas that need to be investigated further include:

1. Future investigations on the mechanisms of safener action should be extended to the molecular level. We must determine whether and how safeners affect gene expression in plants and whether these regulatory effects of safeners are related to the biochemical and physiological effects exerted by the antagonized herbicides.
2. Future studies on the physiological, biochemical, or molecular interactions between herbicides and safeners on plants should establish which effects of the safeners and herbicides are of primary or secondary importance. For example, are the reported effects of oxime ether safeners on sorghum seedling respiration of any significance for the protective action of these safeners (Ketchersid and Merkle, 1983) or are they just secondary effects totally unrelated to the safening function?
3. What is the physiological or biochemical significance of hormonal interactions with selected safeners? Do safeners (e.g., NA) mimic the effects of endogenous plant hormones (e.g., auxins) in protecting grass crops from herbicide injury?
4. What is the role of glutathione (GSSG) reductase in the action of safeners which enhance the metabolic conjugation of selected herbicides with GSH? Is this enzyme activated or induced by these safeners?
5. Do safeners other than flurazole and possibly dichlormid conjugate with reduced glutathione in the protected grass crops? Are such conjugations of safeners with GSH catalyzed by specific GSTs or by the same GST enzymes that catalyze the conjugation of chloroacetanilide and thiocarbamate herbicides to GSH?
6. What is the sensitivity of GST enzymes to stimulation, inhibition, or feedback regulation by the safener–GS conjugates?
7. How stable are the safener–GS conjugates? Are they further catabolized in the same way as GS-conjugates of chloroacetanilide and thiocarbamate herbicides?

8. What is the role of mixed-function oxidases in the metabolism of herbicides whose effects are counteracted by safeners? Are they induced by safeners?
9. Are other oxidative enzymes such as peroxygenases, peroxidases, polyphenol oxidases, lipoxygenases, etc., involved in the metabolism or action of selected herbicides? Do such enzymes play a role in the mechanisms of action of herbicide safeners which antagonize the effects of these herbicides?
10. If safeners act by more than one mechanism, what determines the sequence of steps or events involved in the safener action and what triggers the function of a particular mechanism?
11. What is the impact of herbicide safeners on herbicide metabolism by soil microorganisms, especially in soils where high herbicide residues may exist?
12. What are the potential interactive effects of safeners with mycoherbicides and other naturally occurring phytotoxins? Do safeners protect crops against plant pathogen attack?
13. How is safening action affected by surfactants, adjuvants, and other additives that are commonly found in herbicide formulations?

Further research is definitely needed to provide us with answers to these and other questions that are currently unanswered or, at best, poorly understood.

IV. DEVELOPMENT OF SAFENERS IN THE FUTURE

The initial phase of the exploitation of the concept of crop protection against herbicide injury revolved around the use of activated charcoal, an adsorbent, which acts as a physical barrier minimizing the contact of crops with nonselective herbicides. The high cost of this protection and the difficulty in maintaining the charcoal barrier around the growing plant are frequently cited as the major reasons for the limited practical success of this adsorbent as a crop protectant for herbicides (Blair *et al.*, 1976; Gray *et al.*, 1982).

The second phase of the commercial development of herbicide safeners enjoyed greater success because chemicals discovered and developed during this period are acting inside the plant and offering protection for the desired length of time. Crop safening provided by such chemicals is the result of their interaction with the biochemical metabolism of the protected plant rather than the result of a physical deactivation of the herbicide in the soil (outside the plant).

Almost all the currently marketed crop safeners for herbicides were developed by random screening and subsequent chemical optimization (Gray *et al.*, 1982; Hatzios, 1983; Hoffman, 1978). This approach has been proven very successful for the development of herbicide safeners and it should not be abandoned in the future. Significant antagonistic interactions with other agrochemicals must be identified and exploited to advantage for the targeted synthesis of future herbicide safeners. Inactive analogs of herbicides, shelved after the selection of active compounds, should be tested for potential crop safening activity. The potential practical use of the preconditioning of susceptible crops with subtoxic levels of marginally selective herbicides should also be exploited in the future. The initial findings of Stephenson and Ezra (1985) demonstrate that the metabolic pathways detoxifying herbicides in higher plants are inducible and preconditioning may enable plants to withstand subsequent exposure to greater doses of specific herbicides.

Future advances in the development of crop safeners for herbicides, however, are likely to be based on a biorational approach driven by a combination of biotechnology–chemistry principles rather than by the current dominant process of empirical chemical synthesis coupled with biological screening (Geissbuhler *et al.*, 1983). Targeted synthesis of crop safeners for herbicides in the future would use knowledge on the characteristics of the target site of herbicide action and detoxication mechanisms of herbicides coupled with computer graphics to guide synthesis of novel safeners to be tested.

So far, efforts to exploit biorational design procedures for the development of herbicides and safeners have been seriously hampered by a significant lack of basic scientific knowledge concerning the physiology and biochemistry of crop and weed plants (Geissbuhler *et al.*, 1987). In addition, in spite of recent advances, basic information on biochemical and molecular aspects of the mechanisms of action of most of the presently used herbicides and safeners is incomplete and insufficient to be systematically exploited by biorational procedures. Molecular models evaluating the potential of substrates, analogs, antagonists, and metabolites of current herbicides as well as compounds of novel structure as crop safeners for herbicides are expected to become more useful in the near future following concomitant advances in our understanding of the biochemical and molecular aspects of safener and herbicide action. Chemical alteration of the target site of action of specific herbicides may also become more feasible.

The use of prosafeners and proherbicides is also likely to increase in the future. A desirable margin of crop selectivity could be generated by exploiting metabolic differences of crop and weed plants to design novel proherbicides or prosafeners. Differential bioactivation of proherbicides to active herbicides in weeds and of prosafeners to effective safeners in crops may increase the margin of crop safety and efficacy of specific herbicides. The promising results observed with *N*-phenylmaleimides and *N*-phenylisomaleamic acids as progenitors of the *N*-phenylmaleamic acid safeners of sorghum against chloroacetanilide herbicide injury illustrate the potential of the prosafener concept (Rubin and Kirino, Chapter 14, this volume). The practical utilization of the proherbicide concept has been exploited commercially for a number of years (Wain and Smith, 1976). Proherbicides such as the phenoxy-butyric acids and the alkyl esters of several arylophenoxyalkanoic acid herbicides are activated by means of metabolism in susceptible weeds and used selectively in tolerant crops (Hatzios and Penner, 1982).

Finally, because of the extensive research efforts in the area of recombinant DNA technology it is reasonable to expect that genetically engineered microorganisms could be used as safeners for specific herbicides. Microbial safeners exploiting the herbicide degradative potential of soil bacteria and fungi which colonize the roots of desirable crops may be developed in the near future (Karns, Chapter 15, this volume). Alternatively, microbial safeners could result from the manipulation of endophytic bacteria colonizing the transport system of higher plants. Such microbes could be manipulated genetically to synthesize appropriate substrates or enzymes known to serve as target sites for specific herbicides.

The search for herbicide safeners in the future would also benefit from advances in screening procedures used for the evaluation of candidate chemicals. Cell or protoplast culture systems have long been accepted as useful means for the study of chemical and genetic mechanisms of plants as well as for the chemical and genetic regulation of such mechanisms (Gressel, 1984). The use of such systems for the evaluation of candidate crop safeners for herbicides is expected to increase (Gray *et al.*, 1982).

Regardless of whether whole plants or cell-free systems constitute the basis of a screening program for herbicide safeners, results and observations from studies conducted under laboratory or greenhouse conditions should always be extended to the field. Many times it is easier to observe the effects of safener/herbicide combinations at the biochemical level, but on some occasions these effects are not subsequently manifested in reduced herbicide effects on crop growth under field conditions (Stephenson and Ezra, 1983). Optimization of chemical crop safening against herbicide injury in the field may be determined by several factors which are not always present in a laboratory setup.

A case illustrating this principle is the development of the corn safener dichlormid. Of more than 500 *N,N*-substituted amides that showed initial promise as safeners of corn against thiocarbamate herbicide injury, dichlormid was the most active and best suited for practical application (Pallos *et al.*, 1975). Structure–activity relationship studies conducted by Stephenson *et al.* (1978, 1979) showed that the most active safeners for a series of thiocarbamate herbicides were the dichloroacetamides most similar to the herbicides with respect to the N,N-disubstituted alkyl groups. These studies, however, were carried out in a soil-free bioassay system. Under field conditions dichlormid was found to be the most active dichloroacetamide corn safener (Pallos *et al.*, 1975).

Future research on herbicide safener development should also concentrate on the use of plant species and model systems that are relevant to the actual agricultural environment, even if these biological systems are more difficult to handle. A good deal of past research on herbicide safeners has been conducted with the use of enzymes, organelles, cellular systems, and seedlings of plant species that are either not protected by safeners or not susceptible to the herbicides antagonized by these safeners (Hatzios, 1983). The contribution of the results of such studies toward the advancement of our understanding of practical crop/herbicide/safener combinations is often difficult to assess.

Further research will also need to assess the impact of changing agricultural practices such as minimum tillage, integrated pest and weed management approaches, cancellations of herbicide registrations, biotechnology advances, etc., on the use of and market for herbicide safeners.

V. CONCLUDING REMARKS

Future research in the area of weed management must adapt to changing needs and requirements related to crop safety, control of problem weeds, circumvention of weed resistance, and selected agronomic practices including crop rotations, narrow widths of crop rows, double cropping systems, and limited and no-till crop production. Novel approaches offering potential promise for improving or extending the useful life of currently used, reliable herbicides should be evaluated as viable alternatives to the costly process of discovering and developing new herbicides.

The recent development and use of crop safeners for herbicides offers an attractive means for achieving the above goal. Discoveries of practical herbicide safeners have been impressive considering the limitations in our understanding of how these chemicals protect crops against herbicide injury. The prospects of monetary gains have motivated the undertaking of a large effort by private enterprise in this area. However, the contributions of researchers at universities or other public institutions have been very significant in the development of the herbicide safener concept and they will continue to be invaluable in the future.

The pesticide industry is currently in a phase of transition. During the last few years, consolidations of manufacturers of crop protection chemicals have been making the news headlines (Anonymous, 1981). As a result, research efforts in several areas of crop protection technology have slowed or were discontinued at some companies. The future of crop safeners for herbicides, however, is expected to be bright. The basis for such optimism stems from two factors. First, the contributions of chemical weed control in modern crop production have been very impressive and they will continue to be important. The volume of herbicide sales in the United States currently exceeds the total volume of insecticide and fungicide sales combined (Schaub, 1985). Second, the biological interest in the concept of chemical manipulation of crop tolerance to herbicides coupled with its practical and economic importance is expected to continue providing the needed impetus for future research and developments in this area.

A better understanding of the mechanisms of action of current safeners and herbicides should enable more positive attempts toward increasing the number of situations in which crop safeners for herbicides could be used. Meanwhile, with increased interest in safeners it is hoped that other substances or approaches will be developed that will provide protection to a broader range of crops against a wider spectrum of herbicides including the photosynthetic inhibitors.

References

Anonymous.(1981). Concentration builds in pesticide industry. *Agrichem. Age* **25**(6), 20–25.

Baughan, P. J., Biddle, A. J., Blackett, J. A., and Toms, A. M. (1985). Symposium on application and biology. *Monog.—Br. Crop Prot. Counc.* **28,** 97.

Blair, A. M., Parker, C., and Kasasian, L. (1976). Herbicide protectants and antidotes—A review. *Pestic. Artic. News Summ..* **22,** 65–74.

Casida, J. E. (1978). Herbicide antidotes: Progress and prospects. *In* "Chemistry and Action of Herbicide Antidotes" (F. M. Pallos and J. E. Casida, eds.), pp. 161–164. Academic Press, New York.

Ezra, G., Flowers, H. M., and Gressel, J. (1983). Rapid multilevel interactions of a thiocarbamate herbicide and its protectant in maize cell culture. *In* "Pesticide Chemistry: Human Welfare and the Environment" (J. Miyamoto and P. C. Kearney, eds.), Vol. 3, pp. 225–231. Pergamon, Oxford.

Fedtke, C., and Trebst, A. (1987). Advances in understanding herbicide modes of action. *In* "Pesticide Science and Biotechnology" (R. Greenhalgh and T. R. Roberts, eds.), pp. 161–168. Blackwell, Oxford.

Fletcher, R. A. (1985). Plant growth regulating properties of sterol inhibiting fungicides. *In* "Hormonal Regulation of Plant Growth and Development" (S. S. Purohit, ed.), Vol. 2, pp. 103–113. Agro Bot. Publ., Old Ginnani, Bikaner, India.

Geissbuhler, H., Muller, U., Pachlatko, J. P., and Waespe, H. R. (1983). Biorational design of chemicals. *In* "Chemistry and World Food Supplies: The New Frontiers," (CHEMRAWN II, L. W. Shemilt, ed.), pp. 643–656. Pergamon, Oxford.

Geissbuhler, H., d'Hondt, C., Kunz, E., Nyfeler, R., and Pfister, K. (1987). Reflections on the future of chemical plant protection research. *In* "Pesticide Science and Biotechnology" (R. Greenhalgh and T. R. Roberts, eds.), pp. 3–14. Blackwell, Oxford.

Gray, R. A., Green, L. L., Hoch, P. E., and Pallos, F. M. (1982). The evolution of practical crop safeners. *Proc. Br. Crop Prot. Conf.—Weeds* **2,** 431–437.

Gressel, J. (1984). Plant tissue culture systems for screening of plant growth regulators: Hormones, herbicides and natural phytotoxins. *Adv. Cell Cult.* **3,** 93–181.

Hatzios, K. K. (1983). Herbicide antidotes: Development, chemistry and mode of action. *Adv. Agron.* **36,** 265–316.

Hatzios, K. K., and Penner, D. (1982). "Metabolism of Herbicides in Higher Plants." Burgess, Minneapolis, Minnesota.

Hoffman, O. L. (1978). Herbicide antidotes: From concept to practice. *In* "Chemistry and Action of Herbicide Antidotes" (F. M. Pallos and J. E. Casida, eds.), pp. 1–13. Academic Press, New York.

Jeffs, K. A. (1978). "Seed Treatment," CIPAC Monogr. No. 2. Colloq. Int. Pestic. Anal. Counc., Harpenden, Herts.

Ketchersid, M. L., and Merkle, M. G. (1983). Effects of protectants on sorghum seedling respiration. *Proc., South. Weed Sci. Soc.* **36,** 384.

Lay, M.-M., and Casida, J. E. (1978). Influence of glutathione and glutathione *S*-transferase in the action of dichloroacetamide antidotes for thiocarbamate herbicides. *In* "Chemistry and Action of Herbicide Antidotes" (F. M. Pallos and J. E. Casida, eds.), pp. 151–160. Academic Press, New York.

Muller, G., and Nyffeler, A. (1981). CONCEP (CGA-43089), a safening agent protecting sorghum (*Sorghum bicolor*) from metolachlor injury. *Acta Phytopathol. Acad. Sci. Hung.* **16,** 245–248.

Pallos, F. M., Gray, R. A., Arneklev, D. R., and Brokke, M. E. (1975). Antidotes protect corn from thiocarbamate herbicide injury. *J. Agric. Food Chem.* **23,** 821–822.

Parker, C. (1983). Herbicide antidotes—A review. *Pestic. Sci.* **14,** 40–48.

Peek, J. W., Dill, T. R., and Turner, W. E. (1981). Use of safeners to protect sorghum from herbicide injury. *Proc. Annu. Corn Sorghum Res. Conf.* **36,** 31–47.

Schaub, J. R. (1985). The economics of agricultural pesticide technology. *Beltsville Symp. Agric. Res.* **8,** 15–26.

Stephenson, G. R., and Chang, F. Y. (1978). Comparative activity and selectivity of herbicide antidotes. *In* "Chemistry and Action of Herbicide Antidotes" (F. M. Pallos and J. E. Casida, eds.), pp. 35–61. Academic Press, New York.

Stephenson, G. R., and Ezra, G. (1983). Herbicide antidotes—A new era in selective chemical weed control. *In* "Plant Growth Regulating Chemicals" (L. G. Nickell, ed.), Vol. 2, pp. 193–212. CRC Press, Boca Raton, Florida.

Stephenson, G. R., and Ezra, G. (1985). Use of subtoxic pretreatments to improve crop tolerance to herbicides. *ACS Symp. Ser.* **276,** 69–84.

Stephenson, G. R., Bunce, J. J., Makowski, R. I., and Curry, J. C. (1978). Structure–activity relationships for *S*-ethyl-*N*,*N*-dipropylthiocarbamate (EPTC) antidotes in corn. *J. Agric. Food Chem.* **26,** 137–140.

Stephenson, G. R., Bunce, J. J., Makowski, R. I., Bergsma, M. D., and Curry, J. C. (1979). Structure–activity relationships for antidotes to thiocarbamate herbicides in corn. *J. Agric. Food Chem.* **27,** 543–547.

Sweetser, P. B. (1985). Safening of sulfonylurea herbicides to cereal crops: Mode of herbicide antidote action. *Proc. Br. Crop Prot. Conf.—Weeds* **3,** 1147–1154.

Theologis, A. (1986). Rapid gene regulation by auxin. *Annu. Rev. Plant Physiol.* **37,** 407–438.

Wain, R. L., and Smith, M. S. (1976). Selectivity in relation to metabolism. *In* "Herbicides: Physiology, Biochemistry, Ecology" (L. J. Audus, ed.), 2nd ed., Vol. 1, pp. 279–302. Academic Press, London.

Wiegand, R. C., Shah, D. M., Mozer, T. J., Harding, E. I., Diaz-Collier, J., Sounders, C., Jaworski, E. G., and Tiemeier, D. C. (1986). Messenger RNA encoding a glutathione-*S*-transferase responsible for herbicide tolerance in maize is induced in response to safener treatment. *Plant Mol. Biol.* **7,** 235–243.

Common or Code Names and Corresponding Chemical Names of Herbicides, Safeners, and Other Agrochemicals Mentioned in the Text

AC-222,164	(±)-2-[4,5-dihydro-4-methyl-4-(1-methylethyl)-5-oxo-1*H*-imidazol-2-yl]-3-benzoic acid
AC-222,293	isomeric mixture of 6-[4-isopropyl-4-methyl-5-oxo-2-imidazolin-2-yl]-*m*-toluate and methyl 2-(4-isopropyl-4-methyl-5-oxo-2-imidazolin-2-yl]-*p*-toluate
Acetochlor	2-chloro-*N*-(ethoxymethyl)-*N*-(2-ethyl-6-methylphenyl)acetamide
Acifluorfen	5-[2-chloro-4-(trifluoromethyl)phenoxy]-2-nitrobenzoic acid
ACQ	acenaphthenequinone
AD-67	*N*-dichloroacetyl-1-oxa-4-aza-spiro-4,5-decane
Alachlor	2-chloro-*N*-(2,6-diethylphenyl)-*N*-(methoxymethyl)acetamide

Aldrin	1,2,3,4,10,10-hexachloro-1,4,4α,5,8,8α-hexahydro-1,4:5,8-dimethanonaphthalene
Allidochlor	2-chloro-*N*,*N*-di-2-propenylacetamide
Ametryn	*N*-ethyl-*N*′-(1-methylethyl)-6-(methylthio)-1,3,5-triazine-2,4-diamine
Amitrole	1*H*-1,2,4-triazol-3-amine
AMO-1618	2′-isopropyl-4′-(trimethylammonium chloride) 5′-methylpiperidine carboxylate
Ancymidol	α-cyclopropyl-α-(*p*-methoxyphenyl)-5-pyrimidine methanol
Asulam	methyl[(4-aminophenyl)sulfonyl]carbamate
Atrazine	6-chloro-*N*-ethyl-*N*′-(1-methylethyl)-1,3,5-triazine-2,4-diamine
Barban	4-chloro-2-butynyl-3-chlorophenylcarbamate
BAS 140 810	*N*-allyl-*N*-2-(2,4,6-trichlorophenoxy)-ethyl-piperidinium-bromide
BAS 145 138	see LAB 145 138
BCS	4-bromophenyl chloromethyl sulfone
BDS	benzenediazo sodium sulfonate
Benefin	*N*-butyl-*N*-ethyl-2,6-dinitro-4-(trifluoromethyl) benzenamine
Bensulfuron	2-[[[[[(4,6-dimethoxy-2-pyrimidinyl)amino]carbonyl]amino]-sulfonyl]methyl]benzoic acid
Bensulide	*O*,*O*-bis(1-methylethyl)-*S*-[2-[(phenylsulfonyl)amino]ethyl] phosphorodithioate
Benthiocarb	see thiobencarb
Bialaphos	L-2-amino-4-((1-hydroxy)(methyl)phosphinoyl)-butyric acid
Bromacil	5-bromo-6-methyl-3-(1-methylpropyl)-2,4(1*H*,3*H*) pyrimidinedione
BSM	see bensulfuron-methyl
Butachlor	*N*-(buthoxymethyl)-2-chloro-*N*-(2,6-diethylphenyl) acetamide

Buthidazole	3-[5-(1,1-dimethylethyl)-1,3,4-thiadiazol-2-yl]-4-hydroxy-1-methyl-2-imidiazolidinone
Butralin	4-(1,1-dimethylethyl)-*N*-(1-methylpropyl)-2,6-dinitrobenzenamine
Bu	*S*-ethyl bis(2-methylpropyl)carbamothioate
dylic acid	dimethyl arsinic acid
arbofuran	2,3-dihydro-2,2-dimethyl-7-benzofuranol methylcarbamate
Carboxin	5,6-dihydro-2-methyl-*N*-phenyl-1,4-oxathiin-3-carboxamide
CBDS	*p*-chloro-benzenediazo sodium sulfonate
CDAA	see allidochlor
CGA-43089	see cyometrinil
CGA-92194	see oxabetrinil
CGA-123407	see fenclorim
CGA-133205	*O*-[1,3-dioxolan-2-yl-methyl]-2,2,2-trifluoro-4′-chloroacetophenone-oxime
CGA-154281	4-(dichloroacetyl)-3,4-dihydro-3-methyl-2*H*-1,4-benzoxazine
CH-83	see isopropylate
CHCH	2-(1-cyclohexenyl)-cyclohexanone
Chloramben	3-amino-2,5-dichlorobenzoic acid
Chloramphenicol	D(−)-threo-2,2-dichloro-*N*-[β-hydroxy-α-(hydroxymethyl)-*p*-nitrophenethyl]acetamide
Chlorbromuron	*N*′-(4-bromo-3-chlorophenyl)-*N*-methoxy-*N*-methylurea
Chlordane	1,2,4,5,6,7,8,8-octachloro-2,3,3α,4,7,7α-hexahydro-4,7-methano-1*H*-indene
Chlorfenvinphos	*O*,*O*-diethyl *O*-2-chloro-1-(2,4-dichlorophenyl)vinyl] phosphate
Chlorimuron	2-[[[[(4-chloro-6-methoxy-2-pyrimidinyl)amino]carbonyl]amino]sulfonyl]benzoic acid
Chlornitrofen	4-nitrophenyl 2,4,6-trichlorophenyl ester

Chlorophthalim	*N'*-(4-chlorophenyl)-3,4,5,6-tetrahydrophthalamide
Chloroxuron	*N'*-[4-(4-chlorophenoxy)phenyl]-*N*,*N*-dimethylurea
Chlorpropham	1-methylethyl-3-chlorophenylcarbamate
Chlorsulfuron	2-chloro-*N*-[[(4-methoxy-6-methyl-1,3,5-triazin-2-yl)amino]carbonyl]benzenesulfonamide
Chlorthiamid	2,6-dichlorobenzenecarbothiamide
Cisanilide	*cis*-2,5-dimethyl-*N*-phenyl-1-pyrrolidine-carboxamide
ClNA	4-chloronaphthalene-1,8-dicarboxylic acid
Clomazone	2-[(2-chlorophenyl)methyl]-4,4-dimethyl-3-isoxazolidinone
Concep I	see cyometrinil
Concep II	see oxabetrinil
Concep III	see CGA-133205
CPIM	*N*-(4-chlorophenyl) isomaleimide
CPMA	*N*-(4-chlorophenyl) maleamic acid
CPMI	*N*-(4-chlorophenyl) maleimide
Cyanazine	2-[[4-chloro-6-(ethylamino)-1,3,5-triazin-2-yl]amino]-2-methylpropanenitrile
Cycloate	*S*-ethyl cyclohexylethylcarbamothioate
Cycloheximide	3-[2-(3,5-dimethyl-2-oxocyclohexyl)-2-hydroxyethyl]-glutarimide
Cyometrinil	(*Z*)-α[(cyanomethoxy)imino]benzeneacetonitrile
2,4-D	(2,4-dichlorophenoxy)acetic acid
2,6-D	(2,6-dichlorophenoxy)acetic acid
2,4-DB	4-(2,4-dichlorophenoxy)butanoic acid
DA	Diphenic anhydride
Dalapon	2,2-dichloropropanoic acid
Daminozide	succinic acid-2,2-dimethyl hydrazine
DCBA	2,6-dichlorobenzaldehyde
DCPA	dimethyl-2,3,5,6-tetrachloro-1,4-benzenedicarboxylate

DDCA	see dichlormid
DDT	1,1′-(2,2,2-trichloroethylidene)bis[4-chlorobenzene]
Dexon	see fenaminosulf
Diallate	*S*-(2,3-dichloro-2-propenyl)bis(1-methylethyl) carbamothioate
Dicamba	3,6-dichloro-2-methoxybenzoic acid
Dichlobenil	2,6-dichlorobenzonitrile
Dichlormate	3,4-dichloro benzenemethanol methylcarbamate
Dichlormid	2,2-dichloro-*N*,*N*-di-2-propenylacetamide
Diclofop	(±)-2-[4-(2,4-dichlorophenoxy)phenoxy]propanoic acid
Dieldrin	3,4,5,6,9,9-hexachloro-1α,2,2α,3,6,6α,7,7α-octahydro-2,7:3,6-dimethanonaphth[2,3-*b*]oxirene
Dimepiperate	*S*-(1-methyl-1-phenethyl-piperidine-1-carbothioate)
Dinoseb	2-(1-methylpropyl)-4,6-dinitrophenol
Diphenamid	*N*,*N*-dimethyl-α-phenyl benzeneacetamide
Diquat	6,7-dihydrodipyrido[1,2-α:2′,1′-*c*] pyrazinediium ion
Diuron	*N*′-(3,4-dichlorophenyl)-*N*,*N*-dimethylurea
DKA-24	*N*,*N*′-diallyl-*N*′-dichloroacetylglycinamide
DNBP	see dinoseb
DNOC	2-methyl-4,6-dinitrophenol
DPX-F5384	see bensulfuron-methyl
DPX-F6025	see chlorimuron-ethyl
DSMA	methylarsonic acid disodium salt
Endothall	7-oxabicyclo[2.2.1]heptane-2,3-dicarboxylic acid
EPTC	*S*-ethyl dipropylcarbamothioate
Eradicane	EPTC + dichlormid
Ethofumesate	(±)-2-ethoxy-2,3-dihydro-3,3-dimethyl-5-benzofuranyl methanesulfonate

Fenac	2,3,6-trichlorobenzeneacetic acid
Fenaminosulf	*p*-dimethylamino-benzenediazo sodium sulfonate
Fenclorim	4,6-dichloro-2-phenyl-pyrimidine
Fenoxaprop	(±)-2-[4-[(6-chloro-2-benzoxazolyl)oxy]phenoxy]propanoic acid
Fluazifop	(±)-2-[4-[[5-(trifluoromethyl)-2-pyridinyl]oxy]phenoxy]-propanoic acid
Fluometuron	*N*,*N*-dimethyl-*N*′-[3-(trifluoromethyl)phenyl]urea
Fluorodifen	2-nitro-1-(4-nitrophenoxy)-4-trifluoromethylbenzene
Flurazole	phenylmethyl 2-chloro-4-(trifluoromethyl)-5-thiazolecarboxylate
FMC-57020	see clomazone

Glufosinate	ammonium (3-amino-3-carboxypropyl)methylphosphinate
Glyphosate	*N*-(phosphonomethyl)glycine
G-36393	2-methylthio-4-isopropylamino-6-(3-methoxypropylamino)-*s*-triazine

Heptachlor	1,4,5,6,7,8,8-heptachloro-3α,4,7,7α-tetrahydro-4,7-methanoindene
HMI	3-hydroxy-5-methyl isoxazole

IAA	3-indoleacetic acid
IBA	3-indolebutyric acid
Imazapyr	(±)-2-[4,5-dihydro-4-methyl-4-(1-methylethyl)-5-oxo-1*H*-imidazol-2-yl]-3-pyridinecarboxylic acid
Imazaquin	2-[4,5-dihydro-4-methyl-4-(1-methylethyl)-5-oxo-1*H*-imidazol-2-yl]-3-quinolinecarboxylic acid
Imazethapyr	(±)-2-[4,5-dihydro-4-methyl-4-(1-methylethyl)-5-oxo-1*H*-imidazol-2-yl]-5-ethyl-3-pyridinecarboxylic acid

Isomethiozin	6-(1,1-dimethylethyl)-4-[(2-methylpropylidene)-amino]-3-(methylthio)-1,2,4-triazin-5-(4*H*)-one
Isopropylate	*S*-isopropyl-hexahydro-1*H*-azepine-1-carbothioate
Karbutilate	3-[[(dimethylamino)carbonyl]aminophenyl]-(1,1-dimethylethyl)carbamate
LAB-145138	1-dichloroacetyl-hexahydro-3,3-8α-trimethyl-pyrrolo-[1,2-α]-pyrimidin-6-(2*H*)-one
LAB-147886	1-dichloroacetyl-hexahydro-8α-methyl-imidazo-[1,2-α]-pyridin-5(1*H*)-one
Linuron	*N*′-(3,4-dichlorophenyl)-*N*-methoxy-*N*-methylurea
M-32988	2,2-dichloro-*N*-(3-methyl-4-thiazolin-2-ylidene)acetamide
Maleic hydrazide	1,2-dihydro-3,6-pyridazinedione
MBDS	*p*-methyl-benzenediazo sodium sulfonate
MCPA	(4-chloro-2-methylphenoxy)acetic acid
MCPB	4-(4-chloro-2-methylphenoxy)butanoic acid
Mepiquat chloride	2-chloroethyltrimethylammonium chloride
Metalaxyl	methyl-*N*-(2-methoxyacetyl)-*N*-(2,6-xylyl)-DL-alaninate
Metazachlor	*N*-(2,6-dimethyl-phenyl)-N-(1-pyrazolyl-methyl)-chloroacetamide
Methoxyphenone	3,3′-dimethyl-4-methoxybenzophenone
Metolachlor	2-chloro-*N*-(2-ethyl-6-methylphenyl)-*N*-(2-methoxy-1-methylethyl)acetamide
Metribuzin	4-amino-6-(1,1-dimethylethyl)-3-(methylthio)-1,2,4-triazin-5(4*H*)-one
Metsulfuron	2-[[[[(4-methoxy-6-methyl-1,3,5,triazin-2-yl)amino]carbonyl]amino]sulfonyl]benzoic acid
MG-191	2-dichloromethyl-2-methyl-1,3-dioxolane

MH	see maleic hydrazide
Molinate	*S*-ethyl hexahydro-1*H*-azepine-1-carbothioate
MON-4606	see flurazole
Monuron	*N'*-(4-chlorophenyl)-*N*,*N*-dimethylurea
MSMA	methylarsonic acid monosodium salt
MY-93	see dimepiperate
NA	see naphthalic anhydride
NAA	naphthaleneacetic acid
Naphthalic anhydride	naphthalene-1,8-dicarboxylic acid anhydride
Napropamide	*N*,*N*-diethyl-2-(1-naphthalenyloxy)propanamide
NDiA	1,8:4,5-naphthalenetetracarboxylic acid dianhydride
Neburon	*N*-butyl-*N'*-(3,4-dichlorophenyl)-*N*-methylurea
NEM	*N*-ethylmaleimide
NHNA	naphthalene-1,8-carboxamide
NH_2NA	4-aminonaphthalene-1,8-dicarboxylic acid anhydride
Nitralin	4-(methylsulfonyl)-2,6-dinitro-*N*,*N*-dipropyl-benzenamine
Nitrofen	2,4-dichloro-1-(4-nitrophenoxy)benzene
Norea	*N*,*N*-dimethyl-*N'*-(octahydro-4,7-methano-1*H*-inden-5-yl)urea-3*a*α,4α,5α,7α,7aα-isomer
Norflurazon	4-chloro-5-(methylamino)-2-(3-(trifluoromethyl)phenyl)-3(2*H*)-pyridazinone
OTC	L-2-oxathiazolidine-4-carboxylic acid
Oxabetrinil	α-(1,3-dioxolan-2-yl-methoxy)-iminobenzeneacetonitrile
Oxadiazon	3-[2,4-dichloro-5-(1-methylethoxy)phenyl]-5-(1,1-dimethylethyl)-1,3,4-oxadiazol-2-(3*H*)-one

PA	phenale-1-one
Paclobutrazole	(2RS,3RS)-1-(4-chlorophenyl)-4,4-dimethyl-2-(1,2,3-triazol-1-yl)-pentan-3-ol
Paraquat	1,1′-dimethyl-4,4′-bipyridinium ion
Parathion	*O,O*-diethyl *O-p*-nitrophenyl phosphorothioate
PC-671	formulated lignin by-product
PC940-C	kraft lignin
PCMBS	*p*-chloromercuribenzene sulfonic acid
Pebulate	*S*-propyl butylethylcarbamothioate
Perfluidone	1,1,1-trifluoro-*N*-[2-methyl-4-(phenylsulfonyl)phenyl]-methanesulfonamide
PGA	2-phenylglutaric anhydride
PHA	phthalic anhydride
Phenamiphos	ethyl 3-methyl-4-(methylthio) phenyl-(1-methylethyl)phosphoramidate
Phenobarbital	5-ethyl-5-phenyl-2,4,6(1H,3H,5H)-pyrimidinetrione
Phorate	*O,O*-dimethyl *S*-(ethylthio)methylphosphorodithioate
Phosphon D	tributyl-2,4-dichlorobenzylphosphonium chloride
Phosphon S	tributyl-2,4-dichlorobenzylammonium chloride
Picloram	4-amino-3,5,6-trichloro-2-pyridinecarboxylic acid
Piperonyl butoxide	5-[[2-(2-butoxyethoxy)ethoxy]methyl]-6-propyl-1,3-benzodioxole
PPG-1292	2,2-dichloro-*N*-[1,3-dioxolan-2-yl-methyl]-*N*′-2-propenyl-acetamide
Pretilachlor	2-chloro-2′,6′-diethyl-*N*-(2-propoxyethyl)acetanilide
Prometon	6-methoxy-*N,N*′-bis(1-methylethyl)-1,3,5-triazine-2,4-diamine
Prometryn	*N,N*′-bis(1-methylethyl)-6-(methylthio)-1,3,5 triazine-2,4-diamine
Pronamide	3,5-dichloro-(*N*-1,1-dimethyl-2-propynyl)benzamide
Propachlor	2-chloro-*N*-(1-methylethyl)-*N*-phenylacetamide

Propanil	*N*-(3,4-dichlorophenyl)propanamide
Propazine	6-chloro-*N*,*N′*-bis(1-methylethyl)-1,3,5-triazine-2,4-diamine
Propham	1-methylethyl phenylcarbamate
Propoxur	2-(1-methylethoxy)phenol methylcarbamate
Protect	see naphthalic anhydride
Pyrazon	5-amino-4-chloro-2-phenyl-3(2*H*)-pyridazinone
Pyriclor	2,3,5-trichloro-4-pyridinol
R-25788	see dichlormid
R-28725	3-(dichloroacetyl)-2,2-dimethyl-1,3-oxazolidine
R-29148	3-(dichloroacetyl)-2,2,5-trimethyl-1,3-oxazolidine
R-30228	chemistry not released
RSW-0411	see triapenthenol
S-449	4-chloro-2-hydroxyiminoacetanilide
SC-0774	chemistry not released
SC-2957	*S*-benzyl-*N*-ethyl-*N*-(1,2-dimethylpropyl)thiolcarbamate
Screen	see flurazole
Sethoxydim	2-[1-(ethoxyimino)butyl]-5-[2-(ethylthio)propyl]-3-hydroxy-2-cyclohexen-1-one
SKF-525A	2-(diethylamino)ethyl-2,2-diphenylpentanoate
SHAM	salicylhydroxamic acid
Siduron	*N*-(2-methylcyclohexyl)-*N′*-phenylurea
Silvex	2-(2,4,5-trichlorophenoxy)propanoic acid
Simazine	6-chloro-*N*,*N′*-diethyl-1,3,5-triazine-2,4-diamine
Simetryn	*N*,*N′*-diethyl-6-(methylthio)-1,3-5-triazine-2,4-diamine
Sofit	pretilachlor + fenclorim

Sulfometuron	2-[[[[(4,6-dimethyl-2-pyrimidinyl)amino]carbonyl]amino] sulfonyl]benzoic acid
Sutan$^+$	butylate + dichlormid
2,4,6-T	(2,4,6-trichlorophenoxy)acetic acid
TC	L-thiazolidine-4-carboxylic acid
TCA	trichloroacetic acid
Tebuthiuron	*N*-[5-(1,1-dimethylethyl)-1,3,4-thiadiazol-2-yl]-*N*,*N*′-dimethylurea
Terbacil	5-chloro-3-(1,1-dimethylethyl)-6-methyl-2,4(1*H*,3*H*)-pyrimidinedione
Terbutol	2,6-bis(1,1-dimethylethyl)-4-methylphenyl methylcarbamate
Tetcyclasis	5-(4-chlorophenyl)-3,4,5,9,10-pentaaza-tetra cyclo-[5,4,$10^{2,6}$, $0^{8,11}$]dodeca-3,9-diene
Thimet	see phorate
Thiobencarb	*S*-[(4-chlorophenyl)methyl] diethylcarbamothioate
Triadimefon	1-(chlorophenoxy)-3,3-dimethyl-1-(1H-1,2,4-triazol-1-yl)-2-butanone
Triallate	*S*-(2,3,3,-trichloro-2-propenyl) bis(1-methylethyl)-carbamothioate
Triapenthenol	β-(cyclohexalmethylene)-gamma-(1,1′-dimethyl-ethyl)-1*H*-1,2,4 triazole-1-ethanol
Triarimol	α-(2,4-dichlorophenyl)-α-phenyl-5-pyrimidinemethanol
Tridiphane	2-(3,5-dichlorophenyl)-2-(2,2,2-trichloroethyl)oxirane
Trifluralin	2,6-dinitro-*N*,*N*-dipropyl-4-(trifluoromethyl)-benzenamine
Vernam$^+$	Vernolate + dichlormid
Vernolate	*S*-propyl dipropylcarbamothioate

Index